Lecture Notes in Mathematics

Edited by J.-M. Morel, F. Takens and B. Teissier

Editorial Policy
for the publication of monographs

1. Lecture Notes aim to report new developments in all areas of mathematics and their applications - quickly, informally and at a high level. Mathematical texts analysing new developments in modelling and numerical simulation are welcome.

   Monograph manuscripts should be reasonably self-contained and rounded off. Thus they may, and often will, present not only results of the author but also related work by other people. They may be based on specialised lecture courses. Furthermore, the manuscripts should provide sufficient motivation, examples and applications. This clearly distinguishes Lecture Notes from journal articles or technical reports which normally are very concise. Articles intended for a journal but too long to be accepted by most journals, usually do not have this „lecture notes" character. For similar reasons it is unusual for doctoral theses to be accepted for the Lecture Notes series, though habilitation theses may be appropriate.

2. Manuscripts should be submitted (preferably in duplicate) either to Springer's mathematics editorial in Heidelberg, or to one of the series editors (with a copy to Springer). In general, manuscripts will be sent out to 2 external referees for evaluation. If a decision cannot yet be reached on the basis of the first 2 reports, further referees may be contacted: The author will be informed of this. A final decision to publish can be made only on the basis of the complete manuscript, however a refereeing process leading to a preliminary decision can be based on a pre-final or incomplete manuscript. The strict minimum amount of material that will be considered should include a detailed outline describing the planned contents of each chapter, a bibliography and several sample chapters.

   Authors should be aware that incomplete or insufficiently close to final manuscripts almost always result in longer refereeing times and nevertheless unclear referees' recommendations, making further refereeing of a final draft necessary.

   Authors should also be aware that parallel submission of their manuscript to another publisher while under consideration for LNM will in general lead to immediate rejection.

3. Manuscripts should in general be submitted in English. Final manuscripts should contain at least 100 pages of mathematical text and should always include

   - a table of contents;
   - an informative introduction, with adequate motivation and perhaps some historical remarks: it should be accessible to a reader not intimately familiar with the topic treated;
   - a subject index: as a rule this is genuinely helpful for the reader.

   For evaluation purposes, manuscripts may be submitted in print or electronic form (print form is still preferred by most referees), in the latter case preferably as pdf- or zipped ps-files. Lecture Notes volumes are, as a rule, printed digitally from the authors' files. To ensure best results, authors are asked to use the LaTeX2e style files available from Springer's web-server at:

   ftp://ftp.springer.de/pub/tex/latex/mathegl/mono/ (for monographs) and

   ftp://ftp.springer.de/pub/tex/latex/mathegl/mult/ (for summer schools/tutorials).

   Additional technical instructions, if necessary, are available on request from lnm@springer-sbm.com.

Continued on inside back-cover

Lecture Notes in Mathematics 1888

J. van der Hoeven

Transseries and Real Differential Algebra

Springer

Author

Joris van der Hoeven
Département de Mathématiques, CNRS
Université Paris-Sud
Bâtiment 425
91405 Orsay CX
France
e-mail: joris@texmacs.org

Library of Congress Control Number: 2006930997

Mathematics Subject Classification (2000): 34E13, 03C65, 68W30, 34M35, 13H05

ISSN print edition: 0075-8434
ISSN electronic edition: 1617-9692
ISBN-10 3-540-35590-1 Springer Berlin Heidelberg New York
ISBN-13 978-3-540-35590-8 Springer Berlin Heidelberg New York

DOI 10.1007/3-540-35590-1

Springer is a part of Springer Science+Business Media
springer.com

Typesetting by the author and SPi using a Springer LaTeX package
Cover design: WMXDesign GmbH, Heidelberg

Printed on acid-free paper SPIN: 11771975 VA41/3100/SPi 5 4 3 2 1 0

Table of Contents

Foreword

Transseries find their origin in at least three different areas of mathematics: analysis, model theory and computer algebra. They play a crucial role in Écalle's proof of Dulac's conjecture, which is closely related to Hilbert's 16-th problem.

I personally became interested in transseries because they provide an excellent framework for automating asymptotic calculus. While developing several algorithms for computing asymptotic expansions of solutions to non-linear differential equations, it turned out that still a lot of theoretical work on transseries had to be done. This led to part A of my thesis. The aim of the present book is to make this work accessible for non-specialists. The book is self-contained and many exercises have been included for further studies. I hope that it will be suitable for both graduate students and professional mathematicians. In the later chapters, a very elementary background in differential algebra may be helpful.

The book focuses on that part of the theory which should be of common interest for mathematicians working in analysis, model theory or computer algebra. In comparison with my thesis, the exposition has been restricted to the theory of grid-based transseries, which is sufficiently general for solving differential equations, but less general than the well-based setting. On the other hand, I included a more systematic theory of "strong linear algebra", which formalizes computations with infinite summations. As an illustration of the different techniques in this book, I also added a proof of the "differential intermediate value theorem".

I have chosen not to include any developments of specific interest to one of the areas mentioned above, even though the exercises occasionally provide some hints. People interested in the accelero-summation of divergent transseries are invited to read Écalle's work. Part B of my thesis contains effective counterparts of the theoretical algorithms in this book and work is in progress on the analytic counterparts. The model theoretical aspects are currently under development in a joint project with Matthias Aschenbrenner and Lou van den Dries.

The book in its present form would not have existed without the help of several people. First of all, I would like to thank Jean Écalle, for his support and many useful discussions. I am also indoubted to Lou van den Dries and Matthias Aschenbrenner for their careful reading of several chapters and their corrections. Last, but not least, I would like to thank Sylvie for her patience and aptitude to put up with an ever working mathematician.

We finally notice that the present book has been written and typeset using the GNU $\TeX_{\text{MACS}}$ scientific text editor. This program can be freely downloaded from `http://www.texmacs.org`.

Joris van der Hoeven
Chevreuse 1999–2006

Introduction

The field with no escape

A *transseries* is a formal object, constructed from the real numbers and an infinitely large variable $x \succ 1$, using infinite summation, exponentiation and logarithm. Examples of transseries are:

$$\frac{1}{1-x^{-1}} = 1+\frac{1}{x}+\frac{1}{x^2}+\frac{1}{x^3}+\cdots \tag{1}$$

$$\frac{1}{1-x^{-1}-\mathrm{e}^{-x}} = 1+\frac{1}{x}+\frac{1}{x^2}+\cdots+\mathrm{e}^{-x}+2\frac{\mathrm{e}^{-x}}{x}+\cdots+\mathrm{e}^{-2x}+\cdots \tag{2}$$

$$\frac{\mathrm{e}^{\frac{x}{1-1/\log x}}}{1-x^{-1}} = \mathrm{e}^{x+\frac{x}{\log x}+\frac{x}{\log^2 x}+\cdots}+\frac{1}{x}\mathrm{e}^{x+\frac{x}{\log x}+\frac{x}{\log^2 x}+\cdots}+\cdots \tag{3}$$

$$-\mathrm{e}^{x}\int\frac{\mathrm{e}^{-x}}{x} = \frac{1}{x}-\frac{1}{x^2}+\frac{2}{x^3}-\frac{6}{x^4}+\frac{24}{x^5}-\frac{120}{x^6}+\cdots \tag{4}$$

$$\Gamma(x) = \frac{\sqrt{2\pi}\,\mathrm{e}^{x(\log x-1)}}{x^{1/2}}+\frac{\sqrt{2\pi}\,\mathrm{e}^{x(\log x-1)}}{12\,x^{3/2}}+\frac{\sqrt{2\pi}\,\mathrm{e}^{x(\log x-1)}}{288\,x^{5/2}}+\cdots \tag{5}$$

$$\zeta(x) = 1+2^{-x}+3^{-x}+4^{-x}+\cdots \tag{6}$$

$$\varphi(x) = \frac{1}{x}+\varphi(x^{\pi})=\frac{1}{x}+\frac{1}{x^{\pi}}+\frac{1}{x^{\pi^2}}+\frac{1}{x^{\pi^3}}+\cdots \tag{7}$$

$$\psi(x) = \frac{1}{x}+\psi(\mathrm{e}^{\log^2 x})=\frac{1}{x}+\frac{1}{\mathrm{e}^{\log^2 x}}+\frac{1}{\mathrm{e}^{\log^4 x}}+\frac{1}{\mathrm{e}^{\log^8 x}}+\cdots \tag{8}$$

As the examples suggest, transseries are naturally encountered as formal asymptotic solutions of differential or more general functional equations. The name "transseries" therefore has a double signification: transseries are generally *transfinite* and they can model the asymptotic behaviour of *transcendental* functions.

Whereas the transseries (1), (2), (3), (6), (7) and (8) are convergent, the other examples (4) and (5) are divergent. Convergent transseries have a clear analytic meaning and they naturally describe the asymptotic behaviour of their sums. These properties surprisingly hold in the divergent case as well. Roughly speaking, given a divergent series

$$f = \sum_{n=1}^{\infty} \frac{f_n}{x^n} = \sum_{n=1}^{\infty} \frac{(-1)^{n-1}(n-1)!}{x^n}$$

like (4), one first applies the formal Borel transformation

$$\hat{f}(\zeta) = (\tilde{\mathcal{B}} f)(\zeta) = \sum_{n=1}^{\infty} \frac{f_n}{(n-1)!} \zeta^n = \frac{1}{1+\zeta}.$$

If this Borel transform $\hat{f}$ can be analytically continued on $[0, +\infty)$, then the inverse Laplace transform can be applied analytically:

$$\bar{f}(x) = (\mathcal{L}\hat{f})(x) = \int_0^{\infty} \hat{f}(\zeta)\, \mathrm{e}^{-x\zeta}\, \mathrm{d}x = \int_0^{\infty} \frac{\mathrm{e}^{-x\zeta}}{1+\zeta} \mathrm{d}x.$$

The analytic function $\bar{f}$ obtained admits f as its asymptotic expansion. Moreover, the association $f \mapsto \bar{f}$ preserves the ring operations and differentiation. In particular, both f and $\bar{f}$ satisfy the differential equation

$$f' - f = -\frac{1}{x}.$$

Consequently, we may consider $\bar{f}$ as an analytic realization of f. Of course, the above example is very simple. Also, the success of the method is indirectly ensured by the fact that the formal series f has a "natural origin" (in our case, f satisfies a differential equation). The general theory of accelero-summation of transseries, as developed by Écalle [É92, É93], is far more complex, and beyond the scope of this book. Nevertheless, it is important to remember that such a theory *exists*: even though the transseries studied in this this book are purely formal, they generally correspond to genuine analytic functions.

The attentive reader may have noticed another interesting property which is satisfied by some of the transseries (1–8) above: we say that a transseries is *grid-based*, if

GB1. There exists a finite number $\mathfrak{m}_1$, ..., $\mathfrak{m}_k$ of infinitesimal "transmonomials", such that f is a multivariate Laurent series in $\mathfrak{m}_1, \ldots, \mathfrak{m}_k$:

$$f = \sum_{\nu_1 \leqslant \alpha_1 \in \mathbb{Z}} \cdots \sum_{\nu_k \leqslant \alpha_k \in \mathbb{Z}} f_{\alpha_1, \ldots, \alpha_k} \mathfrak{m}_1^{\alpha_1} \cdots \mathfrak{m}_k^{\alpha_k}.$$

GB2. The property **GB1** is recursively satisfied when replacing f by the logarithm of one of the $\mathfrak{m}_i$.

The examples (1–5) are grid-based. For instance, for (2), we may take $\mathfrak{m}_1 = x^{-1}$ and $\mathfrak{m}_2 = \mathrm{e}^{-x}$. The examples (6–8) are not grid-based, but only *well-based*. The last example even cannot be expanded w.r.t. a finitely generated asymptotic scale with powers in $\mathbb{R}$. As we will see in this book, transseries solutions to algebraic differential equations with grid-based coefficients are necessarily grid-based as well. This immediately implies that the examples (6–8) are differentially transcendental over $\mathbb{R}$ (see also [GS91]). The fact that grid-based transseries may be considered as multivariate Laurent series also makes them particularly useful for effective computations. For these reasons, we will mainly study grid-based transseries in this book, although generalizations to the well-based setting will be indicated in the exercises.

The resolution of differential and more general equations using transseries presupposes that the set of transseries has a rich structure. Indeed, the transseries form a totally ordered field $\mathbb{T}$ (chapter 4), which is real closed (chapter 3), and stable under differentiation, integration, composition and functional inversion (chapter 5). More remarkably, it also satisfies the differential intermediate value property:

> Given a differential polynomial $P \in \mathbb{T}\{F\}$ and transseries $f < g \in \mathbb{T}$ with $P(f)\, P(g) < 0$, there exists a transseries $h \in \mathbb{T}$ with $f < h < g$ and $P(h) = 0$.

In particular, any algebraic differential equation of odd degree over $\mathbb{T}$, like

$$f^3\,(f')^2\,(f''')^4 + \mathrm{e}^{\mathrm{e}^x}\, f^7 - \Gamma(\Gamma(x \log x))\, f^3\, f' = \log \log x$$

admits a solution in $\mathbb{T}$. In other words, the field of transseries is the first concrete example of what one might call a "real differentially closed field".

The above closure properties make the field of transseries ideal as a framework for many branches of mathematics. In a sense, it has a similar status as the field of real or complex numbers. In analysis, it has served in Écalle's proof of Dulac's conjecture — the best currently known result on Hilbert's 16-th problem. In model theory, it can be used as a natural model for many theories (reals with exponentiation, ordered differential fields, etc.). In computer algebra, it provides a sufficiently general formal framework for doing asymptotic computations. Furthermore, transseries admit a rich non-archimedean geometry and surprising connections exist with Conway's "field" of surreal numbers.

Historical perspectives

Historically speaking, transseries have their origin in several branches of mathematics, like analysis, model theory, computer algebra and non-archimedean geometry. Let us summarize some of the highlights of this interesting history.

1 Resolution of differential equations by means of power series

It was already recognized by Newton that formal power series are a powerful tool for the resolution of differential equations [New71]. For the resolution of algebraic equations, he already introduced Puiseux series and the Newton polygon method, which will play an important role in this book. During the 18-th century, formal power series were used more and more systematically as a tool for the resolution of differential equations, especially by Euler.

However, the analytic meaning of a formal power series is not always clear. On the one hand side, convergent power series give rise to germs which can usually be continued analytically into multi-valued functions on a Riemann surface. Secondly, formal power series can be divergent and it is not clear *a priori* how to attach reasonable sums to them, even though several recipes for doing this were already known at the time of Euler [Har63, Chapter 1].

With the rigorous formalization of analysis in the 19-th century, criteria for convergence of power series were studied in a more systematic way. In particular, Cauchy and Kovalevskaya developed the well-known majorant method for proving the convergence of power series solutions to certain partial differential equations [vK75]. The analytic continuation of solutions to algebraic and differential equations were also studied in detail [Pui50, BB56] and the Newton polygon method was generalized to differential equations [Fin89].

However, as remarked by Stieltjes [Sti86] and Poincaré [Poi93, Chapître 8], even though divergent power series did not fit well in the spirit of "rigorous mathematics" of that time, they remained very useful from a practical point of view. This raised the problem of developing rigorous analytic methods to attach plausible sums to divergent series. The modern theory of resummation started with Stieltjes, Borel and Hardy [Sti94, Sti95, Bor28], who insisted on the development of summation methods which are stable under the common operations of analysis. Although the topic of divergent series was an active subject of research in the early 20-th century [Har63], it went out of fashion later on.

2 Generalized asymptotic scales

Another approach to the problem of divergence is to attach only an asymptotic meaning to series expansions. The foundations of modern asymptotic calculus were laid by Dubois-Raymond, Poincaré and Hardy.

More general asymptotic scales than those of the form $x^{\mathbb{Z}}$, $x^{\mathbb{Q}}$ or $x^{\mathbb{R}}$ were introduced by Dubois-Raymond [dBR75, dBR77], who also used "Cantor's" diagonal argument in order to construct functions which cannot be expanded with respect to a given scale. Nevertheless, most asymptotic scales occurring in practice consist of so called L-functions, which are constructed from algebraic functions, using the field operations, exponentiation and logarithm. The asymptotic properties of L-functions were investigated in detail by Hardy [Har10, Har11] and form the start of the theory of Hardy fields [Bou61, Ros80, Ros83a, Ros83b, Ros87, Bos81, Bos82, Bos87].

Poincaré [Poi90] also established the equivalence between computations with formal power series and asymptotic expansions. Generalized power series with real exponents [LC93] or monomials in an abstract monomial group [Hah07] were introduced about the same time. However, except in the case of linear differential equations [Fab85, Poi86, Bir09], it seems that nobody had the idea to use such generalized power series in analysis, for instance by using a monomial group consisting of L-functions.

Newton, Borel and Hardy were all aware of the systematic aspects of their theories and they consciously tried to complete their framework so as to capture as much of analysis as possible. The great unifying theory nevertheless had to wait until the late 20-th century and Écalle's work on transseries and Dulac's conjecture [É85, É92, É93, Bra91, Bra92, CNP93].

His theory of accelero-summation filled the last remaining source of instability in Borel's theory. Similarly, the "closure" of Hardy's theory of L-functions under infinite summation removes its instability under functional inversion (see exercise 5.20) and the resolution of differential equations. In other words, the field of accelero-summable transseries seems to correspond to the "framework-with-no-escape" about which Borel and Hardy may have dreamed.

3 Model theory

Despite the importance of transseries in analysis, the first introduction of the formal field of transseries appeared in model theory [Dah84, DG86]. Its roots go back to another major challenge of 20-th century mathematics: proving the completeness and decidability of various mathematical theories.

Gödel's undecidability theorem and the undecidability of arithmetic are well-known results in this direction. More encouraging were the results on the theory of the field of real numbers by Artin-Schreier and later Tarski-Seidenberg [AS26, Tar31, Tar51, Sei54]. Indeed, this theory is complete, decidable and quantifier elimination can be carried out effectively. Tarski also raised the question how to axiomatize the theory of the real numbers with exponentiation and to determine its decidability. This motivated the model-theoretical introduction of the field of transseries as a good candidate of a non-standard model of this theory, and new remarkable properties of the real exponential function were stated.

The model theory of the field of real numbers with the exponential function has been developed a lot in the nineties. An important highlight is Wilkie's theorem [Wil96], which states that the real numbers with exponentiation form an o-minimal structure [vdD98, vdD99]. In these further developments, the field of transseries proved to be interesting for understanding the singularities of real functions which involve exponentiation.

After the encouraging results about the exponential function, it is tempting to generalize the results to more general solutions of differential equations. Several results are known for Pfaffian functions [Kho91, Spe99], but the thing we are really after is a real and/or asymptotic analogue of Ritt-Seidenberg's elimination theory for differential algebra [Rit50, Sei56, Kol73]. Again, it can be expected that a better understanding of differential fields of transseries will lead to results in that direction; see [AvdD02, AvdD01, AvdD04, AvdDvdH05, AvdDvdH] for ongoing work.

4 Computer algebra and automatic asymptotics

We personally became interested in transseries during our work on automatic asymptotics. The aim of this subject is to effectively compute asymptotic expansions for certain explicit functions (such as "exp-log" function) or solutions to algebraic, differential, or more general equations.

In early work on the subject [GG88, Sha90, GG92, Sal91, Gru96, Sha04], considerable effort was needed in order to establish an appropriate framework and to prove the asymptotic relevance of results. Using formal transseries as the privileged framework leads to considerable simplifications: henceforth, with Écalle's accelero-summation theory in the background, one can concentrate on the computationally relevant aspects of the problem. Moreover, the consideration of transfinite expansions allows for the development of a formally exact calculus. This is not possible when asymptotic expansions are restricted to have at most ω terms and difficult in the framework of nested expansions [Sha04].

However, while developing algorithms for the computation of asymptotic expansions, it turned out that the mathematical theory of transseries still had to be further developed. Our results in this direction were finally regrouped in part A of our thesis, which has served as a basis for this book. Even though this book targets a wider public than the computer algebra community, its effective origins remain present at several places: Cartesian representations, the incomplete transbasis theorem, the Newton polygon method, etc.

5 Non-archimedean geometry

Last but not least, the theory of transseries has a strong geometric appeal. Since the field of transseries is a model for the theory of real numbers with exponentiation, it is natural to regard it as a non-standard version of the real line. However, contrary to the real numbers, the transseries also come with a non-trivial derivation and composition. Therefore, it is an interesting challenge to study the geometric properties of differential polynomials, or more general "functions" constructed using the derivation and composition. The differential intermediate value theorem can be thought of as one of the first results in this direction.

An even deeper subject for further study is the analogy with Conway's construction of the "field" of surreal numbers [Con76]. Whereas the surreal numbers come with the important notion of "earliness", transseries can be differentiated and composed. We expect that it is actually possible to construct isomorphisms between the class of surreal numbers and the class of generalized transseries of the reals with so called transfinite iterators of the exponential function and nested transseries. A start of this project has been carried out in collaboration with my former student M. Schmeling [Sch01]. If this project could be completed, this would lead to a remarkable correspondence between growth-rate functions and numbers.

Outline of the contents

Orderings occur in at least two ways in the theory of transseries. On the one hand, the terms in the expansion of a transseries are naturally ordered by their asymptotic magnitude. On the other hand, we have a natural ordering on the field $\mathbb{T}$ of transseries, which extends the ordering on $\mathbb{R}$. In chapter 1, we recall some basic facts about well-quasi-orderings and ordered fields. We also introduce the concept of "asymptotic dominance relations" $\preccurlyeq$, which can be considered as generalizations of valuations. In analysis, $f \preccurlyeq g$ and $f \prec g$ are alternative notations for $f = O(g)$ and $f = o(g)$.

In chapter 2, we introduce the "strong C-algebra of grid-based series" $C[[\mathfrak{M}]]$, where $\mathfrak{M}$ is a so called monomial monoid with a partial quasi-ordering $\preccurlyeq$. Polynomials, ordinary power series, Laurent series, Puiseux series and multivariate power series are all special types of grid-based series. In general, grid-based series carry a transfinite number of terms (even though the order is always bounded by ω^{ω}) and we study the asymptotic properties of $C[[\mathfrak{M}]]$.

We also lay the foundations for linear algebra with an infinitary summation operator, called "strong linear algebra". Grid-based algebras of the form $C[[\mathfrak{M}]]$, Banach algebras and completions with respect to a valuation are all examples of strong algebras, but we notice that not all strong "serial" algebras are of a topological nature. One important technique in the area of strong linear algebra is to make the infinite sums as large as possible while preserving summability. Different regroupings of terms in such "large sums" can then be used in order to prove identities, using the axiom of "strong associativity". The terms in "large sums" are often indexed by partially ordered grid-based sets. For this reason, it is convenient to develop the theory of grid-based series in the partially ordered setting, even though the ordering $\preccurlyeq$ on transmonomials will be total.

The Newton polygon method is a classical technique for the resolution of algebraic equations with power series coefficients. In chapter 3, we will give a presentation of this method in the grid-based setting. Our exposition is based on the systematic consideration of "asymptotic equations", which are equations with asymptotic side-conditions. This has the advantage that we may associate invariants to the equation like the Newton degree, which simplifies the method from a technical point of view. We also systematically consider derivatives of the equation, so as to quickly separate almost multiple roots.

Chapter 3 also contains a digression on Cartesian representations, which are both useful from a computational point of view and for the definition of convergence. However, they will rarely be used in the sequel, so this part may be skipped at a first reading.

In chapter 4, we construct the field $\mathbb{T} = C[[[x]]]$ of grid-based transseries in x over an "ordered exp-log field" of constants C. Axioms for such constant fields and elementary properties are given in section 4.1. In practice, one usually takes $C = \mathbb{R}$. In computer algebra, one often takes the countable subfield of all "real elementary constants" [Ric97]. It will be shown that $\mathbb{T}$ is again an ordered exp-log field, so it is also possible to take $C = \mathbb{T}$ and construct fields like $\mathbb{R}[[[x]]][[[y]]]$. Notice that our formalism allows for partially defined exponential functions. This is both useful during the construction of $\mathbb{T}$ and for generalizations to the multivariate case.

The construction of $\mathbb{T}$ proceeds by the successive closure of $C[[x^{\mathbb{R}}]]$ under logarithm and exponentiation. Alternatively, one may first close under exponentiation and next under logarithm, following Dahn and Göring or Écalle [DG86, É92]. However, from a model-theoretical point of view, it is more convenient to first close under logarithm, so as to facilitate generalizations of the construction [Sch01]. A consequence of the finiteness properties which underlie grid-based transseries is that they can always be expanded with respect to finite "transbases". Such representations, which will be studied in section 4.4, are very useful from a computational point of view.

In chapter 5, we will define the operations ∂, $\int$, $\circ$ and $\cdot^{\mathrm{inv}}$ on $\mathbb{T}$ and prove that they satisfy the usual rules from calculus. In addition, they satisfy several compatibility properties with the ordering, the asymptotic relations and infinite summation, which are interesting from a model-theoretical point of view. In section 5.4.2, we also prove the Translagrange theorem due to Écalle, which generalizes Lagrange's well-known inversion formula for power series.

Before going on with the study of differential equations, it is convenient to extend the theory from chapter 2 and temporarily return to the general setting of grid-based series. In chapter 6, we develop a "functional analysis" for grid-based series, based on the concept of "grid-based operators". Strongly multilinear operators are special cases of grid-based operators. In particular, multiplication, differentiation and integration of transseries are grid-based operators. General grid-based operators are of the form

$$\Phi(f) = \Phi_0 + \Phi_1(f) + \Phi_2(f, f) + \cdots,$$

where each Φ_i is a strongly i-linear operator. The set $\mathscr{G}(C[[\mathfrak{M}]], C[[\mathfrak{N}]])$ of grid-based operators from $C[[\mathfrak{M}]]$ into $C[[\mathfrak{N}]]$ forms a strong C-vector space, which admits a natural basis of so called "atomic operators". At the end of chapter 6, we prove several implicit function theorems, which will be useful for the resolution of differential equations.

In chapter 7, we study linear differential equations with transseries coefficients. A well-known theorem [Fab85] states that any linear differential equation over $\mathbb{C}[[z]]$ admits a basis of formal solutions of the form

$$(f_0(\sqrt[p]{z}) + \cdots + f_d(\sqrt[p]{z}) \log^d z)\, z^{\alpha}\, \mathrm{e}^{P(1/\sqrt[p]{z})},$$

with $f_0, ..., f_d \in \mathbb{C}[[z]]$, $\alpha \in \mathbb{C}$, $P \in \mathbb{C}[X]$ and $p, d \in \mathbb{N}^>$. We will present a natural generalization of this theorem to the transseries case. Our method is based on a deformation of the algebraic Newton polygon method from chapter 3.

Since the only transseries solution to $f'' + f = 0$ is 0, the solution space of an equation of order r does not necessarily have dimension r. Nevertheless, as will be shown in section 7.7, one does obtain a solution space of dimension r by considering an oscillatory extension of the field of transseries. A remarkable consequence is that linear differential operators can be factored into first order operators in this extension. It will also be shown that operators in $\mathbb{T}[\partial]$ can be factored into first and second order operators.

It should also be noticed that the theory from chapter 7 is compatible with the strong summation and asymptotic relations on $\mathbb{T}$. First of all, the trace T_L of a linear differential operator $L \in \mathbb{T}[\partial]$, which describes the dominant asymptotic behaviour of L, satisfies several remarkable properties (see section 7.3.3). Secondly, any operator $L \in \mathbb{T}[\partial]$ admits a so called distinguished strong right-inverse L^{-1}, with the property that $(L^{-1}\, g)_{\mathfrak{h}} = 0$ when $\mathfrak{h}$ is the dominant monomial of a solution to $Lh = 0$. Similarly, we will construct distinguished bases of solutions and distinguished factorizations.

Non-linear differential equations are studied in chapter 8. For simplicity, we restrict our attention to asymptotic algebraic differential equations like

$$P(f) = 0 \qquad (f \prec \mathfrak{v}),$$

with $P \in \mathbb{T}\{F\} = \mathbb{T}[F, F', ...]$, but similar techniques apply in more general cases. The generalization of the Newton polygon method to the differential setting contains two major difficulties. First, the "slopes" which lead to the first terms of solutions cannot directly be read off from the Newton polygon. Moreover, such slopes may be due to cancellations of terms of different degrees (like in the usual case) or terms of the same degree. Secondly, it is much harder to "unravel" almost multiple solutions.

In order to circumvent the first problem, we first define the differential Newton polynomial $N_P \in C\{F\}$ associated to the "horizontal slope" (it actually turns out that N_P is always of the form $N_P = Q\,(F')^\nu$ with $Q \in C[F]$). Then the slope which corresponds to solutions of the form $f = c\,\mathfrak{m} + \cdots$ is "admissible" if and only if $N_{P_{\times\mathfrak{m}}}$ admits a non-zero root in C. Here $P_{\times\mathfrak{m}}$ is the unique differential polynomial with $P_{\times\mathfrak{m}}(f) = P(\mathfrak{m}\, f)$ for all f. In section 8.4, we next give a procedure for determining the admissible slopes. The second problem is more pathological, because one has to ensure the absence of iterated logarithms $\log_l = \log \circ \overset{l\times}{\cdots} \circ \log$ with arbitrarily high l in the expansions of solutions. This problem is treated in detail in section 8.6.

The suitably adapted Newton polygon methods allows us to prove several structure theorems about the occurrence of exponentials and logarithms into solutions of algebraic differential equation. We also give a theoretical algorithm for the determination of all solutions.

The last chapter of this book is devoted to the proof the intermediate value theorem for differential polynomials $P \in \mathbb{T}\{F\}$. This theorem ensures the existence of a solution to $P(f) = 0$ on an interval $I = [g, h]$ under the simple hypothesis that P admits a sign-change on I. The main part of the chapter contains a detailed study of the non-archimedean geometry of $\mathbb{T}$. This comprises a classification of its "cuts" and a description of the behaviour of differential polynomials in cuts. In the last section, this theory is combined with the results of chapter 8, and the interval on which a sign-change occurs is shrunk further and further until we hit a root of P.

Notations

A few remarks about the notations used in this book will be appropriate. Notice that a glossary can be found at the end.

1. Given a mapping $f: A_1 \times \cdots \times A_n \to B$ and $S_1 \subseteq A_1, \ldots, S_n \subseteq A_n$, we write
$$f(S_1, \ldots, S_n) = \{f(a_1, \ldots, a_n): a_1 \in S_1, \ldots, a_n \in S_n\}.$$
Similarly, given a set S, we will write $S > 0$ or $S \prec 1$ if $a > 0$ resp. $a \prec 1$ for all $a \in S$. These and other classical notations for sets are extended to families in section 2.4.1.
2. We systematically use the double index convention $(f_i)_j = f_{i,j}$. Given a set $\mathfrak{S}$ of monomials, we also denote $f_{\mathfrak{S}} = \sum_{\mathfrak{m} \in \mathfrak{S}} f_{\mathfrak{m}} \mathfrak{m}$ (this is an exception to the above notation).
3. Given a set S, we will denote by $S^{>}$ its subset of strictly positive elements, $S^{\preccurlyeq}$ its subset of bounded elements, $S^{<,\prec}$ of negative infinitesimal elements, etc. If $S \subseteq C[[\mathfrak{M}]]$ is a set of series, then we also denote $S_{\succ} = \{f_{\succ}: f \in S\}$, where $f_{\succ} = f_{\mathfrak{M}^{\succ}}$, and similarly for $S_{\succcurlyeq}$, $S_{\prec}$, etc. Notice that this is really a special case of notations 1 and 2.
4. Intervals are denoted by (f, g), $(f, g]$, $[f, g)$ or $[f, g]$ depending on whether the left and right sides are open or closed.
5. We systematically denote monomials $\mathfrak{m}, \mathfrak{n}, \ldots$ in the fraktur font and families $\mathcal{F}, \mathcal{G}, \ldots$ using calligraphic characters.

Those readers who are familiar with my thesis should be aware of the following notational changes which occurred during the past years:

Former	$\underline{\prec\!\!\prec}$	$\prec\!\!\prec$	$\asymp$	$\sim$	$\underline{\prec\!\!\prec\!\!\prec}$	$\prec\!\!\prec\!\!\prec$	$\stackrel{\asymp}{\asymp}$		$f^{\uparrow}$	f^{c}	$f^{\downarrow}$
New	$\preccurlyeq$	$\prec$	$\asymp$	$\sim$	$\underline{\prec\!\!\prec}$	$\prec\!\!\prec$	$\stackrel{\asymp}{\asymp}$	$\approxeq$	$f_{\succ}$	$f_{\asymp}$	$f_{\prec}$

There are also a few changes in terminology:

Former	New
normal basis	transbasis
purely exponential transseries	exponential transseries
potential dominant —	starting —
privileged refinement	≈ unravelling

1

Orderings

In this chapter, we will introduce some order-theoretical concepts, which prepare the study of generalized power series in the next chapter. Orderings occur in at least two important ways in this study.

First, the terms of a series are naturally ordered according to their asymptotic magnitudes. For instance, the support of $1 + z + z^2 + \cdots \in \mathbb{R}[[z]]$, considered as an ordered set, is isomorphic to $\mathbb{N}$. More interesting examples are

$$1+z_1+z_1^2+\cdots+z_2+z_1 z_2+z_1^2 z_2+\cdots+z_2^2+z_1 z_2^2+\cdots \in \mathbb{R}[[z_1]][[z_2]]$$

and

$$\begin{array}{ccccccccc} 1 & + & z_1 & + & z_1^2 & + & \cdots & + & \\ z_2 & + & z_1 z_2 & + & z_1^2 z_2 & + & \cdots & + & \\ z_2^2 & + & z_1 z_2^2 & + & z_1^2 z_2^2 & + & \cdots & + & \\ \vdots & + & \vdots & + & \vdots & + & \cdots & & \end{array} \in \mathbb{R}[[z_1, z_2]],$$

whose supports are isomorphic to $\mathbb{N} \dot{\times} \mathbb{N}$ and $\mathbb{N} \times \mathbb{N}$ respectively. Here $\mathbb{N} \dot{\times} \mathbb{N}$ denotes the set $\mathbb{N}^2$ with the total anti-lexicographical ordering

$$(m,n) \leqslant (m',n') \Leftrightarrow ((n<n') \vee (m \leqslant m' \wedge n=n'))$$

and $\mathbb{N} \times \mathbb{N}$ denotes the set $\mathbb{N}^2$ with the partial product ordering

$$(m,n) \leqslant (m',n') \Leftrightarrow (m \leqslant m' \wedge n \leqslant n').$$

In general, when the support is totally ordered, it is natural to require the support to be well-ordered. If we want to be able to multiply series, this condition is also necessary, as shown by the example

$$(1+z+z^2+\cdots)(1+z^{-1}+z^{-2}+\cdots).$$

For convenience, we recall some classical results about well-ordered sets and ordinal numbers in section 1.2. In what follows, our treatment will be based on well-quasi-orderings, which are the analogue of well-orderings in the context of partial quasi-orderings. In sections 1.3 and 1.4, we will prove some classical results about well-quasi-orderings.

A second important occurrence of orderings is when we consider an algebra of generalized power series as an ordered structure. For instance $\mathbb{R}[[z]]$ is naturally ordered by declaring a non-zero series $f_n z^n + f_{n+1} z^{n+1} + \cdots$ with $f_n \neq 0$ to be positive if and only if $f_n > 0$. This gives $\mathbb{R}[[z]]$ the structure of a so called totally ordered $\mathbb{R}$-algebra.

In section 1.5, we recall the definitions of several types of ordered algebraic structures. In section 1.6, we will then show how a certain number of typical asymptotic relations, like $\prec$, $\preccurlyeq$, $\asymp$ and $\sim$, can be introduced in a purely algebraic way. In section 1.8, we define groups and fields with generalized exponentiations, and the asymptotic relations $\prec\!\!\prec$, $\preccurlyeq\!\!\preccurlyeq$, $\asymp\!\!\asymp$ and $\approxeq$. Roughly speaking, for infinitely large f and g, we have $f \prec\!\!\prec g$, if $f^\lambda \prec g$ for all λ. For instance, $x \prec\!\!\prec \mathrm{e}^x$, but $x \asymp\!\!\asymp x^{1000}$, for $x \to \infty$.

1.1 Quasi-orderings

Let E be a set. In all what follows, a *quasi-ordering* on E is reflexive and transitive relation $\leqslant$ on E; in other words, for all $x, y, z \in E$ we have

O1. $x \leqslant x$;

O2. $x \leqslant y \wedge y \leqslant z \Rightarrow x \leqslant z$.

An *ordering* is a quasi-ordering which is also antisymmetric:

O3. $x \leqslant y \wedge y \leqslant x \Rightarrow x = y$.

We sometimes write $\leqslant_E$ instead of $\leqslant$ in order to avoid confusion. A mapping $\varphi: E \to F$ between two quasi-ordered sets is said to be *increasing* (or a morphism of quasi-ordered sets), if $x \leqslant y \Rightarrow \varphi(x) \leqslant \varphi(y)$, for all $x, y \in E$.

Given a quasi-ordering E, we say that x, $y \in E$ are *comparable* if $x \leqslant y$ or $y \leqslant x$. If every two elements in E are comparable, then the quasi-ordering is said to be *total*. Two elements x, $y \in E$ are said to be *equivalent*, and we write $x \equiv y$, if $x \leqslant y$ and $y \leqslant x$. If $x \leqslant y$ and $y \not\equiv x$, then we write $x < y$ (see also exercise 1.1(a) below). The quasi-ordering on E induces a natural ordering on the quotient set $E/\equiv$ by $X \leqslant Y \Leftrightarrow (\forall x \in X, \forall y \in Y, x \leqslant y)$ and the corresponding projection $E \to E/\equiv$ is increasing. In other words, we do not really gain in generality by considering quasi-orderings instead of orderings, but it is sometimes more convenient to deal with quasi-orderings.

Some simple examples of totally ordered sets are $\varnothing, \{0\}, \{0,1\}, \ldots$ and $\mathbb{N}$. Any set E can be trivially quasi-ordered both by the finest ordering, for which $x \leqslant y \Leftrightarrow x = y$, and by the roughest quasi-ordering, for which all $x, y \in E$ satisfy $x \leqslant y$. In general, a quasi-ordering $\leqslant$ on E is said to be *finer* than a second quasi-ordering $\leqslant'$ on E if $x \leqslant y \Rightarrow x \leqslant' y$ for all $x, y \in E$. Given quasi-ordered sets E and F, we can construct other quasi-ordered sets as follows:

1. The disjoint union $E \amalg F$ is naturally quasi-ordered, by taking the quasi-orderings on E and F on each summand, and by taking E and F mutually incomparable. In other words,

$$x \leqslant_{E \amalg F} y \Leftrightarrow (x \in E \wedge y \in E \wedge x \leqslant_E y) \vee (x \in F \wedge y \in F \wedge x \leqslant_F y).$$

2. Alternatively, we can quasi-order $E \amalg F$, by postulating any element in E to be strictly smaller than any element in F. This quasi-ordered set is called the ordered union of E and F, and we denote it by $E \dot{\amalg} F$. In other words,

$$x \leqslant_{E \dot{\amalg} F} y \Leftrightarrow x \leqslant_{E \amalg F} y \vee (x \in E \wedge y \in F).$$

3. The Cartesian product $E \times F$ is naturally quasi-ordered by

$$(x, y) \leqslant_{E \times F} (x', y') \Leftrightarrow x \leqslant x' \wedge y \leqslant y'.$$

4. Alternatively, we can quasi-order $E \times F$ *anti-lexicographically* by

$$(x, y) \leqslant_{E \dot{\times} F} (x', y') \Leftrightarrow (x, y) \leqslant_{E \times F} (x', y') \vee y < y'.$$

We write $E \dot{\times} F$ for the corresponding quasi-ordered set.

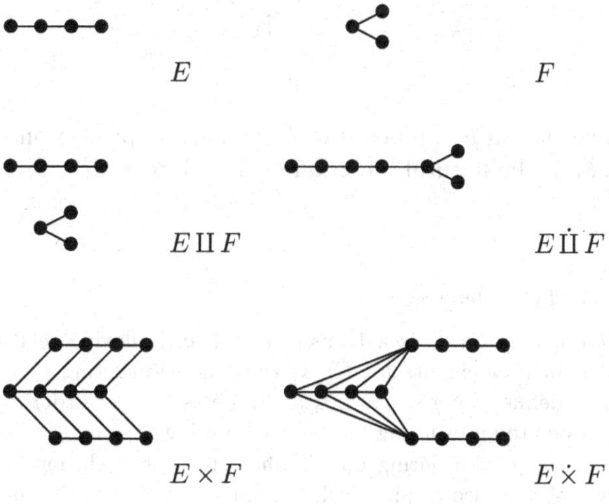

Fig. 1.1. Examples of some basic constructions on ordered sets.

5. Let E^* be the set of *words* over E. Such words are denoted by sequences $x_1 \cdots x_n$ (with $x_1, ..., x_n \in E$) or $[x_1, ..., x_n]$ if confusion may arise. The empty word is denoted by ε and we define $E^+ = E^* \setminus \{\varepsilon\}$. The *embeddability quasi-ordering* on E^* is defined by $x_1 \cdots x_n \leqslant y_1 \cdots y_m$, if and only if there exists a strictly increasing mapping $\varphi: \{1, ..., n\} \to \{1, ..., m\}$, such that $x_i \leqslant y_{\varphi(i)}$ for all i. For instance,

$$\begin{aligned} [2, 31, 15, 7] &\leqslant_{\mathbb{N}^*} [2, 8, 35, 17, 3, 7, 1]; \\ [2, 31, 15, 7] &\nleqslant_{\mathbb{N}^*} [2, 8, 35, 17, 3, 2, 1]. \end{aligned}$$

6. An equivalence relation $\sim$ on E is said to be *compatible* with the quasi-ordering if

$$x \leqslant y \wedge x \sim x' \wedge y \sim y' \Rightarrow x' \leqslant y'$$

for all $x, y, x', y' \in E$. In that case, $E/\sim$ is naturally quasi-ordered by

$$X \leqslant_{E/\sim} Y \Leftrightarrow (\forall x \in X, \forall y \in Y, x \leqslant_E y),$$

and the canonical projection $\pi\colon E \to E/\sim$ is increasing.

If E and F are ordered sets, then it can be verified that the quasi-orderings defined in 1–6 above are actually orderings.

Let $\varphi\colon E \to F$ be an increasing mapping between quasi-ordered sets $(E, \leqslant)$ and $(F, \leqslant)$. Consider the quasi-ordering $\preccurlyeq$ on E defined by

$$x \preccurlyeq y \Leftrightarrow \varphi(x) \leqslant \varphi(y).$$

Then $\leqslant$ is finer than $\preccurlyeq$ and the mapping φ admits a natural factorization

$$\begin{array}{ccc} (E, \leqslant) & \xrightarrow{\varphi} & (F, \leqslant) \\ \downarrow \pi & & \uparrow \iota \\ (E, \preccurlyeq)/\equiv_{\preccurlyeq} & \xrightarrow{\bar{\varphi}} & (\mathrm{Im}\,\varphi, \leqslant) \end{array} . \tag{1.1}$$

Here π is the identity on E composed with the natural projection from $(E, \preccurlyeq)$ on $(E, \preccurlyeq)/\equiv_{\preccurlyeq}$, ι is the natural inclusion of $\mathrm{Im}\,\varphi$ into F and $\bar{\varphi}$ is an isomorphism.

Exercise 1.1. Let E be a set.

a) A *strict ordering* on E is a transitive and antireflexive relation $<$ on E (i.e. $x < x$ for no elements $x \in E$). Given a quasi-ordering $\leqslant$ show that the relation $<$ defined by $x < y \Leftrightarrow x \leqslant y \wedge y \not\equiv x$ is a strict ordering. Show also how to associate an ordering to a strict ordering.
b) Let $\leqslant$ be a quasi-ordering on E. Show that the relation $\geqslant$ defined by $x \geqslant y \Leftrightarrow y \leqslant x$ is also a quasi-ordering on E; we call it the *opposite* quasi-ordering of $\leqslant$.
c) Let $\leqslant$ be a quasi-ordering on E. Show that $x \leqslant^! y \Leftrightarrow x = y \vee x < y$ defines an ordering on E. Show that $\leqslant^!$ is the roughest ordering which is finer than $\leqslant$.

Exercise 1.2. Two quasi-ordered sets E and F are said to be *isomorphic*, and we write $E \cong F$, if there is an increasing bijection between E and F, whose inverse is also increasing. Prove the following:

a) $\amalg$ and $\times$ are commutative modulo $\cong$ (i.e. $E \amalg F \cong F \amalg E$), but not $\dot{\amalg}$ and $\dot{\times}$.
b) $\amalg, \times, \dot{\amalg}$ and $\dot{\times}$ are associative modulo $\cong$.
c) $\amalg$ is distributive w.r.t. $\times$ modulo $\cong$.
d) $\dot{\amalg}$ is right (but not left) distributive w.r.t. $\dot{\times}$ modulo $\cong$ (in other words $E \dot{\times} (F \dot{\amalg} G) \cong (E \dot{\times} F) \dot{\amalg} (E \dot{\times} G)$).

Exercise 1.3. Let E be a quasi-ordered set. We define an equivalence relation on E^*, by taking two words to be equivalent if they are obtained one from another by a permutation of letters. We call $E^\diamond = E^*/\sim$ the set of *commutative words* over E. Show that:

a) We define a quasi-ordering $\preccurlyeq$ on E by $u \preccurlyeq v \Leftrightarrow \exists w \in E, u \leqslant w \wedge v \sim w$.
b) For all $x_1 \cdots x_m, y_1 \cdots y_m \in E^*$, we have $x_1 \cdots x_m \preccurlyeq y_1 \cdots y_n$ if and only if there exists an injection $\varphi: \{1, ..., m\} \to \{1, ..., n\}$ with $x_i \leqslant y_{\varphi(i)}$ for all i.
c) The equivalence relation $\sim$ is compatible with $\preccurlyeq$, so that we may order $E^\diamond$ by the quotient quasi-ordering induced by $\preccurlyeq$.
d) The quasi-ordering $\leqslant$ is finer than $\preccurlyeq$ and we have a natural increasing surjection $E^* \to E^\diamond$.
e) For all ordered sets E, F, prove that $(E \amalg F)^\diamond \cong E^\diamond \times F^\diamond$.
f) For all ordered sets E, F prove that there exists an increasing bijection $(E \dot{\amalg} F)^\diamond \to E^\diamond \dot{\times} F^\diamond$, whose inverse is not increasing, in general.

Exercise 1.4. Let E and F be ordered sets and denote by $\mathscr{F}(E, F)$ the set of mappings from E into F. For $\varphi, \psi \in \mathscr{F}(E, F)$, we define

$$\varphi \leqslant \psi \iff \forall x \in E, \varphi(x) \nleqslant \psi(x) \Rightarrow$$
$$(\exists y > x, \varphi(y) < \psi(y) \wedge (\forall z \geqslant y, \varphi(z) \leqslant \psi(z))).$$

Prove that $\leqslant$ defines an ordering on $\mathscr{F}(E, F)$. Also prove the following properties:

a) If $A = \{0\} \amalg \{0\}$, then $\mathscr{F}(A, B) \cong B \times B$.
b) If $A = \{0, 1\}$, then $\mathscr{F}(A, B) \cong B \dot{\times} B$.
c) $\mathscr{F}(E \amalg F, G) \cong \mathscr{F}(E, G) \times \mathscr{F}(F, G)$.
d) $\mathscr{F}(E \dot{\amalg} F, G) \cong \mathscr{F}(E, G) \dot{\times} \mathscr{F}(F, G)$.

Exercise 1.5. Show that the category of quasi-ordered sets admits direct sums and products, pull-backs, push-outs, direct and inverse limits and free objects (i.e. the forgetful functor to the category of sets admits a right adjoint).

1.2 Ordinal numbers

Let E be a quasi-ordered set. The quasi-ordering on E is said to be *well-founded*, if there is no infinite strictly decreasing sequence in E. A total well-founded ordering is called a *well-ordering*. A total ordering is a well-ordering if and only if each of its non-empty subsets has a least element. The following classical theorems are implied by the axiom of choice [Bou70, Mal79]:

Theorem 1.1. *Every set can be well-ordered.* □

Theorem 1.2. (Zorn's lemma) *Let E be a non-empty ordered set, such that each non-empty totally ordered subset of E has an upper bound. Then E admits a maximal element.* □

An *ordinal number* or *ordinal* is a set α, such that the relation $\in$ forms a strict well-ordering on α. In particular, the natural numbers can "be defined to be" ordinal numbers: $0 = \varnothing$, $1 = \{0\}$, $2 = 1 \cup \{1\}$, $3 = 2 \cup \{2\}$, ... The set $\omega = \{0, 1, 2, ...\}$ of natural numbers is also an ordinal. More generally, if α is an ordinal, then so is $\alpha \cup \{\alpha\}$. For all ordinals α, its elements are also ordinals.

Fig. 1.2. Some examples of ordinal numbers.

It is classical [Mal79] that the class of all ordinal numbers has all the properties of an ordinal number: if α, β and γ are ordinal numbers, then $\alpha \notin \alpha, \alpha \in \beta \Rightarrow \beta \notin \alpha, \alpha \in \beta \wedge \beta \in \gamma \Rightarrow \alpha \in \gamma, \alpha \in \beta \vee \beta \in \alpha \vee \alpha = \beta$ and each non-empty set of ordinals admits a least element for $\in$. The following classification theorem is also classical [Mal79]:

Theorem 1.3. *Each well-ordered set is isomorphic to a unique ordinal.* □

The usual induction process for natural numbers admits an analogue for ordinal numbers. For this purpose, we distinguish between successor ordinals and limit ordinals: an ordinal α is called a *successor ordinal* if $\alpha = \beta \cup \{\beta\}$ for some ordinal β (and we write $\alpha = \beta + 1$) and a *limit ordinal* if not (in which case $\alpha = \bigcup_{\beta \in \alpha} \beta$). For example, the inductive definitions for addition, multiplication and exponentiation can now be extended to ordinal numbers as follows:

Table 1.1. Basic arithmetic on ordinal numbers.

	0	Successor ordinals $\beta+1$	Limit ordinals $\lambda > 0$
$+$	$\alpha + 0 = 0$	$\alpha + (\beta + 1) = (\alpha + \beta) + 1$	$\alpha + \lambda = \bigcup_{\beta \in \lambda} \alpha + \beta$
$\times$	$\alpha \cdot 0 = 0$	$\alpha \cdot (\beta + 1) = (\alpha \cdot \beta) + \alpha$	$\alpha \cdot \lambda = \bigcup_{\beta \in \lambda} \alpha \cdot \beta$
^	$\alpha^0 = 1$	$\alpha^{\beta+1} = \alpha^\beta \cdot \alpha$	$\alpha^\lambda = \bigcup_{\beta \in \lambda} \alpha^\beta$

Similarly, one has the *transfinite induction* principle: assume that a property P for ordinals satisfies $P(\alpha) \Rightarrow P(\alpha + 1)$ for all α and $(\forall \alpha \in \lambda, P(\alpha)) \Rightarrow P(\lambda)$ for all limit ordinals λ. Then $P(\alpha)$ holds for all ordinals α.

The following theorem classifies all countable ordinals smaller than ω^ω, and is due to Cantor [Can99]:

Theorem 1.4. *Let $\alpha < \omega^\omega$ be a countable ordinal. Then there exists a unique sequence of natural numbers $n_d, ..., n_0$ (with $n_d > 0$ if $d > 0$), such that*

$$\alpha = \omega^d \cdot n_d + \cdots + \omega \cdot n_1 + n_0.$$ □

Exercise 1.6. Prove the transfinite induction principle.

Exercise 1.7. For any two ordinals α, β, show that

a) $\alpha + \beta \cong \alpha \dot{\amalg} \beta$;
b) $\alpha \cdot \beta \cong \alpha \dot{\times} \beta$.

In particular, $+$ and $\cdot$ are associative and $+$ is right distributive w.r.t. $\cdot$, by exercise 1.2.

Exercise 1.8. For all ordinals α, β and γ, prove that

a) $(\alpha^\beta)^\gamma = \alpha^{\beta \cdot \gamma}$;
b) $\alpha^{\beta+\gamma} = \alpha^\beta \cdot \alpha^\gamma$.

Do we also have $(\alpha \cdot \beta)^\gamma = \alpha^\gamma \cdot \beta^\gamma$?

1.3 Well-quasi-orderings

Let E be a quasi-ordered set. A *chain* in E is a subset of E which is totally ordered for the induced quasi-ordering. An *anti-chain* is a subset of E of pairwise incomparable elements. A *well-quasi-ordering* is a well-founded quasi-ordering without infinite anti-chains.

A *final segment* is a subset F of E, such that $x \in F \wedge x \leqslant y \Rightarrow y \in F$, for all $x, y \in E$. Given an arbitrary subset A of E, we denote by

$$\operatorname{fin}(A) = \{y \in E : \exists x \in A, x \leqslant y\}$$

the final segment *generated* by A. Dually, an *initial segment* is a subset I of E, such that $y \in I \wedge x \leqslant y \Rightarrow x \in I$, for all $x, y \in E$. We denote by

$$\operatorname{in}(A) = \{y \in E : \exists x \in A, y \leqslant x\}$$

the initial segment generated by A.

Proposition 1.5. *Let E be a quasi-ordered set. Then the following are equivalent:*

a) E is well-quasi-ordered.
b) Any final segment of E is finitely generated.
c) The ascending chain condition w.r.t. inclusion holds for final segments of E.
d) Each sequence $x_1, x_2, \ldots \in E$ admits an increasing subsequence.
e) Any extension of the quasi-ordering on E to a total quasi-ordering on E yields a well-founded quasi-ordering.

Proof. Assume (a) and let F be a final segment of E and $G \subseteq F$ the subset of minimal elements of F. Then G is an anti-chain, whence finite. We claim that G generates F. Indeed, in the contrary case, let $x_1 \in F \setminus \operatorname{fin}(G)$. Since x_1 is not minimal in F, there exists an $x_2 \in F \setminus \operatorname{fin}(G)$ with $x_1 > x_2$. Repeating this argument, we obtain an infinite decreasing sequence $x_1 > x_2 > \cdots$. This proves (b). Conversely, if x_1, x_2, ... is an infinite anti-chain or an infinite strictly decreasing sequence, then the final segment generated by $\{x_1, x_2, \ldots\}$ is not finitely generated. This proves $(a) \Leftrightarrow (b)$.

Now let $F_1 \subseteq F_2 \subseteq \cdots$ be an ascending chain of final segments. If the final segment $F = \bigcup_n F_n$ is finitely generated, say by G, then we must have $G \subseteq F_n$, for some n. This shows that $(b) \Rightarrow (c)$. Conversely, let G be the set of minimal elements of a final segment F. If $x_1, x_2, \ldots$ are pairwise distinct elements of G, then $\operatorname{fin}(x_1) \subsetneq \operatorname{fin}(x_1, x_2) \subsetneq \cdots$ forms an infinite strictly ascending chain of final segments.

Now consider a sequence $x_1, x_2, \ldots$ of elements in E, and assume that $\leqslant$ is a well-quasi-ordering. We extract an increasing sequence $x_{i_1}, x_{i_2}, \ldots$ from it by the following procedure: Let F_n be the final segment generated by the x_k, with $k > i_n$ and $x_k \geqslant x_{i_n}$ ($F_0 = E$ by convention) and assume by induction that the sequence $x_1, x_2, \ldots$ contains infinitely many terms in F_n. Since F_n is finitely generated by (b), we can select a generator $x_{i_{n+1}}$, with $i_{n+1} > i_n$ and such that the sequence $x_1, x_2, \ldots$ contains infinitely many terms in F_{n+1}. This implies (d). On the other hand, it is clear that it is not possible to extract an increasing sequence from an infinite strictly decreasing sequence or from a sequence of pairwise incomparable elements.

Let us finally prove $(a) \Leftrightarrow (e)$. An ordering containing an infinite anti-chain or an infinite strictly decreasing sequence can always be extended to a total quasi-ordering which contains a copy of $-\mathbb{N}$, by a straightforward application of Zorn's lemma. Inversely, any extension of a well-quasi-ordering is a well-quasi-ordering. □

The most elementary examples of well-quasi-orderings are well-orderings and quasi-orderings on finite sets. Other well-quasi-orderings can be constructed as follows.

Proposition 1.6. *Assume that E and F are well-quasi-ordered sets. Then*

a) Any subset of E with the induced ordering is well-quasi-ordered.
b) Let $\varphi \colon E \to F$ be a morphism of ordered sets. Then $\operatorname{Im} \varphi$ *is well-quasi-ordered.*
c) Any ordering on E which extends $\leqslant_E$ is a well-quasi-ordering.
d) $E/\sim$ is well-quasi-ordered, for any compatible equivalence relation $\sim$ on E.
e) $E \amalg F$ and $E \dot{\amalg} F$ are well-quasi-ordered.
f) $E \times F$ and $E \dot{\times} F$ are well-quasi-ordered.

Proof. Properties (a), (b), (e) and (f) follow from proposition 1.5(d). The properties (c) and (d) are special cases of (b). □

Corollary 1.7. (Dickson's lemma) *For each $n \in \mathbb{N}$, the set $\mathbb{N}^n$ with the partial, componentwise ordering is a well-quasi-ordering.* □

Theorem 1.8. (Higman) *Let E is be a well-quasi-ordered set. Then E^* is a well-quasi-ordered set.*

Proof. Our proof is due to Nash-Williams [NW63]. If $\leqslant$ denotes any ordering, then we say that $(x_1, x_2, \ldots)$ is a *bad sequence*, if there do not exist $i < j$ with $x_i \leqslant x_j$. A quasi-ordering is a well-quasi-ordering, if and only if there are no bad sequences.

Now assume for contradiction that $s = (w_1, w_2, ...)$ is a bad sequence for $\leqslant_{E^*}$. Without loss of generality, we may assume that each w_i was chosen such that the length (as a word) of w_i were minimal, under the condition that

$$w_i \in E^* \setminus \mathrm{fin}(w_1, ..., w_{i-1}).$$

We say that $(w_1, w_2, ...)$ is a *minimal bad sequence*.

Now for all i, we must have $w_i \neq \varepsilon$, so we can factor $w_i = x_i\, u_i$, where x_i is the first letter of w_i. By proposition 1.5(d), we can extract an increasing sequence $x_{i_1}, x_{i_2}, ...$ from $x_1, x_2, ...$ Now consider the sequence

$$s' = (w_1, ..., w_{i_1 - 1}, u_{i_1}, u_{i_2}, ...).$$

By the minimality of s, this sequence is good. Hence, there exist $j < k$ with $u_{i_j} \leqslant_{E^*} u_{i_k}$. But then,

$$w_{i_j} = x_{i_j}\, u_{i_j} \leqslant_{E^*} x_{i_k}\, u_{i_k} = w_{i_k},$$

which contradicts the badness of s. □

Exercise 1.9. Show that E is a well-quasi-ordering if and only if the ordering on $E/\equiv$ is a well-quasi-ordering.

Exercise 1.10. Prove the principle of *Noetherian induction*: let P be a property for well-quasi-ordered sets, such that $P(E)$ holds, whenever P holds for all proper initial segments of E. Then P holds for all well-quasi-ordered sets.

Exercise 1.11. Let E and F be well-quasi-ordered sets. With $\mathscr{F}(E, F)$ as in exercise 1.4, when is $\mathscr{F}(E, F)$ also well-quasi-ordered?

Exercise 1.12. Let E be a well-quasi-ordered set. The set $\mathrm{In}(E)$ of initial segments of E is naturally ordered by inclusion. Show that $\mathrm{In}(E)$ is not necessarily well-quasi-ordered. We define E to be a strongly well-quasi-ordered set if $\mathrm{In}(E)$ is also well-quasi-ordered. Which properties from proposition 1.6 generalize to strongly well-quasi-ordered sets?

Exercise 1.13. A *limit well-quasi-ordered set* is a well-quasi-ordered set E, such that there are no final segments of cardinality 1. Given two well-quasi-ordered sets E and F, we define E and F to be equivalent if there exists an increasing injection from E into F and vice versa. Prove that a limit well-quasi-ordered set is equivalent to a unique limit ordinal.

1.4 Kruskal's theorem

An *unoriented tree* is a finite set T of *nodes* with a partial ordering $\leqslant_T$, such that T admits a minimal element $\mathrm{root}(T)$, called the root of T, and such that each other node admits a predecessor. Given $a, b \in T$, we recall that a is a *predecessor* of b (and b a *successor* of a) if $a <_T b$ and $c \leqslant_T a$ for any $c \in T$ with $c <_T b$. A node without successors is called a *leaf*. Any node $a \in T$ naturally induces a *subtree* $T_a = \{b \in T: b \geqslant_T a\}$ with root a. Since T is finite, an easy induction shows that any two nodes a, b of T admit an *infimum* $a \wedge b$ w.r.t. $\leqslant_T$, for which $a \wedge b \leqslant_T a$, $a \wedge b \leqslant_T b$ and $c \leqslant_T a \wedge b$ for all $c \in T$ with $c \leqslant_T a$ and $c \leqslant_T b$.

An *oriented tree* (or simply *tree*) is an unoriented tree T, together with a total ordering $\trianglelefteq_T$ which extends $\leqslant_T$ and which satisfies the condition

$$a \trianglelefteq_T b \wedge a \nleqslant_T b \wedge a \leqslant_T a' \wedge b \leqslant_T b' \Rightarrow a' \trianglelefteq_T b'.$$

It is not hard to see that such a total ordering $\trianglelefteq_T$ is uniquely determined by its restrictions to the sets of $\leqslant_T$-successors for each node a.

Two unoriented or oriented trees T and U will be understood to be equal if there exists a bijection $\varphi: T \to U$ which preserves $\leqslant$ resp. $\leqslant$ and $\trianglelefteq$. In particular, under this identification, the sets of unoriented and oriented trees are countable.

Given a set E, an *E-labeled tree* is a tree T together with a labeling $l: T \to E$. We denote by $E^{\top}$ the set of such trees. An E-labeled tree T may be represented graphically by

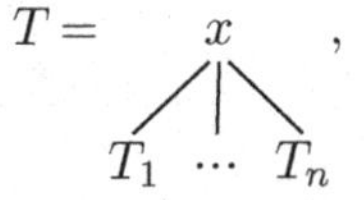

(1.2)

where $x = l(\mathrm{root}(T))$ and $T_1 = T_{a_1}, \ldots, T_n = T_{a_n} \in E^{\top}$ are the subtrees associated to the successors $a_1 \lhd_T \cdots \lhd_T a_n$ of $\mathrm{root}(T)$. We call $T_1, \ldots, T_n$ the *children* of the root and n its *arity*. Notice that we may have $n = 0$.

Example 1.9. We may see usual trees as $\{\bullet\}$-labeled trees, where $\{\bullet\}$ is the set with one symbolic element $\bullet$. The difference between unoriented and oriented trees is that the ordering on the branches is important. For instance, the two trees below are different as oriented trees, but the same as unoriented trees:

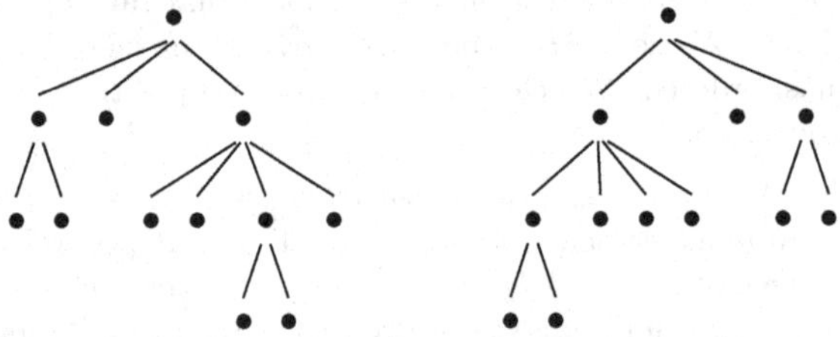

If E is a quasi-ordered set, then the *embeddability quasi-ordering* on $E^{\top}$ is defined by $T \leqslant_{E^{\top}} T'$, if and only if there exists a strictly increasing mapping $\varphi: T \to T'$ for $\trianglelefteq_T$, such that $\varphi(a \wedge b) = \varphi(a) \wedge \varphi(b)$, and $l(a) \leqslant_E l(\varphi(a))$, for all $a, b \in T$. An example of a tree which embeds into another tree is given by

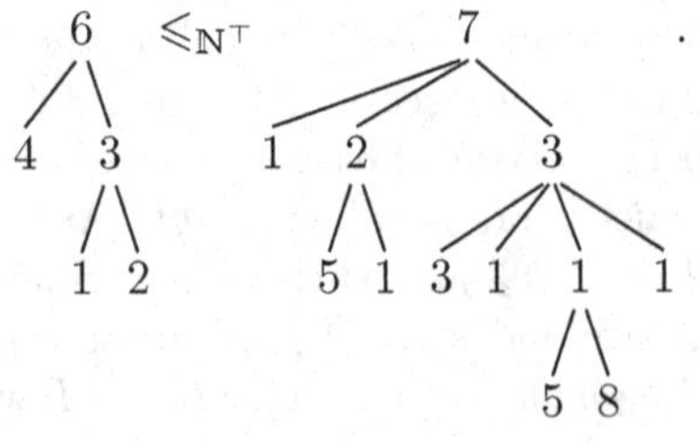

The following theorem is known as Kruskal's theorem:

Theorem 1.10. *If E is a well-quasi-ordered set, then so is $E^\top$.*

Proof. Assume that there exists a bad sequence $T_1, T_2, \ldots$. We may assume that we have chosen each

$$T_i = \underset{\overbrace{T_{i,1} \;\cdots\; T_{i,n_i}}}{x_i}$$

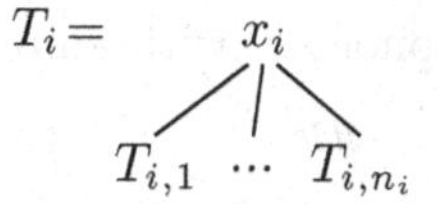

of minimal cardinality (assuming that $T_1, \ldots, T_{i-1}$ have already been fixed), i.e. $T_1, T_2, \ldots$ is a "minimal bad sequence". We claim that the induced quasi-ordering on $S = \{T_{i,j} : j \leqslant n_i\}$ is a well-quasi-ordering. Indeed, suppose the contrary, and let

$$T_{i_1, j_1}, T_{i_2, j_2}, \ldots$$

be a bad sequence. Let k be such that i_k is minimal. Then the sequence

$$T_1, \ldots, T_{i_k - 1}, T_{i_k, j_k}, T_{i_{k+1}, j_{k+1}}, \ldots$$

is also bad, which contradicts the minimality of $T_1, T_2, \ldots$. Hence, S is well-quasi-ordered, and so is $E \times S^*$, by Higman's theorem and proposition 1.6(f). But each tree T_i can be interpreted as an element of $E \times S^*$. Hence, $\{T_1, T_2, \ldots\}$ is a well-quasi-ordered subset of $E^\top$, which contradicts our assumption that $T_1, T_2, \ldots$ is a bad sequence. $\square$

Remark 1.11. In the case when we restrict ourselves to trees of bounded arity, the above theorem was already due to Higman. The general theorem was first conjectured by Vázsonyi. The proof we have given here is due to Nash-Williams.

Exercise 1.14. Let X be a quasi-ordered set and let Ω be an ordered set of operations on X. That is, the elements of Ω are mappings $f : X^{n_f} \to X$. We say that such an operation f is *extensive*, if for all $x \in X^{n_f}$ and $1 \leqslant i \leqslant n_f$, we have

$$x_i \leqslant_X f(x_1, \ldots, x_{n_f})$$

We say that the orderings of X and Ω are *compatible*, if for all $f \leqslant_\Omega g$, $x \in X^{n_f}$ and $y \in X^{n_g}$, we have

$$f(x_1, \ldots, x_{n_f}) \leqslant_X g(y_1, \ldots, y_{g_n}),$$

whenever there exists an increasing mapping $\varphi : \{1, \ldots, n_f\} \to \{1, \ldots, n_g\}$ with $x_i \leqslant_X y_{\varphi(i)}$ for all $1 \leqslant i \leqslant n_f$.

Assume that these conditions are satisfied and let G be a subset of X. The smallest subset of X which contains G and which is stable under Ω is said to be the subset of X *generated* by G w.r.t. Ω, and will be denoted by $(G)_\Omega$. If G is a well-quasi-ordered subset of X and the ordering on Ω is well-quasi-ordered, then prove that $(G)_\Omega$ is well-quasi-ordered.

1.5 Ordered structures

In what follows, all monoids, groups and rings will be commutative and all rings unitary. The following ordered structures will be encountered frequently throughout this book. Recall that we systematically understand all orderings to be partial (contrary to what is customary for certain structures).

- An *ordered monoid* is a monoid X with an ordering $\leqslant$ such that

  OM. $x \leqslant y \wedge x' \leqslant y' \Rightarrow x\, x' \leqslant y\, y'$

  for all $x, y, x', y' \in X$. If X is rather an additive monoid (in which case X is assumed to be abelian), then **OM** becomes

  OA. $x \leqslant y \wedge x' \leqslant y' \Rightarrow x + x' \leqslant y + y'$.

- An *ordered ring* is a ring R with an ordering $\leqslant$ with the following properties:

  OR1. $0 \leqslant 1$;
  OR2. $x \leqslant y \wedge x' \leqslant y' \Rightarrow x + x' \leqslant y + y'$;
  OR3. $0 \leqslant x \wedge 0 \leqslant y \Rightarrow 0 \leqslant x\, y$,

  for all $x, y, x', y' \in R$.
- An *ordered field* is a field K with an ordering $\leqslant$ which makes K an ordered ring and such that $0 < x \Rightarrow 0 < x^{-1}$ for all $x \in K$. Notice that this latter condition is automatically satisfied if $\leqslant$ is total.
- An *ordered R-module* over an ordered ring R is an R-module M with an ordering $\leqslant$ which satisfies

  OM1. $x \leqslant y \wedge x' \leqslant y' \Rightarrow x + x' \leqslant y + y'$;
  OM2. $0 \leqslant \lambda \wedge 0 \leqslant x \Rightarrow 0 \leqslant \lambda x$,

  for all $\lambda \in R$ and $x, y, x', y' \in M$. Any abelian group is trivially an ordered $\mathbb{Z}$-module.
- An *ordered R-algebra* is a morphism $\varphi: R \to A$ of ordered rings, i.e. an increasing ring morphism of an ordered ring R into an ordered ring A. As usual, we denote $\lambda\, x = \varphi(\lambda)\, x$, for $\lambda \in R$ and $x \in A$. Notice that A is in particular an ordered R-module. Any ordered ring R is trivially an ordered $\mathbb{Z}$-algebra.

Let S be an ordered abelian group, ring, R-module or R-algebra. We denote

$$\begin{aligned} S^{>} &= \{x \in S : x > 0\}; \\ S^{\geqslant} &= \{x \in S : x \geqslant 0\}; \\ S^{\neq} &= \{x \in S : x \neq 0\}; \\ S^{\leqslant} &= \{x \in S : x \leqslant 0\}; \\ S^{<} &= \{x \in S : x < 0\}. \end{aligned}$$

We observe that the ordering $\leqslant$ is characterized by $S^{\geqslant}$. If S is totally ordered, then we define the *absolute value* of $x \in S$ by $|x| = x$ if $x \geqslant 0$ and $|x| = -x$, if $x \leqslant 0$.

Example 1.12. $\mathbb{Q}$ and $\mathbb{R}$ are the most common examples of totally ordered fields. $\mathbb{N}$ and $\mathbb{Z}$ are respectively a totally ordered monoid and a totally ordered group. The complex numbers form an ordered abelian group when setting $u < v \Leftrightarrow \operatorname{Re} u < \operatorname{Re} v$. However, this ordering is partial and not compatible with the multiplication. Notice that u and $u + y\,\mathrm{i}$ are incomparable for $u \in \mathbb{C}$ and $y \in \mathbb{R}^{\neq}$.

Example 1.13. The ring of germs at $+\infty$ of infinitely differentiable real valued functions on intervals $(a, +\infty)$ with $a \in \mathbb{R}$ can be ordered by $f \leqslant g$, if there exists an $x_0 \in \mathbb{R}$, such that $f(x) \leqslant g(x)$ for all $x \geqslant x_0$. A totally ordered subfield of this ring is called a *Hardy field.*

Example 1.14. The above definitions naturally generalize to the case of quasi-orderings instead of orderings. If A is a quasi-ordered abelian group, then $A/\equiv$ is an ordered abelian group, and similarly for quasi-ordered rings, R-modules, etc.

Example 1.15. Let A and B be two quasi-ordered abelian groups, rings, R-modules or R-algebras. Their *direct sum* $A \oplus B := A \times B$ is naturally quasi-ordered by the product quasi-ordering

$$(x, y) \leqslant (x', y') \Leftrightarrow x \leqslant x' \wedge y \leqslant y'.$$

Similarly, the *anti-lexicographical direct sum* $A \dot{\oplus} B := A \dot{\times} B$ of A and B is $A \times B$ with the anti-lexicographical quasi-ordering

$$(x, y) \leqslant (x', y') \Leftrightarrow (x \leqslant x' \wedge y = y') \vee y < y'.$$

If A and B are ordered, then so are $A \oplus B$ and $A \dot{\oplus} B$.

Example 1.16. Let A and B be two quasi-ordered abelian groups, rings, R-modules or R-algebras. Their *tensor product* $A \otimes B$ is naturally quasi-ordered, by declaring an element of $A \otimes B$ to be positive if it is a sum of elements of the form $x \otimes y$ with $x \geqslant 0$ and $y \geqslant 0$. Similarly, we define the *anti-lexicographical tensor product* $A \dot{\otimes} B$: its set of positive elements is additively generated by elements in $A \otimes B$ of the form $x \otimes y + x_1 \otimes y_1 + \cdots + x_n \otimes y_n$, with $x, y \geqslant 0$ and $y_1 R + \cdots + y_n R < y$. If A and B are ordered, then the same does not necessarily hold for

Exercise 1.15. Let R be a totally ordered integral domain and let K be its quotient field.

a) Show that $x >_R 0 \wedge y >_R 0 \Rightarrow x y >_R 0$, for all $x, y \in R$.
b) If $\leqslant_R$ is a total ordering, then show that there exists a unique total ordering on K, which extends $\leqslant_R$, and for which K is an ordered field.

Exercise 1.16. Let R be a totally ordered ring.

a) Show that $x\, y = 0 \Rightarrow (x^2 = 0 \vee y^2 = 0)$, for all $x, y \in R$. In particular, if R contains no nilpotent elements, then R is an integral domain.
b) Show that R may contain nilpotent elements.

c) Show that R may contain zero divisors which are not nilpotent.
d) Show that positive non-nilpotent elements are larger than any nilpotent element in R.

Exercise 1.17. Let A, B and C be quasi-ordered rings. Prove the following properties:

a) $(A\oplus B)\oplus C\cong A\oplus(B\oplus C)$ and $(A\dot{\oplus} B)\dot{\oplus} C\cong A\dot{\oplus}(B\dot{\oplus} C)$;
b) $(A\otimes B)\otimes C\cong A\otimes(B\otimes C)$ and $(A\dot{\otimes} B)\dot{\otimes} C\cong A\dot{\otimes}(B\dot{\otimes} C)$;
c) $A\otimes(B\oplus C)\cong(A\otimes B)\oplus(A\otimes C)$ and $(A\oplus B)\otimes C\cong(A\otimes C)\oplus(B\otimes C)$;
d) $A\dot{\otimes}(B\dot{\oplus} C)\cong(A\dot{\otimes} B)\dot{\oplus}(A\dot{\oplus} C)$, but not always $(A\dot{\oplus} B)\dot{\otimes} C\cong(A\dot{\otimes} C)\dot{\oplus}(B\dot{\otimes} C)$.

Exercise 1.18.

a) Show that the categories of ordered abelian groups, rings, R-modules and R-algebras (its morphisms are increasing morphisms of abelian groups, rings, etc.) admit direct sums and products, pull-backs, push-outs, direct and inverse limits and free objects (i.e. the forgetful functor to the category of sets admits a right adjoint).
b) Show that the same thing holds for the categories of ordered torsion free groups, rings without nilpotent elements, torsion free R-modules and ordered R-algebras A without nilpotent elements, and such that the mapping $R\rightarrow A$, $\lambda\mapsto\lambda\cdot 1$ is injective.
c) What can be said about the operations $\oplus$ and $\otimes$ introduced above?

Exercise 1.19. Let S be an ordered abelian group, ring, R-module or R-algebra. We wish to investigate under which circumstances the ordering $\leqslant$ can be extended into a total ordering.

a) If S is an ordered abelian monoid, prove that $\leqslant$ can be extended into a total ordering if and only if S is torsion free (i.e. $n x=0\Rightarrow x=0$, for all $n>0$ and $x\in S$). Hint: use Zorn's lemma.
b) If S is an ordered ring without nilpotent elements, prove that $\leqslant$ can be extended into a total ordering if and only if S is an integral domain, such that
$$a_1^2+\cdots+a_n^2+(b_1^2+\cdots+b_m^2)\,x=0\Rightarrow a_1=0,$$
for all $a_1,\ldots,a_n,b_1,\ldots,b_m,x\in S$, such that $x\geqslant 0$. Hint: first reduce the problem to the case when all squares in S are positive. Next reduce the problem to the case when $a>0\wedge b>0\wedge a x=b\Rightarrow x>0$, for all $a,b,x\in S$.
c) Generalize b to the case when S is an ordered ring, which may contain nilpotent elements.
d) Give conditions in the cases when S is an ordered R-module or an ordered R-algebra without nilpotent elements.

Exercise 1.20. Let S be an ordered group, ring, R-module or R-algebra. For each morphism $\varphi: S\rightarrow T$ of S into a totally ordered structure T of the same kind as S, we define a relation $\preccurlyeq_\varphi$ on S by $x\preccurlyeq_\varphi y\Leftrightarrow\varphi(x)\leqslant\varphi(y)$. Let E be the set of all such relations $\preccurlyeq_\varphi$ on S.

a) Prove that $\hat{\leqslant}=\bigcap_{\leqslant'\in E}\leqslant'$ is a quasi-ordering.
b) Show that $\hat{\leqslant}$ is an ordering, if and only if $\leqslant$ can be extended into a total ordering on S.

c) Let $\hat{=}$ the equivalence relation associated to $\hat{\leqslant}$ and let $\hat{S} = S/\hat{=}$. Show that the ordered set $\hat{S}$ can be given the same kind of ordered algebraic structure as S, in such a way that the natural projection $\pi: S \to \hat{S}$ is a morphism. We call $\hat{S}$ the *closure* of S.

d) S is said to be *perfect* if π is a bijection. Prove that the closure of S is perfect.

e) Show that an ordered abelian group S is perfect if and only if $n x \geqslant 0 \Rightarrow x \geqslant 0$, for all $n > 0$ and $x \in S$.

f) Show that an ordered ring without nilpotent elements is perfect, if and only if $x^2 \geqslant 0$, for all $x \in S$ and $a x = b \wedge a > 0 \wedge b > 0 \Rightarrow x > 0$, for all $a, b, x \in S$.

g) Under which conditions is an ordered R-module perfect? And an ordered R-algebra without nilpotent elements?

1.6 Asymptotic relations

Let f and g be two germs of real valued functions at infinity. Then we have the following classical definitions of the domination and neglection relations $\preccurlyeq$ resp. $\prec$:

$$\begin{aligned} f \preccurlyeq g &\Leftrightarrow f = O(g) \Leftrightarrow \exists C \in \mathbb{R}, \exists x_0 \in \mathbb{R}, \forall x \geqslant x_0, |f(x)| \leqslant C\,|g(x)| \\ f \prec g &\Leftrightarrow f = o(g) \Leftrightarrow \forall \varepsilon > 0, \exists x_0 \in \mathbb{R}, \forall x \geqslant x_0, |f(x)| < \varepsilon\,|g(x)|. \end{aligned}$$

Considered as relations on the $\mathbb{R}$-algebra of germs of real valued functions at infinity, $\preccurlyeq$ and $\prec$ satisfy a certain number of easy to prove algebraic properties. In this section, we will take these properties as the axioms of abstract domination and neglection relations on more general modules and algebras.

Let R be a ring and M an R-module. In all what follows, we denote by R^* the set of non-zero-divisors in R. A *dominance relation* is a quasi-ordering $\preccurlyeq$ on M, such that for all $\lambda \in R$, $\mu \in R^*$ and $x, y, z \in M$, we have

D1. $(x \preccurlyeq z \wedge y \preccurlyeq z) \Rightarrow x - y \preccurlyeq z$;
D2. $\lambda x \preccurlyeq x$ and $y \preccurlyeq \mu y$.

Notice that **D1** and **D2** imply that $O_y = \{x \in M : x \preccurlyeq y\}$ is a submodule of M for each $y \in M$. If $x \preccurlyeq y$, then we say that x is *dominated* by y, and we also write $x = O(y)$. If $x \preccurlyeq y$ and $y \preccurlyeq x$, then we say that x and y are *asymptotic*, and we also write $x \asymp y$. We say that $\preccurlyeq$ is *total*, if $x \preccurlyeq y$ or $y \preccurlyeq x$ for all $x, y \in M$.

A *neglection relation* is a strict ordering $\prec$ on M (i.e. an anti-reflexive, transitive relation), such that for all $\lambda \in R$ and $\mu \in R^*$ and $x, y, z \in M$, we have

N1. $(x \prec z \wedge y \prec z) \Rightarrow x - y \prec z$;
N2. $x \prec y \Rightarrow \lambda x \prec \mu y$ and $\mu y \prec \lambda x \Rightarrow y \prec x$.
N3. $(x \prec z \wedge y \prec z) \Rightarrow x \prec y + z$.

Notice that $o_y = \{x \in M : x \prec y\}$ is a submodule of M if $0 \in o_y$. However, this is not always the case, since $0 \not\prec 0$. If $x \prec y$, then we say that x can be neglected w.r.t. y, and we also write $x = o(y)$. If $x - y \prec x$, then we also say that x and y are *equivalent*, and we write $x \sim y$. Indeed, $\sim$ is an equivalence relation:

$$x \sim y \Rightarrow (x - y \prec x \wedge y - x \prec x) \Rightarrow x - y \prec y \Rightarrow y - x \prec y \Rightarrow y \sim x.$$

Similarly,

$$(x \sim y \wedge y \sim z) \Rightarrow (x - y \prec y \wedge y - z \prec y) \Rightarrow y - z \prec (x - y) + y = x,$$

whence

$$(x \sim y \wedge y \sim z) \Rightarrow (x - y \prec x \wedge y - z \prec x) \Rightarrow x - z \prec x \Rightarrow x \sim z.$$

We say that $\prec$ is *compatible* with a dominance relation $\preccurlyeq$, if $x \prec y \Rightarrow x \preccurlyeq y$ and $x \sim y \Rightarrow x \asymp y$, for all $x, y \in M$. In that case, we call M an *asymptotic R-module*. We say that $\preccurlyeq$ and $\prec$ are *associated*, if $\prec$ is the strict ordering associated to $\preccurlyeq$, i.e. $x \prec y \Leftrightarrow (x \preccurlyeq y \wedge y \not\preccurlyeq x)$ for all $x, y \in M$.

Proposition 1.17.

a) Let $\preccurlyeq$ be a dominance relation such that the strict ordering $\prec$ associated to $\preccurlyeq$ satisfies **N1** *and* **N2**. *Then $\prec$ also satisfies* **N3**.

b) Let $\preccurlyeq$ and $\prec$ be a dominance and a neglection relation. If $\preccurlyeq$ and $\prec$ are associated, then they are compatible.

Proof. Assume that $\preccurlyeq$ satisfies the condition in (a), and let $x, y, z \in M$ be such that $x \prec z$ and $y \prec z$. If $z \not\preccurlyeq y + z$, then $y + z \preccurlyeq z$ implies $y + z \prec z$ and $z \prec z$: contradiction. Hence, we have $z \preccurlyeq y + z$ and $x \prec z \preccurlyeq y + z$.

As to (b), assume that $\preccurlyeq$ and $\prec$ are associated. Then we clearly have $x \prec y \Rightarrow x \preccurlyeq y$. Furthermore, $x \sim y \Rightarrow x - y \prec x \Rightarrow x - y \preccurlyeq x \Rightarrow y \preccurlyeq x$. Similarly, $x \sim y \Rightarrow y \sim x \Rightarrow x \preccurlyeq y$. Hence, $x \sim y \Rightarrow x \asymp y$. □

Proposition 1.18. *Let K be a totally ordered field and V an ordered K-vector space. Then V is an asymptotic K-vector space for the relations $\preccurlyeq$ and $\prec$ defined by*

$$\begin{aligned} x \preccurlyeq y &\Leftrightarrow \forall \lambda \in K, \exists \mu \in K, \lambda x \leqslant \mu y; \\ x \prec y &\Leftrightarrow \exists \mu \in K, \forall \lambda \in K, \lambda x < \mu y. \end{aligned}$$

Moreover, if V is totally ordered, then $\prec$ is associated to $\preccurlyeq$.

Proof. Let us first show that $\preccurlyeq$ is a quasi-ordering. We clearly have $x \preccurlyeq x$ for all $x \in V$, since $\lambda x \leqslant \lambda x$ for all $\lambda \in K$. If $x \preccurlyeq y \preccurlyeq z$ and $\lambda \in K$, then there exists a $\mu \in K$ with $\lambda x \leqslant \mu y$ and a $\nu \in K$ with $\lambda x \leqslant \mu y \leqslant \nu z$. Let us next prove **D1**. Assume that $x \preccurlyeq z$ and $y \preccurlyeq z$ and let $\lambda \in K$. Then there exist $\mu, \nu \in K$ with $\lambda x \leqslant \mu z$ and $-\lambda y \leqslant \nu z$, whence $\lambda (x - y) \leqslant (\mu + \nu) z$. As to **D2**, let $x \in V$, $\alpha \in K$ and $\beta \in K^*$. Then for all $\lambda \in K$, we have $\lambda \alpha x \leqslant \lambda \alpha x$ and $\lambda x \leqslant (\lambda / \beta) \beta x$.

In order to prove the remaining relations, we first notice that

$$x \prec y \Leftrightarrow ((0 < y \vee 0 > y) \wedge \forall \lambda \in K, \lambda x < |y|).$$

Indeed, if $x \prec y$, then there exists a $\mu \in K$ with $\lambda x < \mu y$ for all λ. In particular, $0 < \mu y$, whence either $0 < y$ (if $\mu > 0$) or $0 < -y$ (if $\mu < 0$). Furthermore, for all $\lambda \in K$, we have $\lambda |\mu| x < \mu y$, whence $\lambda x < |y|$. Let us show that $\prec$ is a strict ordering. We cannot have $x \prec x$, since $|x| \not< |x|$. If $x \prec y \prec z$, then we have $\lambda x < |y| < |z|$ for all $\lambda \in K$.

Let us now prove **N1**. If $x \prec z$, $y \prec z$ and $\lambda \in K$, then $2\,\lambda\,x < |z|$ and $-2\,\lambda\,y < |z|$, whence $\lambda(x - y) < |z|$. As to **N2**, let $\alpha \in K$, $\beta \in K^*$ and $\lambda \in K$. If $x \prec y$, then $(\lambda\,\alpha/|\beta|)\,x < |y|$, whence $\lambda\,\alpha < |\beta\,y|$. If $\beta\,x \prec \alpha\,y$, then $\alpha \neq 0$ and $(\lambda\,|\alpha|/\beta)\,\beta\,x < |\alpha\,y|$, whence $\lambda\,x < |y|$. Let us finally prove **N3**. Assume that $x \prec z$, $y \prec z$ and $\lambda \in K$. Then $2\,y < |z|$ implies $\frac{1}{2}\,|z| < |y + z|$. From $2\,\lambda\,x < |z|$ it thus follows that $\lambda\,x < \frac{1}{2}\,|z| < |y + z|$.

Assuming that V is totally ordered, the relation $\prec$ is associated to $\preccurlyeq$, since $x \prec y \Leftrightarrow y \not\preccurlyeq x$. In general, we clearly have $x \prec y \Rightarrow x \preccurlyeq y$. Furthermore, if $x \sim y$, then both $y - x \leqslant |x|$ and $x \leqslant |x|$, whence $y \leqslant 2\,|x|$. Similarly, $x \leqslant 2\,|y|$, so that $x \asymp y$. □

If R is a totally ordered ring, then R cannot have zero-divisors, so its ring of quotients $\mathscr{Q}(R) = (R^*)^{-1}\,R$ is a totally ordered field. Moreover, for any ordered, torsion-free R-module, the natural map $M \to \mathscr{Q}(R) \otimes M$ is an embedding. This allows us to generalize proposition 1.18 to the case of totally ordered rings.

Corollary 1.19. *Let R be a totally ordered ring and M an ordered, torsion-free R-module. Then M is an asymptotic R-module for the restrictions to M of the relations $\preccurlyeq$ and $\prec$ on $\mathscr{Q}(R) \otimes M$. Moreover, if M is totally ordered, then $\prec$ is associated to $\preccurlyeq$.* □

Assume now that A is an R-algebra. A dominance relation on A is defined to be a quasi-ordering $\preccurlyeq$, which satisfies **D1**, **D2** and for all $x, y, z \in A$:

D3. $x \preccurlyeq y \Rightarrow x\,z \preccurlyeq y\,z$.

A neglection relation on A is a strict ordering $\prec$, which satisfies **N1**, **N2**, **N3**, and for all $x, y \in A$ and $z \in A^*$:

N4. $x \prec y \Rightarrow x\,z \prec y\,z$.

An element $x \in A$ is said to be *infinitesimal*, if $x \prec 1$. We say that x is *bounded*, if $x \preccurlyeq 1$ (and *unbounded* if not). Elements with $x \asymp 1$ are called *archimedean*. If all non-zero elements of A are archimedean, then A is said to be *archimedean* itself. In particular, a totally ordered ring said to be archimedean, if it is archimedean as an ordered $\mathbb{Z}$-algebra. If $\preccurlyeq$ and $\prec$ are compatible, then we call A an *asymptotic R-algebra*.

Proposition 1.20. *Let R be a totally ordered ring and A a non-trivial totally ordered R-algebra. Define the relations $\preccurlyeq$ and $\prec$ on A as in corollary 1.19. Then A is an asymptotic R-algebra and $\prec$ is associated to $\preccurlyeq$.*

Proof. Let $x, y, z \in A$ be such that $x \preccurlyeq y$, and let $\lambda \in \mathscr{Q}(R)$. Then there exists a $\mu \in \mathscr{Q}(R)$ with $\lambda\,x < \mu\,y$. If $z \geqslant 0$, then we infer that $\lambda\,x\,z < \mu\,y\,z$, whence $x\,z \preccurlyeq y\,z$. If $z \leqslant 0$, then we obtain $-x\,z \preccurlyeq -y\,z$, whence again $x\,z \preccurlyeq y\,z$, by **D2**. This proves **D3**. As to **N4**, let x, y, $z \in A$ be such that $x \prec y$. Then for all $\lambda \in \mathscr{Q}(R)$, we have $(\lambda\,z/|z|)\,x < |y|$, whence $\lambda\,x\,z = (\lambda\,z/|z|)\,x\,|z| < |y\,z|$. □

Example 1.21. Let A be a totally ordered R-algebra. We may totally order the polynomial extension $A[\varepsilon]$ of A by an infinitesimal element ε by setting $a_0 + a_1 \varepsilon + \cdots + a_d \varepsilon^d > 0$, if and only if there exists an index i with $a_0 = \cdots = a_{i-1} = 0 < a_i$. This algebra is non-archimedean, since $1 \succ \varepsilon \succ \varepsilon^2 \succ \cdots$. Similarly, one may construct an extension $A[\omega]$ with an infinitely large element ω, in which $1 \prec \omega \prec \omega^2 \prec \cdots$.

Exercise 1.21.

a) Given a totally ordered vector space V over a totally ordered field K, show that

$$x \preccurlyeq y \Leftrightarrow \exists \lambda \in K, |x| \leqslant \lambda y;$$
$$x \prec y \Leftrightarrow \forall \lambda \in K, \lambda x < |y|.$$

b) Given a totally ordered module M over a totally ordered ring R, show that

$$x \preccurlyeq y \Leftrightarrow \exists \lambda \in R, \exists \mu \in R^*, |\mu x| \leqslant \lambda y;$$
$$x \prec y \Leftrightarrow \forall \lambda \in R, \forall \mu \in R^*, \lambda x < |\mu y|.$$

Exercise 1.22. Let A be a totally ordered ring. Is it true that the relations $\prec$ and $\preccurlyeq$ are totally determined by the sets of infinitesimal resp. bounded elements of A?

Exercise 1.23. Prove that the sets of infinitesimal and bounded elements in a totally ordered ring A are both convex (a subset B of A is *convex* if for all x, $z \in B$ and $y \in A$, we have $x < y < z \Rightarrow y \in B$). Prove that the set of archimedean elements has two "convex components", provided that $0 < 1$.

Exercise 1.24. Show that the nilpotent elements of a totally ordered ring A are infinitesimal. Does the same thing hold for zero divisors?

Exercise 1.25. Let K be a field. We recall that a *valuation* on K is a mapping $v: K^* \rightarrow \Gamma$ of K^* into a totally ordered additive group, such that

V1. $v(x y) = v(x) + v(y)$ for all $x, y \in K^*$.
V2. $v(x + y) \geqslant \min(v(x), v(y))$, for all $x, y \in K^*$ with $x + y \in K^*$.

Show that the valuations on K correspond to total dominance relations.

Exercise 1.26.

a) Let R be any ring and define $x \preccurlyeq y$, if and only if $\forall z \in R, y z = 0 \Rightarrow x z = 0$, for all $x, y \in R$. Show that $\preccurlyeq$ is a domination relation, for which R is the set of bounded elements, and R^* the set of archimedean elements.
b) Assume that R is a ring with a compatible dominance relation and neglection relation. Show that we may generalize the theory of this section, by replacing all quantifications over $\lambda \in R$ resp. $\mu \in R^*$ by quantifications over $\lambda \preccurlyeq 1$ resp. $\mu \asymp 1$. For instance, the condition **D2** becomes $x \prec y \Rightarrow \lambda x \preccurlyeq \mu y$ for all $x, y \in M$, $\lambda \preccurlyeq 1$ and $\mu \asymp 1$.

Exercise 1.27. Let R be a perfect totally ordered ring and M a perfect ordered R-module. Given $x, y \in M$, we define $x \preccurlyeq y$ resp. $x \prec y$, if $\varphi(x) \preccurlyeq \varphi(y)$ resp. $\varphi(x) \preccurlyeq \varphi(y)$ for all morphisms $\varphi: M \rightarrow N$ of M into a totally ordered R-module N. Prove that $\preccurlyeq$ and $\prec$ compatible domination and neglection relations. Prove that the same thing holds, if we take a perfect ordered R-algebra A instead of M.

Exercise 1.28. Let M be an R-module with a dominance relation $\preccurlyeq$. Let D be the set of total dominance relations $\preccurlyeq'$ on M, with $\preccurlyeq' \supseteq \preccurlyeq$. Prove that $\preccurlyeq = \bigcap_{\preccurlyeq' \in D} \preccurlyeq'$.

1.7 Hahn spaces

Let K be a totally ordered field and V a totally ordered K-vector space. We say that V is a *Hahn space*, if for each $x, y \in V$ with $x \asymp y$, there exists a $\lambda \in K$, with $x \sim \lambda y$.

Proposition 1.22. *Let K be a totally ordered field and V a finite dimensional Hahn space over K. Then V admits a basis $b_1, \ldots, b_n$ with $b_1 \prec \cdots \prec b_n$.*

Proof. We prove the proposition by induction over the dimension n of V. If $n = 0$, then we have nothing to prove. So assume that $n > 0$, and let H be a hyperplane in V of dimension $n - 1$. By the induction hypothesis, H admits a basis $a_1 \prec \cdots \prec a_{n-1}$.

We claim that there exists an $x \in V \setminus H$, such that x is asymptotic to none of the a_i. Indeed, if not, let i be minimal such that there exists an $x \in V \setminus H$ with $x \asymp a_i$. Since V is saturated, there exists a $\lambda \in K$ with $x \sim \lambda a_i$. Then $x - \lambda a_i \prec a_i$, whence $x - \lambda a_i \asymp a_j$ with $j < i$, since $x - \lambda a_i \in V \setminus H$. This contradicts the minimality of i.

So let $x \in V \setminus H$ be such that x is asymptotic to none of the a_i. Since $x \prec a_i \vee x \asymp a_i \vee x \succ a_i$ for all i, the set $\{a_1, \ldots, a_{n-1}, x\}$ is totally ordered w.r.t. $\prec$. □

Exercise 1.29. Show that any totally ordered $\mathbb{R}$-vector space is a Hahn space. Do there exist other totally ordered fields with this property?

Exercise 1.30. Let K be a totally ordered field and V a finite dimensional Hahn space over K. Assume that $b_1 \prec \cdots \prec b_n$ and $b_1' \prec \ldots \prec b_n'$ are both bases of K and denote by B resp. B' the column matrices with entries $b_1, \ldots, b_n$ resp. $b_1', \ldots, b_n'$. Show that $B' = TB$ for some lower triangular matrix T.

Exercise 1.31.

a) Prove that each Hahn space of countable dimension admits a basis which is totally ordered w.r.t. $\prec$.
b) Prove that there exist infinite dimensional Hahn spaces, which do not admit bases of pairwise comparable elements for $\prec$.

1.8 Groups and rings with generalized powers

Let G be a multiplicative group. For any $x \in G$ and $n \in \mathbb{Z}$, we can take the n-th power x^n of x in G. We say that G is a group with $\mathbb{Z}$-powers. More generally, given a ring R, a *group with R-powers* is an R-module G, such that R acts on G through exponentiation. We also say that G is an *exponential R-module*. If R and G are ordered, then we say that G is an *ordered group with R-powers* if $1 \leqslant x \wedge 0 \leqslant \alpha \Rightarrow 1 \leqslant x^\alpha$, for all $x \in G$ and $\alpha \in R$.

Example 1.23. Let G be any group with R-powers and let S be an R-algebra. Then we may form the group $G^{S|R}$ with S-powers, by tensoring the R-modules G and S. However, there is no canonical way to order $G^{S|R}$, if G, R and S are ordered.

A *ring with R-powers* is a ring A, such that a certain multiplicative subgroup $A^\times$ of A carries the structure of a group with R-powers. Any ring A is a ring with $\mathbb{Z}$-powers by taking the group of units of A for $A^\times$. If A is an ordered ring, then we say that the ordering is compatible with the R-power structure if

$$\forall x \in A^\times, \forall \lambda \in R, x \geqslant 0 \Rightarrow x^\lambda \geqslant 0.$$

An *ordered field with R-powers* is an ordered field K, such that the ordered group $K^\times = K^>$ of strictly positive elements in K has R-powers.

Example 1.24. The field $\mathbb{C}(z)$ is a field with $\mathbb{Z}$-powers by taking $\mathbb{C}(z)^\times = \mathbb{C}(z)^{\neq}$. The field $\mathbb{R}(x)$ is a totally ordered field with $\mathbb{Z}$-powers for the ordering

$$f > 0 \Leftrightarrow \exists x_0 \in \mathbb{R}, \forall x \geqslant x_0, f(x) > 0.$$

from example 1.13.

Let A be an *asymptotic ring with R-powers*, i.e. A is both an asymptotic ring and a ring with R-powers, and $1 \preccurlyeq x$ or $x \prec 1$ for all $x \in A^\times$. Given $x \in A^\times$, we denote $\|x\| = x$ if $x \succcurlyeq 1$ and $\|x\| = x^{-1}$ otherwise. Then, given $x, y \in A^\times$, we define

$$\begin{aligned} x \preceq\!\!\!\prec y &\Leftrightarrow \exists \lambda \in R, \exists \mu \in R^*, \|x^\mu\| \preccurlyeq y^\lambda; \\ x \prec\!\!\!\prec y &\Leftrightarrow \forall \lambda \in R, \forall \mu \in R^*, x^\lambda \prec \|y^\mu\|, \end{aligned}$$

and we say that x is *flatter* than y resp. *flatter than or as flat as* y. If $x \preceq\!\!\!\prec y \preceq\!\!\!\prec x$, then we say that x is *as flat as* y and we write $x \asymp\!\!\!\asymp y$. Given $x \in K^\times$, the set of $y \in K^\times$ with $y \asymp\!\!\!\asymp x$ is also called the *comparability class* of x. Finally, if $y/x \prec\!\!\!\prec x$, then we say that x and y are *similar modulo flatness*, and we write $x \approxeq y$.

Example 1.25. Consider the totally ordered field $\mathbb{R}(x)(\mathrm{e}^x)$ with $\mathbb{Z}$-powers and the natural asymptotic relations $\preccurlyeq$ and $\prec$ for $x \to \infty$. Then we have $x \prec\!\!\!\prec \mathrm{e}^x$, $\mathrm{e}^x \asymp\!\!\!\asymp x\,\mathrm{e}^{1000x}$ and $\mathrm{e}^x \not\approxeq x\,\mathrm{e}^{1000x}$.

Let A be an asymptotic ring with R-powers and consider a subring $A^\flat$ with R-powers such that $A^{\flat,\times} = A^\flat \cap A^\times$. The subring $A^\flat$ is said to be *flat* if

$$\forall x \in A^\times, \forall y \in A^{\flat,\times}, x \preccurlyeq\!\!\!\!\prec y \Rightarrow (\exists x' \in A^{\flat,\times}, x' \asymp x).$$

In that case, we define

$$\begin{aligned} x \preccurlyeq^\sharp y &\Leftrightarrow \exists \varphi \in A^{\flat,\times}, x \preccurlyeq \varphi y; \\ x \prec^\sharp y &\Leftrightarrow \forall \varphi \in A^{\flat,\times}, x \prec \varphi y, \end{aligned}$$

for $x, y \in A$. In virtue of the next proposition, we call $\preccurlyeq^\sharp$ a *flattened dominance relation* and $\prec^\sharp$ a *flattened neglection relation.*

Proposition 1.26.

a) A is an asymptotic ring with R-powers for $\preccurlyeq^\sharp$ and $\prec^\sharp$.
b) If for all $x, y \in A$ and $\varphi \in A^{\flat,\times}$ we have

$$x \preccurlyeq y \wedge \varphi y \not\preccurlyeq x \Rightarrow x \prec \varphi y, \tag{1.3}$$

then $\preccurlyeq^\sharp$ and $\prec^\sharp$ are associated.

Proof. Assume that $x \preccurlyeq^\sharp z$ and $y \preccurlyeq^\sharp z$ so that $x \preccurlyeq \varphi z$ and $y \preccurlyeq \psi z$ for certain $\varphi, \psi \in A^{\flat,\times}$. We also have $\varphi \preccurlyeq \psi$ or $\psi \preccurlyeq \varphi$, so, by symmetry, we may assume that $\varphi \preccurlyeq \psi$. Now $x \preccurlyeq \psi z$, whence $x - y \preccurlyeq \psi z$, which proves **D1**. We trivially have **D2**, since $x \preccurlyeq y \Rightarrow x \preccurlyeq^\sharp y$ for all $x, y \in A$. The properties **D3**, **N1**, **N2**, **N3**, **N4** and the quasi-ordering properties directly follow from the corresponding properties for $\preccurlyeq$ and $\prec$.

Assume now that $x \prec^\sharp y$. Then in particular $x \prec y$, whence $x \preccurlyeq y$ and $x \preccurlyeq^\sharp y$. Furthermore, if we had $y \preccurlyeq^\sharp x$, then we would both have $x \preccurlyeq \varphi y$ and $y \prec \varphi^{-1} x$ for some $\varphi \in A^{\flat,\times}$, which is impossible. This proves that $x \prec^\sharp y \Rightarrow x \preccurlyeq^\sharp y \wedge y \not\preccurlyeq^\sharp x$.

Conversely, assume that we have $x \preccurlyeq^\sharp y$ and $y \not\preccurlyeq^\sharp x$, together with (1.3). Then $x \preccurlyeq \varphi y$ for some $\varphi \in A^{\flat,\times}$ and $y \not\preccurlyeq \psi x$ for all $\psi \in A^{\flat,\times}$. Given $\psi \in A^{\flat,\times}$, we then have $\psi y \not\preccurlyeq x$, since otherwise $y \preccurlyeq \psi^{-1} x$. Applying (1.3) to x, φy and $\psi \varphi^{-1}$, we conclude that $x \prec \psi y$ and $x \prec^\sharp y$. □

Example 1.27. Given an element $\varphi \in A$, we may take $A^\flat = \langle x \in A^\times : x \prec\!\!\!\prec \varphi \rangle$ to be the ring generated by all $x \in A^\times$ with $x \preccurlyeq\!\!\!\!\prec \varphi$. Then we define

$$\begin{aligned} \preccurlyeq_\varphi &= \preccurlyeq^\sharp \\ \prec_\varphi &= \prec^\sharp. \end{aligned}$$

We may also take $A^\flat = \langle x \in A^\times : x \preccurlyeq\!\!\!\!\prec \varphi \rangle$, in which case we define

$$\begin{aligned} \preccurlyeq_\varphi^* &= \preccurlyeq^\sharp \\ \prec_\varphi^* &= \prec^\sharp. \end{aligned}$$

For instance, if $A = \mathbb{R}(x)(\mathrm{e}^x)$, then $x^{10} \mathrm{e}^x \prec_{\mathrm{e}^x} \mathrm{e}^{2x}$, $x^{10} \mathrm{e}^x \preccurlyeq_{\mathrm{e}^x} \mathrm{e}^x$ and $f \asymp_{\mathrm{e}^x}^* g$ for all $f, g \in A^{\neq}$.

Exercise 1.32. Let K be a totally ordered field with R-powers and let L be its smallest subfield with R-powers.

a) Show that K has a natural asymptotic L-algebra structure with R-powers.
b) Show that $\preceq\!\!\!\preceq$ and $\prec\!\!\!\prec$ are characterized by

$$\begin{aligned} x \preceq\!\!\!\preceq y &\Leftrightarrow \exists \lambda \in R, \exists \mu \in R^*, \|x^\mu\| \leqslant y^\lambda; \\ x \prec\!\!\!\prec y &\Leftrightarrow \forall \lambda \in R, \forall \mu \in R^*, x^\lambda < \|y^\mu\|. \end{aligned}$$

Exercise 1.33. Consider $A^\times$ as a "quasi-ordered vector space" for $\preccurlyeq$ and the R-power operation. Show that we may quotient this vector space by $\asymp$ and that $\preceq\!\!\!\preceq$ and $\prec\!\!\!\prec$ correspond to the natural dominance and neglection relations on this quotient.

2

Grid-based series

Let C be a commutative ring, and $\mathfrak{M}$ a quasi-ordered monomial monoid. In this chapter, we will introduce the ring $C[[\mathfrak{M}]]$ of generalized power series in $\mathfrak{M}$ over C. For the purpose of this book, we have chosen to limit ourselves to the study of grid-based series, whose supports satisfy a strong finiteness property. On the other hand, we allow $\mathfrak{M}$ to be partially ordered, so that multivariate power series naturally fit into our context. Let us briefly discuss these choices.

In order to define a multiplication on $C[[\mathfrak{M}]]$, we have already noticed in the previous chapter that the supports of generalized power series have to satisfy an ordering condition. One of the weakest possible conditions is that the supports be well-based and one of the strongest conditions is that the supports be grid-based. But there is a wide range of alternative conditions, which correspond to the natural origins of the series we want to consider (see exercises 2.1 and 2.7). For instance, a series like

$$f=\frac{1}{x}+\frac{1}{x^{\pi}}+\frac{1}{x^{\pi^2}}+\cdots$$

is the natural solution to the functional equation

$$f(x)=x^{-1}+f(x^{\pi}).$$

However, f is not grid-based, whence it does not satisfy any algebraic differential equation with power series coefficients (as will be seen in chapter 8).

Actually, the setting of grid-based power series suffices for the resolution of differential equations and that is the main reason why we have restricted ourselves to this setting. Furthermore, the loss of generality is compensated by the additional structure of grid-based series. For example, they are very similar to multivariate Laurent series (as we will see in the next chapter) and therefore particularly suitable for effective purposes [vdH97]. In chapter 4, we will also show that grid-based "transseries" satisfy a useful structure theorem.

Although we might have proved most results in this book for series with totally ordered supports only, we have chosen to develop theory in a partially ordered setting, whenever this does not require much additional effort. First of all, this lays the basis for further generalizations of our results to multivariate and oscillating transseries [vdH97, vdH01a]. Secondly, we will frequently have to "fully expand" expressions for generalized series. This naturally leads to the concepts of grid-based families and strong linear algebra (see sections 2.4, 2.5.3 and 2.6), which have a very "partially-ordered" flavour. Actually, certain proofs greatly simplify when we allow ourselves to use series with partially ordered supports.

Let us illustrate the last point with a simple but characteristic example. Given a classical power series f and an "infinitesimal" generalized power series g, we will define their composition $f \circ g$. In particular, when taking $f(z) = \sum_{i=0}^{\infty} z^n/n!$, this yields a definition for the exponential $\mathrm{e}^g = f \circ g$ of g. Now given two infinitesimal series g_1 and g_2, the proof of the equality $\mathrm{e}^{g_1+g_2} = \mathrm{e}^{g_1}\mathrm{e}^{g_2}$ is quite long in the totally ordered context. In the partially ordered context, on the contrary, this identity trivially follows from the fact that $\mathrm{e}^{z_1+z_2} = \mathrm{e}^{z_1}\mathrm{e}^{z_2}$ in the ring $\mathbb{Q}[[z_1, z_2]]$ of multivariate power series.

2.1 Grid-based sets

Let $\mathfrak{M}$ be a commutative, multiplicative monoid of *monomials*, quasi-ordered by $\preccurlyeq$. A subset $\mathfrak{G} \subseteq \mathfrak{M}$ is said to be *grid-based*, if there exist $\mathfrak{m}_1, \ldots, \mathfrak{m}_m, \mathfrak{n}_1, \ldots, \mathfrak{n}_n \in \mathfrak{M}$, with $\mathfrak{m}_1, \ldots, \mathfrak{m}_m \prec 1$, and such that

$$\mathfrak{G} \subseteq \{\mathfrak{m}_1, \ldots, \mathfrak{m}_m\}^* \{\mathfrak{n}_1, \ldots, \mathfrak{n}_n\}. \tag{2.1}$$

In other words, for each monomial $\mathfrak{v} \in \mathfrak{G}$, there exist $k_1, \ldots, k_m \in \mathbb{N}$ and l with

$$\mathfrak{v} = \mathfrak{m}_1^{k_1} \cdots \mathfrak{m}_m^{k_m} \mathfrak{n}_l.$$

Notice that we can always take $n = 1$ if the ordering on $\mathfrak{M}$ is total.

By Dickson's lemma, grid-based sets are well-quasi-ordered for the opposite quasi-ordering of $\preccurlyeq$ (carefully notice the fact that this is true for the *opposite* quasi-ordering of $\preccurlyeq$ and not for $\preccurlyeq$ itself). Actually, a grid-based set is even well-quasi-ordered for the opposite ordering of $\preccurlyeq^!$ (recall that $x \preccurlyeq^! y \Leftrightarrow x = y \vee x \prec y$). More generally, a subset of $\mathfrak{M}$ which has this latter property is said to be *well-based*.

Proposition 2.1. *Let $\mathfrak{G}$ and $\mathfrak{H}$ be grid-based subsets of $\mathfrak{M}$. Then*

a) Each finite set is grid-based.
b) $\mathfrak{G} \cup \mathfrak{H}$ is grid-based.
c) $\mathfrak{G}\,\mathfrak{H}$ is grid-based.
d) If $\mathfrak{G} \prec 1$, then $\mathfrak{G}^*$ is grid-based.

Proof. The first three assertions are trivial. As to the last one, we will prove that $\mathfrak{G} \prec 1$ implies that there exist elements $\mathfrak{v}_1, \ldots, \mathfrak{v}_v \prec 1$ in $\mathfrak{M}$, with

$$\mathfrak{G} \subseteq \{\mathfrak{v}_1, \ldots, \mathfrak{v}_v\}^*.$$

This clearly implies the last assertion. So assume that we have $\mathfrak{G} \prec 1$ and (2.1). For each l, the set

$$\{(k_1, \ldots, k_m) \in \mathbb{N}^m : \mathfrak{m}_1^{k_1} \cdots \mathfrak{m}_m^{k_m} \mathfrak{n}_l \prec 1\}$$

is a final segment of $\mathbb{N}^m$. Let F_l be a finite set of generators of this final segment and let

$$\mathfrak{V}_l = \{\mathfrak{m}_1^{k_1} \cdots \mathfrak{m}_m^{k_m} \mathfrak{n}_l : (k_1, \ldots, k_m) \in F_l\}.$$

Then $\{\mathfrak{v}_1, \ldots, \mathfrak{v}_v\} = \mathfrak{V}_1 \cup \cdots \cup \mathfrak{V}_n \cup \{\mathfrak{m}_1, \ldots, \mathfrak{m}_m\}$ fulfills our requirements. □

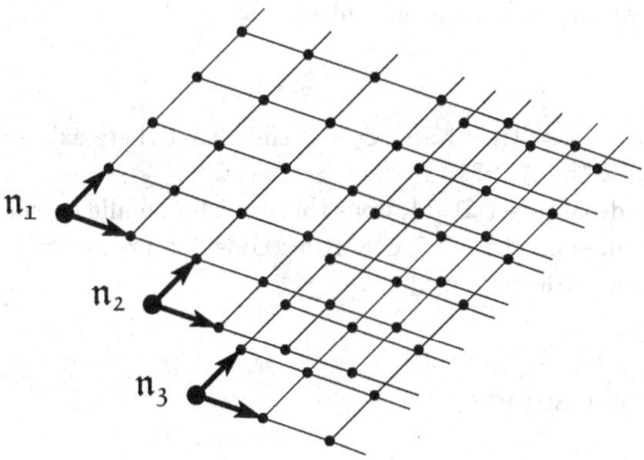

Fig. 2.1. Illustration of a grid-based set with three base points $\mathfrak{n}_1$, $\mathfrak{n}_2$, $\mathfrak{n}_3$ and two infinitesimal generators $\mathfrak{m}_1$ and $\mathfrak{m}_2$. Notice that we used "logarithmic paper" in the sense that multiplication by $\mathfrak{m}_1$ or $\mathfrak{m}_2$ corresponds to a translation via one of the vectors in the picture. Alternatively, one may write $\mathfrak{M} = z^\Gamma$, where z is a formal variable and Γ is a formal ordered additive "value group" which is "anti-isomorphic" to $\mathfrak{M}$. Instead of representing monomials $\mathfrak{M}$, one may then represent their values in Γ.

Exercise 2.1. Show that proposition 2.1 also holds for the following types of subsets of $\mathfrak{M}$:

a) Well-based subsets;
b) Countable well-based subsets;
c) R-finite subsets, when $\mathfrak{M}$ is an ordered group with R-powers. Here an *R-finite subset of* $\mathfrak{M}$ is a well-based subset, which is contained in a finitely generated subgroup with R-powers of $\mathfrak{M}$;
d) Accumulation-free subsets, when $\mathfrak{M}$ is an ordered group with $\mathbb{R}$-powers. Here an *accumulation-free subset of* $\mathfrak{M}$ is a subset $\mathfrak{S}$, such that for all $\mathfrak{m}, \mathfrak{n} \in \mathfrak{M}$ with $\mathfrak{n} \prec 1$, there exists an $\varepsilon > 0$, such that

$$\forall \mathfrak{v} \in \mathfrak{S}, (\mathfrak{v}\, \mathfrak{n}^\varepsilon \prec \mathfrak{m} \Rightarrow (\forall \delta > 0, \mathfrak{v}\, \mathfrak{n}^\delta \prec \mathfrak{m})).$$

Exercise 2.2. Assume that $\mathfrak{M}$ is a group. Show that $\mathbb{Z}$-finite subsets of $\mathfrak{M}$ are not necessarily grid-based.

Exercise 2.3. If $\mathfrak{M}=z^{\mathbb{R}}=\{z^\alpha:\alpha\in\mathbb{R}\}$, with $z^\alpha\succcurlyeq z^\beta\Leftrightarrow\alpha\leqslant\beta$, then accumulation-free subsets of $\mathfrak{M}$ are also called Levi-Civitian subsets. Show that infinite Levi-Civitian subsets of $\mathfrak{M}$ are of the form $\{z^{\alpha_1},z^{\alpha_2},\ldots\}$, with $\lim_{n\to\infty}\alpha_n=+\infty$.

Exercise 2.4. Assume that $\mathfrak{M}$ is a partially ordered monomial group with $\mathbb{Q}$-powers. A subset $\mathfrak{S}$ of $\mathfrak{M}$ is said to be *weakly based*, if for each injective morphism $\varphi\colon\mathfrak{M}\to\mathfrak{N}$ of $\mathfrak{M}$ into a totally ordered monomial group $\mathfrak{N}$ with $\mathbb{Q}$-powers we have:

1. The image $\varphi(\mathfrak{S})$ is well-ordered.
2. For every $\mathfrak{n}\in\mathfrak{N}$, the set $\{\mathfrak{m}\in\mathfrak{S}\colon\varphi(\mathfrak{m})=\mathfrak{n}\}$ is finite.

Show that proposition 2.1 also holds for weakly based subsets and give an example of a weakly based subset which is not well-based.

Exercise 2.5.

a) For grid-based sets $\mathfrak{E}_1\prec 1$ and $\mathfrak{E}_2\prec 1$, show that there exists a grid-based set $\mathfrak{D}\prec 1$ with $\mathfrak{D}^*=\mathfrak{E}_1^*\cap\mathfrak{E}_2^*$.
b) Given a grid-based set $\mathfrak{D}\prec 1$, does there exist a smallest grid-based set $\mathfrak{E}\prec 1$ for inclusion, such that $\mathfrak{D}\subseteq\mathfrak{E}^*$? Hint: consider $\{z_1 z_2^{-2}, z^2\}^*\cap\{z_1^2 z_2^{-1}, z_2^3\}^*$ for a suitable ordering on $z_1^{\mathbb{Z}} z_2^{\mathbb{Z}}$.

2.2 Grid-based series

Let C be a commutative, unitary ring of *coefficients* and $\mathfrak{M}$ a commutative, multiplicative monoid of *monomials*. The *support* of a mapping $f\colon\mathfrak{M}\to C$ is defined by

$$\operatorname{supp} f=\{\mathfrak{m}\in\mathfrak{M}\colon f(\mathfrak{m})\neq 0\}.$$

If $\operatorname{supp} f$ is grid-based, then we call f a *grid-based series*. We denote the set of all grid-based series with coefficients in C and monomials in $\mathfrak{M}$ by $C[\![\mathfrak{M}]\!]$. We also write $f_{\mathfrak{m}}=f(\mathfrak{m})$ for the *coefficient* of $\mathfrak{m}\in\mathfrak{M}$ in such a series and $\sum_{\mathfrak{m}\in\mathfrak{M}} f_{\mathfrak{m}}\mathfrak{m}$ for f. Each $f_{\mathfrak{m}}\mathfrak{m}$ with $\mathfrak{m}\in\operatorname{supp} f$ is called a *term* occurring in f.

Let $(f_i)_{i\in I}$ be a family of grid-based series in $C[\![\mathfrak{M}]\!]$. We say that $(f_i)_{i\in I}$ is a *grid-based family*, if $\bigcup_{i\in I}\operatorname{supp} f_i$ is grid-based and for each $\mathfrak{m}\in\mathfrak{M}$ there exist only a finite number of $i\in I$ with $\mathfrak{m}\in\operatorname{supp} f_i$. In that case, we define its sum by

$$\sum_{i\in I} f_i=\sum_{\mathfrak{m}\in\mathfrak{M}}\left(\sum_{i\in I} f_{i,\mathfrak{m}}\right)\mathfrak{m}. \tag{2.2}$$

This sum is again a grid-based series. In particular, given a grid-based series f, the family $(f_{\mathfrak{m}}\mathfrak{m})_{\mathfrak{m}\in\mathfrak{M}}$ is grid-based and we have $f=\sum_{\mathfrak{m}\in\mathfrak{M}} f_{\mathfrak{m}}\mathfrak{m}$.

Let us now give $C[[\mathfrak{M}]]$ the structure of a C-algebra; we will say that $C[[\mathfrak{M}]]$ is a *grid-based algebra*. C and $\mathfrak{M}$ are clearly contained in $C[[\mathfrak{M}]]$ via $c \mapsto c \cdot 1$ resp. $\mathfrak{m} \mapsto 1 \cdot \mathfrak{m}$. Let $f, g \in C[[\mathfrak{M}]]$. We define

$$f + g = \sum_{\mathfrak{m} \in \operatorname{supp} f \cup \operatorname{supp} g} (f_{\mathfrak{m}} + g_{\mathfrak{m}})\, \mathfrak{m}$$

and

$$f g = \sum_{(\mathfrak{m}, \mathfrak{n}) \in \operatorname{supp} f \times \operatorname{supp} g} f_{\mathfrak{m}}\, g_{\mathfrak{n}}\, \mathfrak{m}\, \mathfrak{n}.$$

By propositions 1.6 and 2.1, $f + g$ and $f g$ are well-defined as sums of grid-based families. It is not hard to show that $C[[\mathfrak{M}]]$ is indeed a C-algebra. For instance, let us prove the associativity of the multiplication. For each $\mathfrak{v} \in \mathfrak{M}$, we have

$$((f g)\, h)_{\mathfrak{v}} = \sum_{\substack{\mathfrak{m} \in \operatorname{supp} f g \\ \mathfrak{n} \in \operatorname{supp} h \\ \mathfrak{m}\mathfrak{n} = \mathfrak{v}}} (f g)_{\mathfrak{m}}\, h_{\mathfrak{n}} = \sum_{\substack{\mathfrak{m}' \in \operatorname{supp} f \\ \mathfrak{m}'' \in \operatorname{supp} g \\ \mathfrak{n} \in \operatorname{supp} h \\ \mathfrak{m}'\mathfrak{m}''\mathfrak{n} = \mathfrak{v}}} f_{\mathfrak{m}'}\, g_{\mathfrak{m}''}\, h_{\mathfrak{n}}.$$

The right hand side of this equation is symmetric in f, g and h and a similar expression is obtained for $(f\,(g h))_{\mathfrak{v}}$.

Let $g \in C[[z]]$ be a power series and $f \in C[[\mathfrak{M}]]$ an infinitesimal grid-based series, i.e. $\mathfrak{m} \prec 1$ for all $\mathfrak{m} \in \operatorname{supp} f$. Then we define

$$g \circ f = \sum_{\mathfrak{m}_1 \cdots \mathfrak{m}_n \in (\operatorname{supp} f)^*} g_n\, f_{\mathfrak{m}_1} \cdots f_{\mathfrak{m}_n}\, \mathfrak{m}_1 \cdots \mathfrak{m}_n,$$

where the sum ranges over all words over the alphabet $\operatorname{supp} f$. The right hand side is indeed the sum of a grid-based family, by Higman's theorem and proposition 2.1. In section 2.5.3, we will consider more general substitutions and we will prove that $(g h) \circ f = (g \circ f)\,(h \circ f)$ and $(h \circ g) \circ f = h \circ (g \circ f)$ for all $g, h \in C[[z]]$.

In particular, we have $((1 + z) \circ f)\;((1 + z)^{-1} \circ f) = 1$ for all f with $\operatorname{supp} f \prec 1$. This yields an inverse for all elements $g \in C[[\mathfrak{M}]]$ of the form $g = 1 + f$ with $\operatorname{supp} f \prec 1$. Assume now that C is a field and that $\mathfrak{M}$ is a totally ordered group. Then we claim that $C[[\mathfrak{M}]]$ is a field. Indeed, let $f \neq 0$ be a series in $C[[\mathfrak{M}]]$ and let $f_{\mathfrak{d}}\, \mathfrak{d}$ be its dominant term (i.e. $\mathfrak{d}$ is maximal for $\prec$ in $\operatorname{supp} f$). Then we have

$$f^{-1} = f_{\mathfrak{d}}^{-1}\, \mathfrak{d}^{-1} \left(\frac{f}{f_{\mathfrak{d}}\, \mathfrak{d}} \right)^{-1}.$$

Example 2.2. Let $\mathfrak{M}$ be any multiplicative monoid with the finest ordering for which no two distinct elements are comparable. Then $C[[\mathfrak{M}]]$ is the polynomial ring $C[\mathfrak{M}]$.

Example 2.3. Let A be any ordered abelian monoid and $z \prec 1$ a formal, infinitely small variable. We will denote by z^A the formal ordered multiplicative monoid of powers z^α with $\alpha \in A$, where $z^\alpha \prec z^\beta \Leftrightarrow \alpha > \beta$ (i.e. A and z^A are *anti-isomorphic*). We call $C[[z^A]]$ the ring of *grid-based series in z over C and along A*. If A is clear from the context, then we also write $C[[z]] = C[[z^A]]$. The following special cases are classical:

a) $C[[z^{\mathbb{N}}]]$ is the ring $C[[z]]$ of *formal power series* in z.
b) $C[[z^{\mathbb{Z}}]]$ is the field $C((z))$ of *Laurent series* in z. Elements of $C[[z^{\mathbb{Z}}]]$ are of the form $\sum_{n \geqslant v} f_n z^n$ with $v \in \mathbb{Z}$.
c) $C[[z^{\mathbb{Q}}]]$ is the field of *Puiseux series* in z. Elements of $C[[z^{\mathbb{Q}}]]$ are of the form $\sum_{n \geqslant v} f_n z^{n/k}$ with $v \in \mathbb{Z}$ and $k \in \mathbb{N}^>$.
d) $C[[z^{\mathbb{N}^n}]]$ is the ring $C[[z_1, ..., z_n]]$ of *multivariate power series*, when $\mathbb{N}^n$ is given the product ordering.
e) $C[[z^{\mathbb{Z}^n}]]$ is the ring $C((z_1, ..., z_n))$ of multivariate Laurent series, when $\mathbb{Z}^n$ is given the product ordering. We recall that a *multivariate Laurent series* $f \in C((z_1, ..., z_n))$ is the product of a series in $C[[z_1, ..., z_n]]$ and a monomial $z_1^{\alpha_1} \cdots z_n^{\alpha_n} \in z^{\mathbb{Z}^n}$. Given $f \in C[[z^{\mathbb{Z}^n}]]^{\neq}$, let $\{z_1^{\beta_{1,j}} \cdots z_n^{\beta_{n,j}} : 1 \leqslant j \leqslant p\}$ be the set of dominant monomials of f. Then we may take $\alpha_i = \min_{1 \leqslant j \leqslant p} \beta_{i,j}$ for each i.

Often, we rather assume that $z \succ 1$ is an infinitely large variable. In that case, z^A is given the opposite ordering $z^\alpha \prec z^\beta \Leftrightarrow \alpha < \beta$.

Example 2.4. There are two ways of explicitly forming rings of multivariate grid-based series: let $z_1, ..., z_n$ be formal variables and $A_1, ..., A_n$ ordered additive monoids. Then we define the rings of *natural grid-based power series* resp. *recursive grid-based power series* in $z_1, ..., z_n$ over C and along $A_1, ..., A_n$ by

$$\begin{aligned} C[[z_1^{A_1}, ..., z_n^{A_n}]] &= C[[z_1^{A_1} \times \cdots \times z_n^{A_n}]]; \\ C[[z_1^{A_1}; ...; z_n^{A_n}]] &= C[[z_1^{A_1} \dot{\times} \cdots \dot{\times} z_n^{A_n}]]. \end{aligned}$$

If $A_1 = \cdots = A_n = A$, where A is clear from the context, then we simply write

$$\begin{aligned} C[[z_1, ..., z_n]] &= C[[z_1^A, ..., z_n^A]]; \\ C[[z_1; ...; z_n]] &= C[[z_1^A; ...; z_n^A]]. \end{aligned}$$

Any series f in $C[[z_1; ...; z_n]]$ may also be considered as a series in $C[[z_1]] \cdots [[z_n]]$ and we may recursively expand f as follows:

$$\begin{aligned} f &= \sum_{\alpha_n \in A} f_{\alpha_n} z_n^{\alpha_n} \\ &\vdots \\ f_{\alpha_n, ..., \alpha_2} &= \sum_{\alpha_1 \in A} f_{\alpha_n, ..., \alpha_1} z_1^{\alpha_1}. \end{aligned}$$

Notice that $C[[z_1; ...; z_n]] \subsetneq C[[z_1]] \cdots [[z_n]]$, in general (see exercise 2.6).

Exercise 2.6. Show that, in general,

$$C[\![z_1,\ldots,z_n]\!] \subsetneq C[\![z_1;\ldots;z_n]\!] \subsetneq C[\![z_1]\!]\cdots[\![z_n]\!]$$

and

$$C[\![z_1;\ldots;z_n]\!] \neq C[\![z_{\sigma(1)};\ldots;z_{\sigma(n)}]\!],$$

for non-trivial permutations σ of $\{1,\ldots,n\}$.

Exercise 2.7. Show that the definitions of this section generalize to the case when, instead of considering grid-based subsets of $\mathfrak{M}$, we consider subsets of one of the types from exercise 2.1 or 2.4. Accordingly, we have the notions of well-based families, well-based series, accumulation-free series, etc. The C-algebra of well-based series in $\mathfrak{M}$ over C will be denoted by $C[[\mathfrak{M}]]$.

Now consider the monomial group

$$\mathfrak{M} = x^{\mathbb{R}} \times \exp^{\mathbb{R}} x \times \exp^{\mathbb{R}} \exp x \times \cdots,$$

where $x, \exp x, \exp\exp x, \ldots \succ 1$. The *order type* of a series is the unique ordinal number which is isomorphic to the support of the series, considered as an ordered set. Determine the order types of the following series in $C[[\mathfrak{M}]]$, as well as their origins (like an equation which is satisfied by the series):

a) $\frac{1}{x} + \frac{1}{\exp x} + \frac{1}{\exp\exp x} + \cdots;$

b) $1 + \frac{1}{x} + \frac{1}{x^2} + \cdots + \frac{1}{e^x} + \frac{1}{x\,e^x} + \frac{1}{x^2\,e^x} + \cdots + \frac{1}{e^{2x}} + \frac{1}{x\,e^{2x}} + \frac{1}{x^2\,e^{2x}} + \cdots;$

c) $1 + 2^{-x} + 3^{-x} + 4^{-x} + \cdots;$

d) $\frac{1}{x} + \frac{1}{x^{\pi}} + \frac{1}{x^{\pi^2}} + \frac{1}{x^{\pi^3}} + \cdots;$

e) $1 + \frac{1}{x} + \frac{1}{x^2} + \frac{1}{x^{e}} + \frac{1}{x^3} + \frac{1}{x^{e+1}} + \frac{1}{x^4} + \frac{1}{x^{e+2}} + \frac{1}{x^5} + \frac{1}{x^{2e}} + \frac{1}{x^{e+3}} + \frac{1}{x^6} + \cdots;$

f) $x + \sqrt{x} + \sqrt{\sqrt{x}} + \sqrt{\sqrt{\sqrt{x}}} + \cdots;$

g) $1 + \frac{1}{x^{1/2}} + \frac{1}{x^{3/4}} + \cdots + \frac{1}{x} + \frac{1}{x^{3/2}} + \frac{1}{x^{7/4}} + \cdots + \frac{1}{x^2} + \frac{1}{x^{5/2}} + \frac{1}{x^{11/4}} + \cdots.$

Also determine the order types of the squares of these series.

Exercise 2.8. Let C be a Noetherian ring and let $\mathfrak{M}$ be a well-based monomial monoid. Show that $C[\![\mathfrak{M}]\!]$ is a Noetherian ring.

Exercise 2.9. For all constant rings C and monomial groups, let $C[[\mathfrak{M}]]$ either denote the ring of well-based, countably well-based, R-finite or accumulation-free series over $\mathfrak{M}$ in C. In which cases do we have $C[[\mathfrak{M} \times \mathfrak{N}]] \cong C[[\mathfrak{M}]][[\mathfrak{N}]]$ for all $\mathfrak{M}$ and $\mathfrak{N}$?

Exercise 2.10. Let $\mathfrak{M}$ be a monomial group and let $\asymp$ be the equivalence relation associated to $\preccurlyeq$ as in exercise 1.1(c) Let $\mathfrak{U} = \{\mathfrak{m} \in \mathfrak{M} : \mathfrak{m} \asymp 1\}$ and let π^{-1} be a right inverse for the projection $\pi : \mathfrak{M} \to \mathfrak{M}/\asymp$. Show that we have natural embeddings

$$\begin{aligned} \nu_1 : C[\![\mathfrak{M}/\asymp]\!][\mathfrak{U}] &\longrightarrow C[\![\mathfrak{M}]\!] \\ \sum_{\mathfrak{n}\in\mathfrak{U}} \sum_{\mathfrak{m}\in\mathfrak{M}/\asymp} f_{\mathfrak{m},\mathfrak{n}}\, \mathfrak{m}\,\mathfrak{n} &\longmapsto \sum_{\mathfrak{n}\in\mathfrak{U}} \sum_{\mathfrak{m}\in\mathfrak{M}/\asymp} f_{\mathfrak{m},\mathfrak{n}}\, \pi^{-1}(\mathfrak{m})\,\mathfrak{n} \end{aligned}$$

and

$$\nu_2 \colon C[\![\mathfrak{M}]\!] \longrightarrow C[\mathfrak{U}][\![\mathfrak{M}/\asymp]\!]$$
$$\sum_{\mathfrak{m} \in \mathfrak{M}/\asymp} \sum_{\mathfrak{n} \in \mathfrak{U}} f_{\mathfrak{m},\mathfrak{n}} \, \pi^{-1}(\mathfrak{m}) \, \mathfrak{n} \longmapsto \sum_{\mathfrak{m} \in \mathfrak{M}/\asymp} \sum_{\mathfrak{n} \in \mathfrak{U}} f_{\mathfrak{m},\mathfrak{n}} \, \mathfrak{m} \, \mathfrak{n}.$$

Show that the embeddings ν_1 and ν_2 are strict, in general.

Exercise 2.11. Let $\mathfrak{M}$ be a quasi-ordered monomial group and $\mathfrak{N}$ an "ideal" of $\mathfrak{M}$ in the sense that $\mathfrak{m}\,\mathfrak{n} \in \mathfrak{N}$, for all $\mathfrak{m} \in \mathfrak{M}$ and $\mathfrak{n} \in \mathfrak{N}$. Define a ring structure on $C[\![\mathfrak{M} \setminus \mathfrak{N}]\!]$, such that $\mathfrak{m}\,\mathfrak{n} = 0$ in $C[\![\mathfrak{M} \setminus \mathfrak{N}]\!]$, for all $\mathfrak{m}, \mathfrak{n} \in \mathfrak{M} \setminus \mathfrak{N}$ with $\mathfrak{m}\,\mathfrak{n} \in \mathfrak{N}$.

2.3 Asymptotic relations

2.3.1 Dominance and neglection relations

Let $f \in C[\![\mathfrak{M}]\!]$ be a grid-based power series. The set of maximal elements for $\preccurlyeq$ in the support of f is called its *set of dominant monomials*. If this set is a singleton, then we say that f is *regular*, we denote by $\mathfrak{d}_f$ or $\mathfrak{d}(f)$ its unique *dominant monomial*, by $c_f = f_{\mathfrak{d}_f}$ its *dominant coefficient*, and by $\tau_f = c_f \, \mathfrak{d}_f$ its *dominant term*. If τ_f is invertible, then we also denote $\delta_f = \frac{f}{\tau_f} - 1$, so that $f = \tau_f (1 + \delta_f)$.

Notice that any grid-based series f can be written as a finite sum of regular series. Indeed, let $\mathfrak{d}_1, \ldots, \mathfrak{d}_n$ be the dominant monomials of f. Then we have

$$f = \sum_{i=1}^{n} \left[\sum_{\mathfrak{m} \in \operatorname{in}(\mathfrak{d}_1, \ldots, \mathfrak{d}_i) \setminus \operatorname{in}(\mathfrak{d}_1, \ldots, \mathfrak{d}_{i-1})} f_{\mathfrak{m}} \, \mathfrak{m} \right],$$

where we recall that $\operatorname{in}(\mathfrak{d}_1, \ldots, \mathfrak{d}_i) = \{\mathfrak{m} \in \mathfrak{M} \colon \mathfrak{m} \preccurlyeq \mathfrak{d}_1 \vee \cdots \vee \mathfrak{m} \preccurlyeq \mathfrak{d}_i\}$.

Assume that C is an ordered ring. We give $C[\![\mathfrak{M}]\!]$ the structure of an ordered C-algebra by setting $f \geqslant 0$, if and only if for each dominant monomial $\mathfrak{d}$ of f, we have $f_{\mathfrak{d}} > 0$ (see exercise 2.12).

Assume now that C and $\mathfrak{M}$ are totally ordered, so that each non-zero series in $C[\![\mathfrak{M}]\!]$ is regular. Then we define a dominance relation $\preccurlyeq$ on $C[\![\mathfrak{M}]\!]$, whose associated strict quasi-ordering $\prec$ is a neglection relation, by

$$f \preccurlyeq g \Leftrightarrow (f = 0 \vee (f \neq 0 \wedge g \neq 0 \wedge \mathfrak{d}_f \preccurlyeq \mathfrak{d}_g)).$$

For non-zero f and g, we have

$$\begin{aligned} f \preccurlyeq g &\Leftrightarrow \mathfrak{d}_f \preccurlyeq \mathfrak{d}_g; \\ f \prec g &\Leftrightarrow \mathfrak{d}_f \prec \mathfrak{d}_g; \\ f \asymp g &\Leftrightarrow \mathfrak{d}_f = \mathfrak{d}_g; \\ f \sim g &\Leftrightarrow \tau_f = \tau_g. \end{aligned}$$

Given $f \in C[\![\mathfrak{M}]\!]$, we define its *canonical decomposition* by

$$f = f_{\succ} + f_{\asymp} + f_{\prec},$$

where $f_{\succ} = \sum_{\mathfrak{m} \succ 1} f_{\mathfrak{m}} \mathfrak{m}$, $f_{\asymp} = f_1$ and $f_{\prec} = \sum_{\mathfrak{m} \prec 1} f_{\mathfrak{m}} \mathfrak{m}$ are respectively the *purely infinite*, *constant* and *infinitesimal parts* of f. We also define $f_{\succcurlyeq} = f_{\succ} + f_{\asymp}$, $f_{\preccurlyeq} = f_{\asymp} + f_{\prec}$ and $f_{\not\asymp} = f - f_{\asymp}$; we call $f_{\preccurlyeq}$ the *bounded part* of f. The canonical decomposition of $C[[\mathfrak{M}]]$ itself is given by:

$$C[[\mathfrak{M}]] = C[[\mathfrak{M}]]_{\succ} \oplus C \oplus C[[\mathfrak{M}]]_{\prec},$$

where

$$\begin{aligned} C[[\mathfrak{M}]]_{\succ} &= C[[\mathfrak{M}^{\succ}]] &= \{f \in C[[\mathfrak{M}]] : f_{\succ} = f\}; \\ C[[\mathfrak{M}]]_{\prec} &= C[[\mathfrak{M}^{\prec}]] &= \{f \in C[[\mathfrak{M}]] : f_{\prec} = f\}. \end{aligned}$$

Similarly, we define $C[[\mathfrak{M}]]_{\succcurlyeq} = C[[\mathfrak{M}^{\succcurlyeq}]] = \{f \in C[[\mathfrak{M}]] : f_{\succcurlyeq} = f\}$ and $C[[\mathfrak{M}]]_{\preccurlyeq} = C[[\mathfrak{M}^{\preccurlyeq}]] = \{f \in C[[\mathfrak{M}]] : f_{\preccurlyeq} = f\}$.

Example 2.5. Let $f = \frac{x^4}{x-1} \in \mathbb{C}[[x^{\mathbb{Z}}]]$ with $x \succ 1$. Then the canonical decomposition of with f is given by

$$\begin{array}{ccccccc} f & = & f_{\succ} & + & f_{\asymp} & + & f_{\prec} \\ & & \| & & \| & & \| \\ & & x^3 + x^2 + x & & 1 & & \dfrac{x^{-1}}{1 - x^{-1}} \end{array}$$

Warning 2.6. One should not confuse $C[[\mathfrak{M}]]_{\succ}$ with $C[[\mathfrak{M}]]^{\succ}$, since $C[[\mathfrak{M}]]_{\succ}$ is strictly contained in $C[[\mathfrak{M}]]^{\succ}$, in general. We always do have $C[[\mathfrak{M}]]^{\prec} = C[[\mathfrak{M}]]_{\prec}$ and $C[[\mathfrak{M}]]^{\preccurlyeq} = C[[\mathfrak{M}]]_{\preccurlyeq}$.

Proposition 2.7. *Assume that C is a totally ordered integral domain and $\mathfrak{M}$ a totally ordered monomial group. Then*

a) $C[[\mathfrak{M}]]$ is a totally ordered C-algebra.
b) The relations $\preccurlyeq$ and $\prec$ coincide with those defined in proposition 1.20.
c) If C is a field, then $C[[\mathfrak{M}]]$ is a Hahn space over C.
d) $C[[\mathfrak{M}]]_{\preccurlyeq}$ is the set of bounded elements in $C[[\mathfrak{M}]]$.
e) $C[[\mathfrak{M}]]_{\prec}$ is the set of infinitesimal elements in $C[[\mathfrak{M}]]$.

Proof. Given f in $C[[\mathfrak{M}]]$, we have either $f = 0$, or $c_f > 0$ (and thus $f > 0$), or $c_f < 0$ (and thus $f < 0$). This proves (a).

Assume that $f \preccurlyeq g$, i.e. $f = 0$ or $f \neq 0 \wedge g \neq 0 \wedge \mathfrak{d}_f \preccurlyeq \mathfrak{d}_g$. If $f = 0$, then clearly $|f| \leqslant |g|$. If $f \neq 0$, then either $\mathfrak{d}_f \prec \mathfrak{d}_g$ and $c_{|g|-|f|} = |c_g| > 0$ implies $|f| < |g|$, or $\mathfrak{d}_f = \mathfrak{d}_g$ and $c_{|2c_f g| - |c_g f|} = |c_f\, c_g| > 0$ implies $|c_g\, f| < |2\, c_f\, g|$. Inversely, assume that $f \not\preccurlyeq g$, i.e. $f \neq 0$ and either $g = 0$ or $\mathfrak{d}_f \succ \mathfrak{d}_g$. If $g = 0$, then clearly $|\lambda\, f| > |\mu\, g| = 0$, for all $\lambda \in C^*$ and $\mu \in C$. Otherwise, $\mathfrak{d}_{|\lambda f|} = \mathfrak{d}_f$ and $|\mu\, g| = 0$ or $\mathfrak{d}_{|\mu g|} = \mathfrak{d}_g$ for all $\lambda \in C^*$ and $\mu \in C$, so that $\mathfrak{d}_{|\lambda f| - |\mu g|} = \mathfrak{d}_f$ and again $|\lambda\, f| > |\mu\, g|$. We conclude that the above definition of $\preccurlyeq$ coincides with the definition in proposition 1.20, using exercise 1.21(b). This proves (b), since for both definitions of $\prec$ we have $f \prec g \Leftrightarrow g \not\preccurlyeq f$.

If C is a field, then for $f, g \in C[[\mathfrak{M}]]^{\neq}$, we have $f \asymp g \Leftrightarrow \mathfrak{d}_f = \mathfrak{d}_g \Rightarrow \tau_f = \tau_{(c_f/c_g)g} \Leftrightarrow f \sim (c_f/c_g)\, g$. This shows (c). If $f \in C[[\mathfrak{M}]]$ is bounded, then either $f = 0$ and clearly $f \in C[[\mathfrak{M}]]_{\preccurlyeq}$, or $f \neq 0 \wedge \mathfrak{d}_f \preccurlyeq 1$ and $\mathfrak{m} \preccurlyeq \mathfrak{d}_f \preccurlyeq 1$ for all $\mathfrak{m} \in \operatorname{supp} f$, whence again $f \in C[[\mathfrak{M}]]_{\preccurlyeq}$. If f is unbounded, then $\mathfrak{d}_f \succ 1$, whence $f \notin C[[\mathfrak{M}]]_{\preccurlyeq}$. This proves (d), and (e) is proved similarly. □

In the case when $\mathfrak{M}$ is not necessarily totally ordered, we may still define the *constant* and *infinitesimal parts* of a series $f \in C[[\mathfrak{M}]]$ by $f_{\asymp} = f_1$ and $f_{\prec} = \sum_{\mathfrak{m} \prec 1} f_{\mathfrak{m}}$. We say that f is *bounded* resp. *infinitesimal*, if $f \in C \oplus C[[\mathfrak{M}]]_{\prec}$ resp. $f \in C[[\mathfrak{M}]]_{\prec}$. In other words, f is bounded resp. infinitesimal, if for all $\mathfrak{m} \in \operatorname{supp} f$, we have $\mathfrak{m} \preccurlyeq 1$, resp. $\mathfrak{m} \prec 1$.

2.3.2 Flatness relations

Assume now that C is both a totally ordered R-module and a totally ordered field with R-powers, for some totally ordered ring R, and assume that $\mathfrak{M}$ is a totally ordered group with R-powers. Let $f \in C[[\mathfrak{M}]]^{>}$ and write $f = c_f\, \mathfrak{d}_f\, (1 + \varepsilon)$ with $\varepsilon \prec 1$. Given $\lambda \in R$, let $\pi_\lambda(z) = (1 + z)^{\lambda \cdot 1} \in C[[z]]$. Then we define

$$f^\lambda = c_f^\lambda\, \mathfrak{d}_f^\lambda\, (\pi_\lambda \circ \varepsilon). \tag{2.3}$$

In this way, we give the field $C[[\mathfrak{M}]]$ the structure of a C-algebra with R-powers, by taking

$$C[[\mathfrak{M}]]^{\times} = \{f \in C[[\mathfrak{M}]]^{\neq} : c_f \in C^{\times}\}.$$

Indeed, $\pi_{\lambda+\mu} \circ \varepsilon = (\pi_\lambda \pi_\mu) \circ \varepsilon = (\pi_\lambda \circ \varepsilon)\, (\pi_\mu \circ \varepsilon)$ for all $\lambda, \mu \in R$ and infinitesimal $\varepsilon \in C[[\mathfrak{M}]]$.

Proposition 2.8. *Let $C, \mathfrak{M}$ and R be as above and let $\prec\!\!\prec$ and $\preccurlyeq\!\!\preccurlyeq$ be defined as in section 1.8. For $\mathfrak{m} \in \mathfrak{M}$, denote $\|\mathfrak{m}\| = \mathfrak{m}$ if $\mathfrak{m} \succcurlyeq 1$ and $\|\mathfrak{m}\| = \mathfrak{m}^{-1}$ otherwise. Then, given $f, g \in C[[\mathfrak{M}]]^{>}$, we have*

a) $f \preccurlyeq\!\!\preccurlyeq g \Leftrightarrow (\exists \lambda \in R, \exists \mu \in R^*, \|\mathfrak{d}_f^\mu\| \preccurlyeq \mathfrak{d}_g^\lambda)$;
b) $f \prec\!\!\prec g \Leftrightarrow (\forall \lambda \in R, \forall \mu \in R^*, \mathfrak{d}_f^\lambda \prec \|\mathfrak{d}_g^\mu\|)$.

Proof. The characterizations of $\preccurlyeq\!\!\preccurlyeq$ and $\prec\!\!\prec$ immediately follow from the fact that $f^\lambda \asymp \mathfrak{d}_f^\lambda$ for all $f \in C[[\mathfrak{M}]]^{>}$. □

2.3.3 Truncations

Let $\mathfrak{M}$ be an arbitrary monomial monoid and $f \in C[[\mathfrak{M}]]$. Given a subset $\mathfrak{S} \subseteq \mathfrak{M}$, we define the *restriction* $f_{\mathfrak{S}} \in C[[\mathfrak{S}]] \subseteq C[[\mathfrak{M}]]$ of f to $\mathfrak{S}$ by

$$f_{\mathfrak{S}} = \sum_{\mathfrak{m} \in \mathfrak{S} \cap \operatorname{supp} f} f_{\mathfrak{m}}\, \mathfrak{m}.$$

For instance, $f_{\succ} = f_{\mathfrak{M}^{\succ}}$, $f_{\asymp} = f_{\{1\}}$, $f_{\prec} = f_{\mathfrak{M}^{\prec}}$ and $f_{\{\mathfrak{m}\}} = f_{\mathfrak{m}}\,\mathfrak{m}$. By our general notations, we recall that $F_{\mathfrak{S}} = \{f_{\mathfrak{S}}: f \in F\}$, for sets $F \subseteq C[[\mathfrak{M}]]$. Notice that $\mathfrak{M}_{\succ} = \mathfrak{M}_{\mathfrak{M}^{\succ}} = \mathfrak{M}^{\succ}$, $\mathfrak{M}_{\prec} = \mathfrak{M}^{\prec}$, etc.

Given two series $f, g \in C[[\mathfrak{M}]]$, we say that f is a *truncation* of g (and we write $f \trianglelefteq g$), if there exists a final segment $\mathfrak{F}$ of supp g, such that $f = g_{\mathfrak{F}}$. The truncation $\trianglelefteq$ is a partial ordering on $C[[\mathfrak{M}]]$.

Let $(f_i)_{i \in I} \in C[[\mathfrak{M}]]^I$ be a non-empty family of series. A *common truncation* of the f_i is a series $g \in C[[\mathfrak{M}]]$, such that $g \trianglelefteq f_i$ for all $i \in I$. A *greatest common truncation* of the f_i is a common truncation, which is greatest for $\trianglelefteq$. Similarly, a *common extension* of the f_i is a series $g \in C[[\mathfrak{M}]]$, such that $f_i \trianglelefteq g$ for all $i \in I$. A *least common extension* of the f_i is a common extension, which is least for $\trianglelefteq$. Greatest common truncations always exist:

Proposition 2.9. *Any non-empty family $(f_i)_{i \in I} \in C[[\mathfrak{M}]]^I$ admits a greatest common truncation.*

Proof. Fix some $j \in I$ and consider the set $\mathscr{F}$ of initial segments $\mathfrak{F}$ of supp f_j, such that $f_{j,\mathfrak{F}} \trianglelefteq f_i$ for all $i \in I$. We observe that arbitrary unions of initial segments of a given ordering are again initial segments. Hence $\mathfrak{F}_{\max} = \bigcup_{\mathfrak{F} \in \mathscr{F}} \mathfrak{F}$ is an initial segment of each supp f_i. Furthermore, for each $\mathfrak{m} \in \mathfrak{F}_{\max}$, there exists an $\mathfrak{F} \in \mathscr{F}$ with $f_{j,\mathfrak{F},\mathfrak{m}} = f_{j,\mathfrak{m}} = f_{i,\mathfrak{m}}$ for all $i \in I$. Hence $f_{j,\mathfrak{F}_{\max}} = f_{i,\mathfrak{F}_{\max}} \trianglelefteq f_i$ for all $i \in I$. This proves that $f_{\mathfrak{F}_{\max}}$ is a common truncation of the f_i. It is also greatest for $\trianglelefteq$, since any common truncation is of the form $f_{j,\mathfrak{F}}$ for some initial segment $\mathfrak{F} \in \mathscr{F}$ of $\mathfrak{F}_{\max}$ with $f_{j,\mathfrak{F}} \trianglelefteq f_{j,\mathfrak{F}_{\max}}$. □

Exercise 2.12. Let C be an ordered ring and $\mathfrak{M}$ a monomial group. Given $\lambda \in C^{\geqslant}$ and series $f, g \in C[[\mathfrak{M}]]^{\geqslant}$, determine the sets of dominant monomials of λf, $f + g$ and $f g$. Show that $C[[\mathfrak{M}]]$ is an ordered C-algebra.

Exercise 2.13. Assume that C is a perfect ordered ring and $\mathfrak{M}$ a perfect ordered monoid.

a) Show that $C[[\mathfrak{M}]]$ is a perfect ordered C-algebra.
b) Let $\prec$ and $\preccurlyeq$ be defined as in exercise 1.27. Show that $z_1^2 + z_2^3 \not\prec z_1 - z_2$ in $C[[z_1, z_2]]$.
c) For $f, g \in C[[\mathfrak{M}]]$ and g regular, show that $f \prec g$, if and only if supp $f \prec \mathfrak{d}_g$.
d) For $f, g \in C[[\mathfrak{M}]]$ and g regular, show that $f \preccurlyeq g$, if and only if supp $f \preccurlyeq \mathfrak{d}_g$.

In other words, there is no satisfactory way to define the relations $\prec$ and $\preccurlyeq$ purely formally, except in the case when the second argument is regular.

Exercise 2.14.

a) Let C be an ordered ring and let $\mathfrak{M}$ be a monomial set, i.e. a set which is ordered by $\preccurlyeq$. Show that the set $C[[\mathfrak{M}]]$ of series $f: \mathfrak{M} \rightarrow C$ with well-based support has the natural structure of an ordered C-module. Show also that this ordering is total if the orderings on C and $\mathfrak{M}$ are both total.
b) Prove Hahn's embedding theorem [Hah07]: let V be a Hahn space over a totally ordered field C. Then $V/\asymp$ is a totally ordered set for $\succ/\asymp$ and V may be embedded into $C[[V/\asymp]]$.

c) If $V \subseteq C[[\mathfrak{M}]]$ in proposition 1.22, then show that V admits a unique basis $(b_1, ..., b_n)$, such that $b_1 \prec \cdots \prec b_n$ and $b_{i,\mathfrak{d}(b_j)} = \delta_{i,j}$ for all $i, j \in \{1, ..., n\}$.

Exercise 2.15.

a) Let $L \supseteq K$ be a field extension and $\mathfrak{M}$ a monomial set. Given a K-subvector space V of $K[[\mathfrak{M}]]$, show that $L \otimes_K C[[\mathfrak{M}]]$ is isomorphic to the L-subvector space of $L[[\mathfrak{M}]]$, which is generated by V.
b) Let $L \supseteq K$ be an extension of totally ordered fields. Given a Hahn space V over K, show that $L \otimes_K V$ has the structure of a Hahn space over L.

Exercise 2.16. Let $\mathfrak{M}$ be a totally ordered monomial group and let $\mathfrak{M}^\flat \subseteq \mathfrak{M}$ be a *flat* subset (i.e. $\forall \mathfrak{m} \in \mathfrak{M}, \forall \mathfrak{n} \in \mathfrak{M}^\flat \colon \mathfrak{m} \preccurlyeq\!\!\!\prec \mathfrak{n} \Rightarrow \mathfrak{m} \in \mathfrak{M}^\flat$).

a) Show that $C[[\mathfrak{M}^\flat]]$ is a flat subring of $C[[\mathfrak{M}]]$.
b) Characterize the relations $\preccurlyeq^\sharp$ and $\prec^\sharp$.

Exercise 2.17. Generalize the notion of truncation to the well-based setting. A directed index set is an ordered set I, such that for any $i, j \in I$, there exist a $k \in I$ with $i \leqslant k$ and $j \leqslant k$. Let $(f_i)_{i \in I}$ be a $\trianglelefteq$-increasing family of series in $C[[\mathfrak{M}]]$, i.e. $f_i \trianglelefteq f_j$ whenever $i \leqslant j$. If $\mathfrak{M}$ is Noetherian or totally ordered, then show that there exists a least common extension of the f_i. Show that this property does not hold in the grid-based setting.

2.4 Strong linear algebra

Just as "absolutely summable series" provide a useful setting for doing analysis on infinite sums (for instance, they provide a context for changing the order of two summations), "grid-based families" provide an analogue setting for formal asymptotics. Actually, there exists an abstract theory for capturing the relevant properties of infinite summation symbols, which can be applied in both cases. In this section, we briefly outline this theory, which we call "strong linear algebra".

2.4.1 Set-like notations for families

It will be convenient to generalize several notations for sets to families. We will denote families by calligraphic characters $\mathcal{F}$, $\mathcal{G}$, ... and write $\mathscr{F}(S)$ for the collection of all families with values in S. Explicit families $(f_i)_{i \in I}$ will sometimes be denoted by $(f_i \colon i \in I)$. Consider two families $\mathcal{F} = (f_i)_{i \in I} \in S^I$ and $\mathcal{G} = (g_j) \in S^J$, where I, J and S are arbitrary sets. Then we define

$$\begin{aligned} \mathcal{F} \amalg \mathcal{G} &= (h_i)_{i \in I \amalg J}, \text{ where } h_i = \begin{cases} f_i & \text{if } i \in I \\ g_i & \text{if } i \in J \end{cases} \\ \mathcal{F} \times \mathcal{G} &= (f_i, g_j)_{(i,j) \in I \times J} \end{aligned}$$

More generally, if $I = \coprod_{j \in J} I_j$, and $\mathcal{G}_j = (f_i)_{i \in I_j}$ for all $j \in J$, then we denote

$$\coprod_{j \in J} \mathcal{G}_j = \mathcal{F}.$$

Given an operation $\varphi: S_1 \times \cdots \times S_n \to T$ and families $\mathcal{F}_k = (f_{k,i})_{i \in I_k} \in S_k^{I_k}$ for $k = 1, \ldots, n$, we define

$$\varphi(\mathcal{F}_1, \ldots, \mathcal{F}_n) = (\varphi(f_{1,i_1}, \ldots, f_{n,i_n}))_{(i_1, \ldots, i_n) \in I_1 \times \cdots \times I_n}. \tag{2.4}$$

It is also convenient to allow bounded variables to run over families. This allows us to rewrite (2.4) as

$$\varphi(\mathcal{F}_1, \ldots, \mathcal{F}_n) = (\varphi(f_1, \ldots, f_n))_{f_1 \in \mathcal{F}_1, \ldots, f_n \in \mathcal{F}_n}$$

Similarly, sums of grid-based families $\mathcal{F} = (f_i)_{i \in I} \in C[[\mathfrak{M}]]^I$ may be denoted by

$$\sum \mathcal{F} = \sum_{f \in \mathcal{F}} f = \sum_{i \in I} f_i$$

We say that $\mathcal{F} = (f_i)_{i \in I}$ and $\mathcal{G} = (g_j)_{j \in J}$ are *equivalent*, and we write $\mathcal{F} \approx \mathcal{G}$, if there exists a bijection $\varphi: I \to J$ with $f_i = g_{\varphi(i)}$ for all $i \in I$. If φ is only injective, then we write $\mathcal{F} \precsim \mathcal{G}$. If $I \subseteq J$ and φ is the natural inclusion, then we simply write $\mathcal{F} \subseteq \mathcal{G}$.

2.4.2 Infinitary operators

The main idea behind strong linear algebra is to consider classical algebraic structures with additional infinitary summation operators $\sum$. These summation symbols are usually only partially defined and they satisfy natural axioms, which will be specified below for a few structures. Most abstract nonsense properties for classical algebraic structures admit natural strong analogues (see exercise 2.20).

A *partial infinitary operator* on a set S is a partial map

$$\Phi: \mathscr{P}(\kappa; S) \rightharpoonup S,$$

where κ is an infinite cardinal number and

$$\mathscr{P}(\kappa; S) = \bigcup_{I \subseteq \kappa} S^I.$$

We call κ the *maximal arity* of the operator Φ. For our purposes, we may usually take $\kappa = \omega$, although higher arities can be considered [vdH97]. The operator $\Phi: \mathscr{P}(\kappa; S) \rightharpoonup S$ is said to be *strongly commutative*, if for all equivalent families $\mathcal{F}$ and $\mathcal{G}$ in $\mathscr{P}(\kappa; S)$, we have $\mathcal{F} \in \operatorname{dom} \Phi \Leftrightarrow \mathcal{G} \in \operatorname{dom} \Phi$ and $\mathcal{F} \in \operatorname{dom} \Phi \Rightarrow \Phi(\mathcal{F}) = \Phi(\mathcal{G})$.

It is convenient to extend commutative operators Φ to arbitrary families $\mathcal{F} = (f_i)_{i \in I} \in S^I$ of cardinality $\operatorname{card} I \leqslant \kappa$. This is done by taking a bijection $\varphi: I \to J$ with $J \subseteq \kappa$ and setting $\Phi(\mathcal{F}) = \Phi((f_{\varphi^{-1}(j)})_{j \in J})$, whenever $(f_{\varphi^{-1}(j)})_{j \in J} \in \operatorname{dom} \Phi$. When extending Φ in this way, we notice that the domain $\operatorname{dom} \Phi$ of Φ really becomes a class (instead of a set) and that Φ is not really a map anymore.

2.4.3 Strong abelian groups

Let A be an abelian group with a partial infinitary operator $\sum: \mathscr{P}(\kappa; A) \rightharpoonup A$. We will denote by $\mathscr{S}(A)$ the domain of $\sum$. We say that A is a *strong abelian group*, if

SA1. $\sum$ is strongly commutative.
SA2. For all $I \subseteq \kappa$ and $\mathcal{O}_I = (0)_{i \in I}$, we have $\sum \mathcal{O}_I = 0$.
SA3. For all $x \in A$ and $\mathcal{S}_x = (x)$, we have $\sum \mathcal{S}_x = x$.
SA4. For all $\mathcal{F}, \mathcal{G} \in \mathscr{S}(A)$, we have $\sum \mathcal{F} \amalg \mathcal{G} = \sum \mathcal{F} + \sum \mathcal{G}$.
SA5. For all $\mathcal{F} \in \mathscr{S}(A)$ and decompositions $\mathcal{F} = \coprod_{j \in J} \mathcal{G}_j$, we have

$$\sum_{j \in J} \sum \mathcal{G}_j = \sum \mathcal{F}.$$

SA6. For all $\mathcal{F} = (n_j f_j)_{i \in J} \in \mathscr{S}(A)$ with $(n_j)_{j \in J} \in (\mathbb{N}^{>})^J$, we have

$$\sum_{\substack{j \in J \\ 1 \leqslant i \leqslant n_j}} f_j = \sum \mathcal{F}.$$

We understand that $\mathcal{F} \in \mathscr{S}(A)$, whenever we use the notation $\sum \mathcal{F}$. For instance, **SA2** should really be read: for all $I \subseteq \kappa$ and $\mathcal{O}_I = (0)_{i \in I}$, we have $\mathcal{O}_I \in \mathscr{S}(A)$ and $\sum \mathcal{O}_I = 0$.

Remark 2.10. Given a strong abelian group A, it is convenient to extend the summation operator $\sum$ to arbitrary families $\mathcal{F} \in \mathscr{F}(A)$: we define $\mathcal{F}$ to be summable in the extended sense if and only if $\mathcal{G} = (f \in \mathcal{F}: f \neq 0)$ is summable in the usual sense; if this is the case, then we set $\sum \mathcal{F} = \sum \mathcal{G}$.

Example 2.11. Any abelian group A carries a *trivial strong structure*, for which $\mathcal{F} \in \mathscr{S}(A)$ if only if $(f \in \mathcal{F}: f \neq 0)$ is a finite family of elements in A.

We call **SA5** the axiom of *strong associativity*. It should be noticed that this axiom can only be applied in one direction: given a large summable family $\mathcal{F}$, we may cut it into pieces $\mathcal{G}_j$, which are all summable and whose sums are summable. On the other hand, given summable families $\mathcal{G}_j$ such that $(\sum \mathcal{G}_j)_{j \in J}$ is again summable, the sum $\sum \coprod_{j \in J} \mathcal{G}_j$ is not necessarily defined: consider $(1-1) + (1-1) + \cdots = 0$. The axiom **SA6** of *strong repetition* aims at providing a partial inverse for **SA5**, in the case when each piece consists of a finite number of repetitions of an element.

Remark 2.12. In **SA5**, we say that the family $\mathcal{F}$ *refines* the family $(\sum \mathcal{G}_j)_{j \in J}$. In order to prove identities of the form $\sum \mathcal{F} = \sum \mathcal{G}$, a common technique is to construct a large summable family $\mathcal{H}$, which refines both $\mathcal{F}$ and $\mathcal{G}$.

2.4.4 Other strong structures

Let R be a ring with a strong summation $\sum$ (which satisfies **SA1–SA6**). We say that R is a *strong ring* if

SR. For all $\mathcal{F},\mathcal{G}\in\mathscr{S}(A)$, we have

$$\sum \mathcal{F}\mathcal{G}=\left(\sum \mathcal{F}\right)\left(\sum \mathcal{G}\right).$$

Let M be a module over such a strong ring R and assume that we also have a strong summation on M. Then M is said to be a *strong R-module* if

SM. For all $\mathcal{F}\in\mathscr{S}(R)$ and $\mathcal{G}\in\mathscr{S}(M)$, we have

$$\sum \mathcal{F}\mathcal{G}=\left(\sum \mathcal{F}\right)\left(\sum \mathcal{G}\right).$$

Notice that **SM** is trivially satisfied when R carries the trivial strong structure. We say that M is an *ultra-strong R-module*, if we also have

UM. For all $(\lambda_i)_{i\in I}\in R^I$ and $(f_i)_{i\in I}\in\mathscr{S}(M)$, we have $(\lambda_i f_i)_{i\in I}\in\mathscr{S}(M)$.

A *strong R-algebra* (resp. an *ultra-strong R-algebra*) is an R-algebra A, together with a strong summation, for which A carries both the structures of a strong ring and a strong R-module (resp. an ultra-strong R-module).

Let M and N be two strong R-modules. A linear mapping $\varphi\colon M\rightarrow N$ is said to be *strong* if it preserves the infinite summation symbols, i.e.

SL. For all $\mathcal{F}\in\mathscr{S}(M)$, we have $\sum\varphi(\mathcal{F})=\varphi(\sum\mathcal{F})$.

In the case of ultra-strong modules, this condition implies

$$\varphi\left(\sum_{i\in I}\lambda_i x_i\right)=\sum_{i\in I}\varphi(\lambda_i x_i)=\sum_{i\in I}\lambda_i\varphi(x_i),$$

whenever $(\lambda_i)_{i\in I}\in R^I$ and $(x_i)_{i\in I}\in\mathscr{S}(M)$. Notice that strong abelian groups and rings can be considered as strong $\mathbb{Z}$-modules resp. $\mathbb{Z}$-algebras, so the definition of strongly linear mappings also applies in these cases.

Exercise 2.18. Let $\mathcal{F}=(f_i)_{i\in I}\in A^I$ and $\mathcal{G}=(g_j)_{j\in J}\in A^J$. Prove that

$$\begin{aligned}\mathcal{F}\approx\mathcal{G} &\Leftrightarrow (\forall x\in A, \operatorname{card}\{i\in I\colon f_i=x\}=\operatorname{card}\{j\in J\colon g_j=x\});\\ \mathcal{F}\precsim\mathcal{G} &\Leftrightarrow (\forall x\in A, \operatorname{card}\{i\in I\colon f_i=x\}\leqslant\operatorname{card}\{j\in J\colon g_j=x\}).\end{aligned}$$

Deduce that $\mathcal{F}\approx\mathcal{G}\Leftrightarrow\mathcal{F}\precsim\mathcal{G}\precsim\mathcal{F}$.

Exercise 2.19.

a) Let $C=\mathbb{R}$, or a more general Banach algebra. Consider the infinite summation operator on C, which associates $\sum_{i\in\mathbb{N}}x_i$ to each absolutely summable family $(x_i)_{i\in\mathbb{N}}$. Show that C is a strong ring for this operator (and the usual finite summation operators).
b) Given a set S, show how to construct the free strong $\mathbb{R}$-module in S.

c) Let $\mathscr{B}$ be a σ-algebra on a set E. We define $M_{\mathscr{B}}$ to be the free strong $\mathbb{R}$-module in $\mathscr{B}$, quotiented by all relations $\sum_{i \in I} U_i = \coprod_{i \in I} U_i$ for at most countable families $(U_i)_{i \in I} \in \mathscr{B}^I$, whose members are mutually disjoint. Show that finite measures can then be interpreted as strongly linear mappings from $M_{\mathscr{B}}$ into $\mathbb{R}$.

Exercise 2.20. Strong abelian groups, rings, modules and algebras form categories, whose morphisms are strongly linear mappings. Show that these categories admit direct sums and products, direct and inverse limits, pull-backs, push-outs and free objects (i.e. the forgetful functor to the category of sets admits a left adjoint).

2.5 Grid-based summation

Let $C[[\mathfrak{M}]]$ be a grid-based algebra as in section 2.2. Given a countable family $\mathcal{F} \in \mathscr{F}(C[[\mathfrak{M}]])$, we define $\mathcal{F}$ to be summable if and only if $\mathcal{F}$ is a grid-based family, in which case its sum is given by formula (2.2). After extension of the strong summation operator to arbitrary families using remark 2.10, it can be checked that the notions of strong summation and summation of grid-based families coincide.

2.5.1 Ultra-strong grid-based algebras

Proposition 2.13. *$C[[\mathfrak{M}]]$ is an ultra-strong C-algebra.*

Proof. The proof does not present any real difficulties. In order to familiarize the reader with strong summability, we will prove **SA5** and **SR** in detail. The proofs of the other properties are left as exercises.

Let $\mathcal{F}$ be a countable grid-based family and $\mathcal{F} = \coprod_{j \in J} \mathcal{G}_j$ a decomposition of $\mathcal{F}$. For each $\mathfrak{m} \in \mathfrak{M}$, let $\mathcal{F}_{;\mathfrak{m}} = (f \in \mathcal{F} \colon f_{\mathfrak{m}} \neq 0)$ and $\mathcal{G}_{j;\mathfrak{m}} = (f \in \mathcal{G}_j \colon f_{\mathfrak{m}} \neq 0)$, so that

$$\mathcal{F}_{;\mathfrak{m}} = \coprod_{j \in J} \mathcal{G}_{j;\mathfrak{m}} \tag{2.5}$$

Now $\mathcal{G}_j$ is a grid-based family for all $j \in J$, since $\bigcup_{f \in \mathcal{G}_j} \operatorname{supp} f \subseteq \bigcup_{f \in \mathcal{F}} \operatorname{supp} f$ and $\mathcal{G}_{j;\mathfrak{m}} \subseteq \mathcal{F}_{;\mathfrak{m}}$ is finite for all $\mathfrak{m} \in \mathfrak{M}$. Furthermore,

$$\bigcup_{j \in J} \operatorname{supp} \sum \mathcal{G}_j \subseteq \bigcup_{j \in J} \bigcup_{f \in \mathcal{G}_j} \operatorname{supp} f = \bigcup_{f \in \mathcal{F}} \operatorname{supp} f,$$

and the set $\{j \in J \colon (\sum \mathcal{G}_j)_{\mathfrak{m}} \neq 0\} \subseteq \{j \in J \colon \mathcal{G}_{j;\mathfrak{m}} \neq \varnothing\}$ is finite for all $\mathfrak{m} \in \mathfrak{M}$, because of (2.5). Hence, the family $(\sum \mathcal{G}_j)_{j \in J}$ is grid-based and for all $\mathfrak{m} \in \mathfrak{M}$, we have

$$\left(\sum_{j \in J} \sum \mathcal{G}_j\right)_{\mathfrak{m}} = \sum_{j \in J} \sum_{f \in \mathcal{G}_{j;\mathfrak{m}}} f_{\mathfrak{m}} = \sum_{f \in \mathcal{F}_{;\mathfrak{m}}} f_{\mathfrak{m}} = \left(\sum \mathcal{F}\right)_{\mathfrak{m}}.$$

This proves **SA5**.

Now let $\mathcal{F}$ and $\mathcal{G}$ be two grid-based families. Then

$$\begin{aligned}\bigcup_{(f,g)\in\mathcal{F}\times\mathcal{G}} \operatorname{supp} f g &\subseteq \bigcup_{(f,g)\in\mathcal{F}\times\mathcal{G}} (\operatorname{supp} f)(\operatorname{supp} g)\\ &= \left(\bigcup_{f\in\mathcal{F}} \operatorname{supp} f\right)\left(\bigcup_{g\in\mathcal{G}} \operatorname{supp} g\right)\end{aligned}$$

is grid-based. Given $\mathfrak{m}\in\mathfrak{M}$, the couples

$$(\mathfrak{v},\mathfrak{w})\in(\bigcup_{f\in\mathcal{F}} \operatorname{supp} f)\times(\bigcup_{g\in\mathcal{G}} \operatorname{supp} g)$$

with $\mathfrak{v}\,\mathfrak{w}=\mathfrak{m}$ form a finite anti-chain; let $(\mathfrak{v}_1,\mathfrak{w}_1)$, ..., $(\mathfrak{v}_n,\mathfrak{w}_n)$ denote those couples. Then

$$\begin{aligned}&((f,g)\in\mathcal{F}\times\mathcal{G}\colon (f g)_{\mathfrak{m}}\neq 0)\\ \subseteq\ &((f,g)\in\mathcal{F}\times\mathcal{G}\colon \exists k\in\{1,\ldots,n\}, f_{\mathfrak{v}_k}\neq 0\wedge g_{\mathfrak{w}_k}\neq 0)\end{aligned}$$

is finite, whence $(f g)_{(f,g)\in\mathcal{F}\times\mathcal{G}}$ is a grid-based family. Given $\mathfrak{m}\in\mathfrak{M}$, and using the above notations, we also have

$$\begin{aligned}\left(\sum_{(f,g)\in\mathcal{F}\times\mathcal{G}} f g\right)_{\mathfrak{m}} &= \sum_{(f,g)\in\mathcal{F}\times\mathcal{G}}\ \sum_{1\leqslant k\leqslant n} f_{\mathfrak{v}_k} g_{\mathfrak{w}_k}\\ &= \sum_{1\leqslant k\leqslant n}\left(\sum \mathcal{F}\right)_{\mathfrak{v}_k}\left(\sum \mathcal{G}\right)_{\mathfrak{w}_k}\\ &= \left(\left(\sum \mathcal{F}\right)\left(\sum \mathcal{G}\right)\right)_{\mathfrak{m}}.\end{aligned}$$

This proves **SR**. □

2.5.2 Properties of grid-based summation

Let $C[[\mathfrak{M}]]$ be a grid-based algebra. Given $\mathcal{F}\in\mathscr{F}(C[[\mathfrak{M}]])$, let

$$\begin{aligned}\operatorname{term}\mathcal{F} &= (f_{\mathfrak{m}}\,\mathfrak{m})_{f\in\mathcal{F},\mathfrak{m}\in\operatorname{supp} f}\\ \operatorname{mon}\mathcal{F} &= (\mathfrak{m})_{f\in\mathcal{F},\mathfrak{m}\in\operatorname{supp} f}\end{aligned}$$

We have

$$\mathcal{F}\in\mathscr{S}(C[[\mathfrak{M}]]) \Leftrightarrow \operatorname{term}\mathcal{F}\in\mathscr{S}(C[[\mathfrak{M}]]) \tag{2.6}$$

$$\Leftrightarrow \operatorname{mon}\mathcal{F}\in\mathscr{S}(C[[\mathfrak{M}]]) \tag{2.7}$$

Indeed,

$$\bigcup_{f\in\mathcal{F}} \operatorname{supp} f = \bigcup_{f\in\operatorname{term}\mathcal{F}} \operatorname{supp} f = \bigcup_{f\in\operatorname{mon}\mathcal{F}} \operatorname{supp} f$$

and for every $\mathfrak{m} \in \mathfrak{M}$,

$$\begin{aligned} \operatorname{card}(f \in \mathcal{F}: f_{\mathfrak{m}} \neq 0) &= \operatorname{card}(f \in \operatorname{term} \mathcal{F}: f_{\mathfrak{m}} \neq 0) \\ &= \operatorname{card}(f \in \operatorname{mon} \mathcal{F}: f_{\mathfrak{m}} \neq 0). \end{aligned}$$

Moreover, if $\mathcal{F}$ is a grid-based, then $\operatorname{term} \mathcal{F}$ refines $\mathcal{F}$.

It is convenient to generalize proposition 2.1 to grid-based families. Given $\mathcal{F} = (f_i)_{i \in I} \in C[[\mathfrak{M}]]^I$, we denote

$$\begin{aligned} \mathcal{F} \prec 1 &\Leftrightarrow (\forall i \in I, f_i \prec 1) \\ \mathcal{F}^* &= (f_{i_1} \cdots f_{i_n})_{i_1 \cdots i_n \in I^*} \end{aligned}$$

Proposition 2.14. *Given grid-based families $\mathcal{F}, \mathcal{G} \in \mathscr{F}(C[[\mathfrak{M}]])$, we have*

a) $\mathcal{F} \amalg \mathcal{G}$ is grid-based.
b) $\mathcal{F}\mathcal{G}$ is grid-based.
c) If $\mathcal{F} \prec 1$, then $\mathcal{F}^*$ is grid-based.

Proof. Properties (a) and (b) follow from **SA4** and **SR**. As to (c), let $\mathfrak{S}$ be the well-based set of pairs $(f, \mathfrak{m})$ with $f \in \mathcal{F}$ and $\mathfrak{m} \in \mathfrak{M}$, for the ordering

$$(f, \mathfrak{m}) \prec (g, \mathfrak{n}) \Leftrightarrow \mathfrak{m} \prec \mathfrak{n}.$$

Now consider the family $\mathcal{T} = (\tau_{\mathfrak{w}})_{\mathfrak{w} \in \mathfrak{S}^*}$ with $\tau_{\mathfrak{w}} = f_{1,\mathfrak{m}_1} \cdots f_{l,\mathfrak{m}_l} \mathfrak{m}_1 \cdots \mathfrak{m}_l$ for each word $\mathfrak{w} = (f_1, \mathfrak{m}_1) \cdots (f_l, \mathfrak{m}_l) \in \mathfrak{S}^*$. This family is well-based, since $\mathfrak{S}^*$ is well-based and the mapping $\mathfrak{w} \mapsto \tau_{\mathfrak{w}}$ increasing. Moreover,

$$\bigcup_{\tau \in \mathcal{T}} \operatorname{supp} \tau \subseteq \left(\bigcup_{f \in \mathcal{F}} \operatorname{supp} f \right)^*,$$

so $\mathcal{T}$ is a grid-based. Hence $\mathcal{F}^*$ is grid-based, since $\mathcal{T}$ refines $\mathcal{F}^*$. □

2.5.3 Extension by strong linearity

Let $C[[\mathfrak{M}]]$ and $C[[\mathfrak{N}]]$ be two grid-based algebras. A mapping $\varphi: \mathfrak{M} \to C[[\mathfrak{N}]]$ is said to be *grid-based* if grid-based subsets $\mathfrak{S} \subseteq \mathfrak{M}$ are mapped to grid-based families $(\varphi(\mathfrak{m}))_{\mathfrak{m} \in \mathfrak{S}}$.

Proposition 2.15. *Let $\varphi: \mathfrak{M} \to C[[\mathfrak{N}]]$ be a grid-based mapping. Then φ extends uniquely to a strongly linear mapping $\hat{\varphi}: C[[\mathfrak{M}]] \to C[[\mathfrak{N}]]$.*

Proof. Let $f \in C[[\mathfrak{M}]]$. Then $(\varphi(\mathfrak{m}))_{\mathfrak{m} \in \operatorname{supp} f}$ is a grid-based family, by definition, and so is $(f_{\mathfrak{m}} \varphi(\mathfrak{m}))_{\mathfrak{m} \in \operatorname{supp} f}$. We will prove that

$$\begin{aligned} \hat{\varphi}: C[[\mathfrak{M}]] &\longrightarrow C[[\mathfrak{N}]] \\ f &\longmapsto \sum_{\mathfrak{m} \in \operatorname{supp} f} f_{\mathfrak{m}} \varphi(\mathfrak{m}) \end{aligned}$$

is the unique strongly linear mapping which coincides with φ on $\mathfrak{M}$.

Given $\lambda \in C$ and $f \in C[[\mathfrak{M}]]$ we clearly have $\hat{\varphi}(\lambda f) = \lambda \hat{\varphi}(f)$, by **SM**. Now let $\mathcal{F} \in \mathscr{S}(C[[\mathfrak{M}]])$ and $\mathfrak{S} = \bigcup_{f \in \mathcal{F}} \operatorname{supp} f$. We claim that

$$(f_{\mathfrak{m}} \varphi(\mathfrak{m}))_{(f, \mathfrak{m}) \in \mathcal{F} \times \mathfrak{S}}$$

is grid-based. Indeed,

$$\bigcup_{(f, \mathfrak{m}) \in \mathcal{F} \times \mathfrak{S}} \operatorname{supp} f_{\mathfrak{m}} \varphi(\mathfrak{m}) \subseteq \bigcup_{\mathfrak{m} \in \mathfrak{S}} \operatorname{supp} \varphi(\mathfrak{m})$$

is grid-based. Secondly, given $\mathfrak{n} \in \mathfrak{N}$, the set $\{\mathfrak{m} \in \mathfrak{S} : \varphi(\mathfrak{m})_{\mathfrak{n}} \neq 0\}$ is finite, since $(\varphi(\mathfrak{m}))_{\mathfrak{m} \in \mathfrak{S}}$ is grid-based. Finally, for each $\mathfrak{m} \in \mathfrak{S}$ with $\varphi(\mathfrak{m})_{\mathfrak{n}} \neq 0$, the family $(f \in \mathcal{F} : f_{\mathfrak{m}} \neq 0)$ is finite. Hence, the family $((f, \mathfrak{m}) \in \mathcal{F} \times \mathfrak{S} : f_{\mathfrak{m}} \varphi(\mathfrak{m})_{\mathfrak{n}} \neq 0)$ is finite, which proves our claim. Now our claim, together with **SA5**, proves that $\hat{\varphi}(\mathcal{F}) = (\sum_{\mathfrak{m} \in \mathfrak{S}} f_{\mathfrak{m}} \varphi(\mathfrak{m}))_{f \in \mathcal{F}}$ is grid-based and

$$\begin{aligned}
\sum \hat{\varphi}(\mathcal{F}) &= \sum_{f \in \mathcal{F}} \sum_{\mathfrak{m} \in \mathfrak{S}} f_{\mathfrak{m}} \varphi(\mathfrak{m}) \\
&= \sum_{(f, \mathfrak{m}) \in \mathcal{F} \times \mathfrak{S}} f_{\mathfrak{m}} \varphi(\mathfrak{m}) \\
&= \sum_{\mathfrak{m} \in \mathfrak{S}} \sum_{f \in \mathcal{F}} f_{\mathfrak{m}} \varphi(\mathfrak{m}) \;=\; \hat{\varphi}\left(\sum \mathcal{F}\right).
\end{aligned}$$

This establishes the strong linearity of $\hat{\varphi}$.

In order to see that $\hat{\varphi}$ is unique with the desired properties, it suffices to observe that for each $f \in C[[\mathfrak{M}]]$, we must have $\hat{\varphi}(f_{\mathfrak{m}} \mathfrak{m}) = f_{\mathfrak{m}} \varphi(\mathfrak{m})$ by linearity and $\hat{\varphi}(f) = \sum_{\mathfrak{m} \in \operatorname{supp} f} f_{\mathfrak{m}} \varphi(\mathfrak{m})$ by strong linearity. □

Proposition 2.16. *Assume, with the notations from the previous proposition that φ preserves multiplication. Then so does $\hat{\varphi}$.*

Proof. This follows directly from the fact that the mappings $(f, g) \mapsto \hat{\varphi}(f g)$ and $(f, g) \mapsto \hat{\varphi}(f)\, \hat{\varphi}(g)$ are both strongly bilinear mappings from $C[[\mathfrak{M}]]^2$ into $C[[\mathfrak{N}]]$, which coincide on $\mathfrak{M}^2$.

Strong bilinearity will be treated in more detail in section 6.2. Translated into terms of strong linearity, the proof runs as follows. Given $\mathfrak{m} \in \mathfrak{M}$, we first consider the mapping $\xi_{\mathfrak{m}} : \mathfrak{n} \mapsto \varphi(\mathfrak{m}\, \mathfrak{n}) = \varphi(\mathfrak{m})\, \varphi(\mathfrak{n})$. Its extension by strong linearity maps $g \in C[[\mathfrak{M}]]$ to

$$\sum_{\mathfrak{n} \in \operatorname{supp} g} g_{\mathfrak{n}} \varphi(\mathfrak{m}\, \mathfrak{n}) = \hat{\varphi}\left(\sum_{\mathfrak{n} \in \operatorname{supp} g} g_{\mathfrak{n}} \mathfrak{m}\, \mathfrak{n}\right) = \hat{\varphi}(\mathfrak{m}\, g),$$

but also to

$$\sum_{\mathfrak{n} \in \operatorname{supp} g} g_{\mathfrak{n}} \varphi(\mathfrak{m})\, \varphi(\mathfrak{n}) = \hat{\varphi}(\mathfrak{m})\, \hat{\varphi}(g).$$

We next consider the mapping $\chi: \mathfrak{m} \mapsto \xi_{\mathfrak{m}}(g)$. Its extension by strong linearity maps $f \in C[[\mathfrak{M}]]$ to

$$\sum_{\mathfrak{m} \in \operatorname{supp} f} f_{\mathfrak{m}} \, \hat{\varphi}(\mathfrak{m}\, g) = \hat{\varphi}\left(\sum_{\mathfrak{m} \in \operatorname{supp} f} f_{\mathfrak{m}} \, \mathfrak{m}\, g \right) = \hat{\varphi}(f g),$$

but also to

$$\sum_{\mathfrak{m} \in \operatorname{supp} f} f_{\mathfrak{m}} \, \hat{\varphi}(\mathfrak{m}) \, \hat{\varphi}(g) = \hat{\varphi}(f) \, \hat{\varphi}(g). \qquad \square$$

Proposition 2.17. *Let $\varphi: \mathfrak{M} \to C[[\mathfrak{N}]]$ and $\psi: \mathfrak{N} \to C[[\mathfrak{V}]]$ be two grid-based mappings. Then*

$$\widehat{\hat{\psi} \circ \varphi} = \hat{\psi} \circ \hat{\varphi}.$$

Proof. This follows directly from the uniqueness of extension by strong linearity, since $\hat{\psi} \circ \varphi$ and $\hat{\psi} \circ \hat{\varphi}$ coincide on $\mathfrak{M}$. $\square$

In section 2.2, we defined the composition $\varphi \circ f$ for $\varphi \in C[[z]]$ and infinitesimal $f \in C[[\mathfrak{M}]]$. We now have a new interpretation of this definition as follows. Consider the mapping $\varphi: z^{\mathbb{N}} \to C[[\mathfrak{M}]]$, which maps z^n to f^n. By proposition 2.1 and Higman's theorem, $(f^n)_{n \in \mathbb{N}}$ is a grid-based family, whence we may extend φ by strong linearity. Given $g \in C[[z]]$, we have

$$\begin{aligned} g \circ f &= \sum_{\mathfrak{m}_1 \cdots \mathfrak{m}_n \in (\operatorname{supp} f)^*} g_n \, f_{\mathfrak{m}_1} \cdots f_{\mathfrak{m}_n} \mathfrak{m}_1 \cdots \mathfrak{m}_n \\ &= \sum_{n \in \mathbb{N}} \sum_{(\mathfrak{m}_1, \ldots, \mathfrak{m}_n) \in (\operatorname{supp} f)^n} g_n \, f_{\mathfrak{m}_1} \cdots f_{\mathfrak{m}_n} \mathfrak{m}_1 \cdots \mathfrak{m}_n \\ &= \sum_{n \in \mathbb{N}} g_n \left[\sum_{(\mathfrak{m}_1, \ldots, \mathfrak{m}_n) \in (\operatorname{supp} f)^n} f_{\mathfrak{m}_1} \cdots f_{\mathfrak{m}_n} \mathfrak{m}_1 \cdots \mathfrak{m}_n \right] \\ &= \sum_{n \in \mathbb{N}} g_n \, f^n = \hat{\varphi}(g). \end{aligned}$$

Now propositions 2.16 and 2.17 respectively imply that

$$(g h) \circ f = (g \circ f) \, (h \circ f)$$

and

$$(h \circ g) \circ f = h \circ (g \circ f)$$

for all $g, h \in C[[z]]$. More generally, we have

Proposition 2.18. *Let $f_1, \ldots, f_k$ be infinitesimal grid-based series in $C[[\mathfrak{M}]]$ and consider the mapping*

$$\begin{aligned} \varphi: z_1^{\mathbb{N}} \cdots z_k^{\mathbb{N}} &\to C[[\mathfrak{M}]] \\ z_1^{n_1} \cdots z_k^{n_k} &\mapsto f_1^{n_1} \cdots f_k^{n_k}. \end{aligned}$$

Given $g \in C[[z_1, \ldots, z_k]]$, *we define* $g \circ (f_1, \ldots, f_k) = \hat{\varphi}(g)$. *Then*

a) For $g, h \in C[[z_1, \ldots, z_k]]$, *we have*

$$(g h) \circ (f_1, \ldots, f_k) = g \circ (f_1, \ldots, f_k)\, h \circ (f_1, \ldots, f_k).$$

b) For $h \in C[[z_1, \ldots, z_l]]$ *and infinitesimal* $g_1, \ldots, g_l \in C[[z_1, \ldots, z_k]]$, *we have*

$$(h \circ (g_1, \ldots, g_l)) \circ (f_1, \ldots, f_k) = h \circ (g_1 \circ (f_1, \ldots, f_k), \ldots, g_l \circ (f_1, \ldots, f_k)).$$

Exercise 2.21. Assume that C is a strong ring and $\mathfrak{M}$ a monomial monoid. A family $\mathcal{F} \in \mathscr{F}(C[[\mathfrak{M}]])$ is said to be *grid-based*, if $\bigcup_{f \in \mathcal{F}} \operatorname{supp} f$ is grid-based and $(f_{\mathfrak{m}})_{f \in \mathcal{F}} \in \mathscr{S}(C)$, for each $\mathfrak{m} \in \mathfrak{M}$. Show that this definition generalizes the usual definition of grid-based families and generalize proposition 2.13.

Exercise 2.22. Give $\mathbb{R}$ the strong field structure from exercise 2.19(a) and $\mathbb{R}[[\mathfrak{M}]]$ the strong ring structure from exercise 2.21. Show that the strong summation on $\mathbb{R}[[\mathfrak{M}]]$ does not necessarily satisfy **US**. Prove that it does satisfy the following axiom:

RS. Let $\mathcal{F} \in \mathscr{S}(\mathbb{R}[[\mathfrak{M}]])$ and $\mathcal{G}_f \in \mathscr{S}(\mathbb{R}^{\geqslant})$ be such that $\sum \mathcal{G}_f = 1$ for all $f \in \mathcal{F}$. Then $(\lambda f)_{f \in \mathcal{F}, \lambda \in \mathcal{G}_f} \in \mathscr{S}(\mathbb{R}[[\mathfrak{M}]])$.

Exercise 2.23. Generalize the results from this section to the case when we consider well-based (or R-finite, accumulation-free series, etc.) series instead of grid-based series.

2.6 Asymptotic scales

Let C both be an R-module and a field with R-powers, for some ring R, and let $\mathfrak{M}$ be an ordered monomial group with R-powers. The the definition of f^λ in (2.3) generalizes to the case when $f \in C[[\mathfrak{M}]]$ is a regular series with $c_f \in C^\times$. As before, the group $C[[\mathfrak{M}]]^\times$ of such f has R-powers.

Proposition 2.19. *Let* $\mathfrak{N}$ *be another ordered monomial group with* R*-powers and let* $\varphi: \mathfrak{M} \to C[[\mathfrak{N}]]$ *be a grid-based mapping such that*

- $\varphi(\mathfrak{m}) \in C[[\mathfrak{N}]]^\times$, *for all* $\mathfrak{m} \in \mathfrak{M}$.
- $\varphi(\mathfrak{m}\,\mathfrak{n}) = \varphi(\mathfrak{m})\,\varphi(\mathfrak{n})$ *and* $\varphi(\mathfrak{m}^\lambda) = \varphi(\mathfrak{m})^\lambda$, *for all* $\mathfrak{m}, \mathfrak{n} \in \mathfrak{M}$ *and* $\lambda \in R$.
- *The mapping* $\mathfrak{d} \circ \varphi: \mathfrak{M} \to \mathfrak{N}, \mathfrak{m} \mapsto \mathfrak{d}_{\varphi(\mathfrak{m})}$ *is increasing.*

Then

a) $\hat{\varphi}(f g) = \hat{\varphi}(f)\,\hat{\varphi}(g)$ *and* $\hat{\varphi}(f^\lambda) = \hat{\varphi}(f)^\lambda$, *for all* $f, g \in C[[\mathfrak{M}]]^\times$ *and* $\lambda \in R$.
b) If $\ker \mathfrak{d} \circ \varphi = 1$, *then* $\hat{\varphi}$ *is injective.*

Proof. By proposition 2.16, $\hat{\varphi}$ preserves multiplication. Let $f = c_f \mathfrak{d}_f (1 + \varepsilon) \in C[[\mathfrak{M}]]^\times$ be a regular series and $\lambda \in R$. Then

$$\hat{\varphi}(f^\lambda) = c_f^\lambda \varphi(\mathfrak{d}_f)^\lambda ((1 + z)^\lambda \circ \hat{\varphi}(\varepsilon)) = c_f^\lambda \varphi(\mathfrak{d}_f)^\lambda (1 + \hat{\varphi}(\varepsilon))^\lambda,$$

by the propositions of the previous section. Furthermore, $\mathfrak{d} \circ \varphi$ is strictly increasing (otherwise, let $\mathfrak{m} \in \mathfrak{M}$ be such that $\mathfrak{m} \prec 1$, but $\mathfrak{d}_{\varphi(\mathfrak{m})} = 1$. Then $(\varphi(\mathfrak{m}^n))_{n \in \mathbb{N}}$ is not grid-based). Hence, $1 + \hat{\varphi}(\varepsilon)$ is in $C[[\mathfrak{N}]]^{\times}$, and so are c_f and $\varphi(\mathfrak{d}_f)$. Therefore,

$$\hat{\varphi}(f^{\lambda}) = c_f^{\lambda} \varphi(\mathfrak{d}_f)^{\lambda} (1 + \hat{\varphi}(\varepsilon))^{\lambda} = (c_f \varphi(\mathfrak{d}_f) (1 + \hat{\varphi}(\varepsilon))^{\lambda} = \hat{\varphi}(f)^{\lambda},$$

since $C[[\mathfrak{N}]]^{\times}$ is a group with R-powers. This proves (a).

Assume now that $\ker \mathfrak{d} \circ \varphi = 1$. Then $\mathfrak{d} \circ \varphi$ is injective and strictly increasing. Given $f \in C[[\mathfrak{M}]]$ with dominant monomials $\mathfrak{d}_1, \ldots, \mathfrak{d}_n$, the monomials $\mathfrak{d}_{\varphi(\mathfrak{d}_1)}, \ldots, \mathfrak{d}_{\varphi(\mathfrak{d}_n)}$ are pairwise distinct. Consequently, the dominant monomials of $\hat{\varphi}(f)$ are precisely the maximal elements for $\preccurlyeq$ among the $\mathfrak{d}_{\varphi(\mathfrak{d}_i)}$. In particular, if $f \neq 0$, then there exists at least one such maximal element, so that $\hat{\varphi}(f) \neq 0$. This proves (b). □

An *asymptotic scale* in $C[[\mathfrak{M}]]$ is a subgroup $\mathfrak{S}$ of $C[[\mathfrak{M}]]^{\times}$ with R-powers, such that $\mathfrak{d}_{|\mathfrak{S}}: \mathfrak{S} \to \mathfrak{M}$ is injective. Then $\mathfrak{S}$ is naturally ordered by $f \succcurlyeq g \Leftrightarrow \mathfrak{d}_f \succcurlyeq \mathfrak{d}_g$, for all $f, g \in \mathfrak{S}$. The previous proposition now shows that we may identify $C[[\mathfrak{S}]]$ with a subset of $C[[\mathfrak{M}]]$ via the strongly linear extension $\hat{\nu}_{\mathfrak{S}}$ of the inclusion $\nu_{\mathfrak{S}}: \mathfrak{S} \to C[[\mathfrak{M}]]$. This identification is coherent in the sense that $\hat{\nu}_{\mathfrak{S}} \circ \hat{\nu}_{\mathfrak{T}} = \hat{\nu}_{\hat{\nu}_{\mathfrak{S}}(\mathfrak{T})}$, for any asymptotic scale $\mathfrak{T}$ in $C[[\mathfrak{S}]]$, by proposition 2.17.

A *basis* of an asymptotic scale $\mathfrak{S}$ is a basis of $\mathfrak{S}$, when considering $\mathfrak{S}$ as an exponential R-module. If $\mathfrak{B}$ is such a basis, then $\mathfrak{d}_{\mathfrak{B}}$ is a basis of $\mathfrak{d}_{\mathfrak{S}}$. In particular, if $\mathfrak{d}_{\mathfrak{S}} = \mathfrak{M}$, then $\mathfrak{d}_{\mathfrak{B}}$ is a basis of $\mathfrak{M}$. In this case, the bijection $\mathfrak{d}_{|\mathfrak{S}}: \mathfrak{S} \to \mathfrak{M}$ is called a *scale change* and its restriction to $\mathfrak{B}$ a *base change*. We also say that $\mathfrak{B}$ is an *asymptotic basis* for $C[[\mathfrak{M}]]$ in this case.

When dealing with finite bases, it will often be convenient to consider them as ordered n-tuples $\mathfrak{B} = (\mathfrak{b}_1, \ldots, \mathfrak{b}_n)$ instead of sets without any ordering.

Exercise 2.24. Generalize the results from this section to the case when we consider well-based series instead of grid-based series. In the definition of asymptotic scales, one should add the requirement that the natural inclusion mapping $\mathfrak{S} \to C[[\mathfrak{M}]]$ be well-based (i.e. well-based subsets of $\mathfrak{S}$ are mapped to well-based families).

Exercise 2.25.

a) Assume that $\mathfrak{M}$ is a perfect monomial group, i.e. $\mathfrak{m}^n \preccurlyeq 1 \Rightarrow \mathfrak{m} \preccurlyeq 1$, for all $\mathfrak{m} \in \mathfrak{M}$ and $n \geqslant 1$. Prove that a series $f \in C[[\mathfrak{M}]]$ is invertible, if and only if f is regular. Hint: show that for each dominant monomial $\mathfrak{m}$ of $f \in C[[\mathfrak{M}]]$, there exists an extension $\preccurlyeq'$ of the ordering on $\mathfrak{M}$, such that $\mathfrak{n} \preccurlyeq' \mathfrak{m}$, for all $\mathfrak{n} \in \operatorname{supp} f$.
b) Prove that the above characterization of invertible series does not hold for general monomial groups.

Exercise 2.26. Let K be a field and $\mathfrak{M}$ be a monomial group with K-powers. Assume that $\mathfrak{M}$ admits a finite basis $\mathfrak{B} = (\mathfrak{b}_1, \ldots, \mathfrak{b}_n)$.

a) Let $\mathfrak{B}' = (\mathfrak{b}'_1, \ldots, \mathfrak{b}'_{n'})$ be another asymptotic basis of $C[[\mathfrak{M}]]$. Show that $n' = n$ and that there exists a square matrix

$$P_{\mathfrak{B}',\mathfrak{B}} = \begin{pmatrix} \lambda_{1,1} & \cdots & \lambda_{1,n} \\ \vdots & & \vdots \\ \lambda_{n,1} & \cdots & \lambda_{n,n} \end{pmatrix},$$

such that $\mathfrak{d}(\mathfrak{B}') = \mathfrak{B}^{P_{\mathfrak{B}',\mathfrak{B}}}$, that is, $\mathfrak{d}(\mathfrak{b}'_i) = \mathfrak{b}_1^{\lambda_{i,1}} \cdots \mathfrak{b}_n^{\lambda_{i,n}}$ for all n.

b) Show that $P_{\mathfrak{B},\mathfrak{B}'} P_{\mathfrak{B}',\mathfrak{B}} = \mathrm{Id}_n$.

c) If $C[[\mathfrak{M}]] = C[[\mathfrak{b}_1, \ldots, \mathfrak{b}_n]] = C[[\mathfrak{b}'_1, \ldots, \mathfrak{b}'_n]]$, then show that the matrix $P_{\mathfrak{B}',\mathfrak{B}}$ is diagonal, modulo a permutation of the elements of $\mathfrak{B}'$.

d) If $C[[\mathfrak{M}]] = C[[\mathfrak{b}_1; \ldots; \mathfrak{b}_n]] = C[[\mathfrak{b}'_1; \ldots; \mathfrak{b}'_n]]$, then show that the matrix $P_{\mathfrak{B}',\mathfrak{B}}$ is lower triangular.

3

The Newton polygon method

Almost all techniques for solving asymptotic systems of equations are explicitly or implicitly based on the Newton polygon method. In this section we explain this technique in the elementary case of algebraic equations over grid-based algebras $C[[\mathfrak{M}]]$, where C is a constant field of characteristic zero and $\mathfrak{M}$ a totally ordered monomial group with $\mathbb{Q}$-powers. In later chapters of this book, the method will be generalized to linear and non-linear differential equations.

In section 3.1, we first illustrate the Newton polygon method by some examples. One important feature of our exposition is that we systematically work with "asymptotic algebraic equations", which are polynomial equations $P(f) = 0$ over $C[[\mathfrak{M}]]$ together with asymptotic side-conditions, like $f \prec \mathfrak{v}$. Asymptotic algebraic equations admit natural invariants, like the "Newton degree", which are useful in the termination proof of the method. Another important ingredient is the consideration of equations $P'(f) = 0$, $P''(f) = 0$, etc. in the case when $P(f) = 0$ admits almost multiple roots.

In section 3.2, we prove a version of the implicit function theorem for grid-based series. Our proof uses a syntactic technique which will be further generalized in chapter 6. The implicit function theorem corresponds to the resolution of asymptotic algebraic equations of Newton degree one. In section 3.3, we show how to compute the solutions to an asymptotic algebraic equation using the Newton polygon method. We also prove that $C[[\mathfrak{M}]]$ is algebraically closed or real closed, if this is the case for C.

The end of this chapter contains a digression on "Cartesian representations", which allow for a finer calculus on grid-based series. This calculus is based on the observation that any grid-based series can be represented by a multivariate Laurent series. By restricting these Laurent series to be of a special form, it is possible to define special types of grid-based series, such as convergent, algebraic or effective grid-based series. In section 3.5, we will show that the Newton polygon method can again be applied to these more special types of grid-based series.

Cartesian representations are essential for the development of effective asymptotics [vdH97], but they will only rarely occur later in this book (the main exceptions being section 4.5 and some of the exercises). Therefore, sections 3.4 and 3.5 may be skipped in a first reading.

3.1 The method illustrated by examples

3.1.1 The Newton polygon and its slopes

Consider the equation

$$P(f)=\sum_{i\geqslant 0} P_i f^i = z^3 f^6 + z^4 f^5 + f^4 - 2 f^3 + f^2 + \frac{z}{1-z^2} f + \frac{z^3}{1-z} = 0 \qquad (3.1)$$

and a Puiseux series $f = c\, z^{\mu} + \cdots \in \mathbb{C}[c]\,[\![z^{\mathbb{Q}}]\!]$, where $c \neq 0$ is a formal parameter. We call $\mu = \operatorname{val} f$ the *dominant exponent* or *valuation* of f. Then

$$\alpha = \min_i \operatorname{val}(P_i\, z^{i\mu}) = \min\{3, \mu+1, 2\,\mu, 3\,\mu, 4\,\mu, 5\,\mu+4, 6\,\mu+3\}$$

is the dominant exponent of $P(f) \in \mathbb{C}[c]\,[\![z^{\mathbb{Q}}]\!]$ and

$$N_{P,z^{\mu}}(c) := P(f)_{z^{\alpha}} = 0 \qquad (3.2)$$

is a non-trivial polynomial equation in c. We call $N_{P,z^{\mu}}$ and (3.2) the *Newton polynomial* resp. *Newton equation* associated to z^{μ}.

Let us now replace c by a non-zero value in C, so that $f = c\, z^{\mu} + \cdots \in \mathbb{C}[\![z^{\mathbb{Q}}]\!]$. If f is a solution to (3.1), then we have in particular $N_{P,z^{\mu}}(c) = 0$. Consequently, $N_{P,z^{\mu}}$ must contain at least two terms, so that α occurs at least twice among the numbers $3, \mu+1, 2\,\mu, 3\,\mu, 4\,\mu, 5\,\mu+4, 6\,\mu+3$. It follows that

$$\mu \in \{2, 1, 0, -\tfrac{3}{2}\}.$$

We call 2, 1, 0 and $-\frac{3}{2}$ the *starting exponents* for (3.1). The corresponding monomials z^2, z, 1 and $z^{-3/2}$ are called *starting monomials* for (3.1).

The starting exponents may be determined graphically from the *Newton polygon* associated to (3.1), which is defined to be the convex hull of all points (i, ν) with $\nu \geqslant \operatorname{val} P_i$. Here points $(i, \nu) \in \mathbb{N} \times \mathbb{Q}$ really encode points $(f^i, z^{\nu}) \in f^{\mathbb{N}} \times z^{\mathbb{Q}}$ (recall the explanations below figure 2.1). The Newton polygon associated to (3.1) is drawn at the left hand side of figure 3.1. The diagonal slopes

$$\begin{array}{lcll} (1, z^3) & \rightarrow & (f, z) & (\mu = 2) \\ (f, z) & \rightarrow & (f^2, 1) & (\mu = 1) \\ (f^2, 1) & \rightarrow & (f^4, 1) & (\mu = 0) \\ (f^4, 1) & \rightarrow & (f^6, z^3) & (\mu = -\tfrac{3}{2}) \end{array}$$

correspond to the starting exponents for (3.1).

Given a starting exponent $\mu \in \mathbb{Q}$ for (3.1), a non-zero solution c of the corresponding Newton equation is called a *starting coefficient* and $c\, z^\mu$ a *starting term*. Below, we listed the starting coefficients c as a function of μ in the case of equation (3.2):

μ	$N_{P,\mu}$	c	multiplicity
2	$c+1$	-1	1
1	c^2+c	-1	1
0	$c^4-2c^3+c^2$	1	2
$\frac{3}{2}$	c^6+c^4	$-\mathrm{i}, \mathrm{i}$	1

Notice that the Newton polynomials can again be read off from the Newton polygon. Indeed, when labeling each point (f^i, z^μ) by the coefficient of z^μ in P_i, the coefficients of N_{P,z^μ} are precisely the coefficients on the edge with slope μ.

Given a starting term $c\, z^\mu \in \mathbb{C}\, z^{\mathbb{Q}}$, we can now consider the equation $\tilde{P}(\tilde{f}) = 0$ which is obtained from (3.1), by substituting $c\, z^\mu + \tilde{f}$ for f, and where $\tilde{f}$ satisfies the asymptotic constraint $\tilde{f} \prec z^\mu$. For instance, if $c z^\mu = 1\, z^0$, then we obtain:

$$\begin{aligned} \tilde{P}(\tilde{f}) \;=\; & z^3 \tilde{f}^6 + (6z^3)\, \tilde{f}^5 + (15\, z^3 + 5\, z^4 + 1)\, \tilde{f}^4 + \\ & (20\, z^3 + 10\, z^4 + 2)\, \tilde{f}^3 + (15\, z^3 + 10\, z^4 + 1)\, \tilde{f}^2 + \\ & \left(6\, z^3 + 5\, z^4 + \tfrac{z}{1-z^2}\right) \tilde{f} + z^4 + z^3 + \tfrac{z^4+z^3+z}{1-z^2} = 0 \quad (\tilde{f} \prec 1) \end{aligned} \tag{3.3}$$

The Newton polygon associated to (3.3) is illustrated at the right hand side of figure 3.1. It remains to be shown that we may solve (3.3) by using the same method in a recursive way.

3.1.2 Equations with asymptotic constraints and refinements

First of all, since the new equation (3.3) comes with the asymptotic side-condition $\tilde{f} \prec 1$, it is convenient to study polynomial equations with asymptotic side-conditions

$$P(f) = 0 \quad (f \prec z^\nu) \tag{3.4}$$

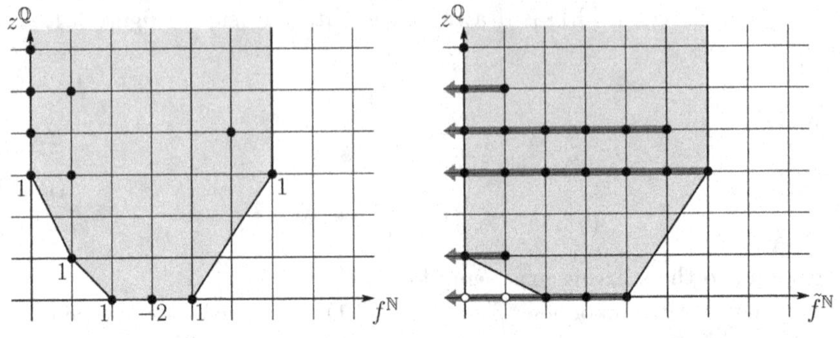

Fig. 3.1. The left-hand side shows the Newton polygon associated to the equation (3.1). The slopes of the four edges correspond to the starting exponents 2, 1, 0 and $-\frac{3}{2}$ (from left to right). After the substitution

$$f \to 1 + \tilde{f} \ (\tilde{f} \prec 1),$$

we obtain the equation (3.3), whose Newton polygon is shown at the right-hand side. Each non-zero coefficient P_{i,z^α} in the equation (3.1) for f induces a "row" of (potentially) non-zero coefficients $\tilde{P}_{\tilde{i},z^\alpha}$ in the equation for $\tilde{f}$, in the direction of the arrows. The horizontal direction of the arrows corresponds to the slope of the starting exponent 0. Moreover, the fact that 1 is a starting term corresponds to the fact that the coefficient of the lowest left-most induced point vanishes.

in a systematic way. The case of usual polynomial equations is recovered by allowing $\nu = -\infty$. In order to solve (3.4), we now only keep those starting monomials z^μ for $P(f)=0$ which satisfy the asymptotic side condition $z^\mu \prec z^\nu$, i.e. $\mu > \nu$.

The highest degree of N_{P,z^μ} for a monomial $z^\mu \prec z^\nu$ is called the Newton degree of (3.4). If $d>0$, then P is either divisible by f (and $f=0$ is a solution to (3.4)), or (3.4) admits a starting monomial (and we can carry out one step of the above resolution procedure). If $d=0$, then (3.4) admits no solutions.

Remark 3.1. Graphically speaking, the starting exponents for (3.4) correspond to sufficiently steep slopes in the Newton polygon (see figure 3.2). Using a substitution $f = z^\nu \tilde{f}$, the equation (3.4) may always be transformed into an equation

$$\tilde{P}(\tilde{f}) = 0 \quad (\tilde{f} \prec 1)$$

with a normalized asymptotic side-condition (the case $\nu = -\infty$ has to be handled with some additional care). Such transformations, called *multiplicative conjugations*, will be useful in chapter 8, and their effect on the Newton polygon is illustrated in figure 3.2.

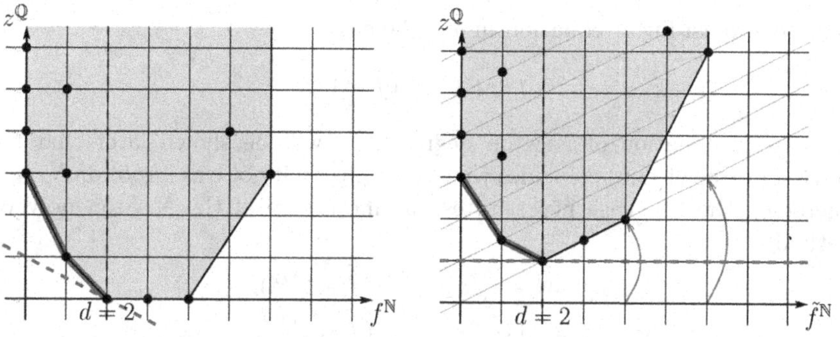

Fig. 3.2. At the left-hand side, we have illustrated the Newton polygon for the asymptotic equation $P(f)=0$ $(f \prec z^{1/2})$. The dashed line corresponds to the slope 1/2 and the edges of the Newton polygon with slope > 1/2 have been highlighted. Notice that the Newton degree $d=2$ corresponds to the first coordinate of the rightmost point on an edge with slope > 1/2. At the right-hand side, we have shown the "pivoting" effect around the origin of the substitution $f=z^{1/2}\tilde{f}$ on the Newton polygon.

Given a starting term $\varphi=\tau=c\,z^{\mu}$ or a more general series $\varphi=c\,z^{\mu}+\cdots\in \mathbb{C}[[z^{\mathbb{Q}}]]$, we next consider the transformation

$$f=\varphi+\tilde{f} \quad (\tilde{f}\prec z^{\tilde{\nu}}), \tag{3.5}$$

with $z^{\tilde{\nu}}\preccurlyeq z^{\mu}$, which transforms (3.4) into a new asymptotic polynomial equation

$$\tilde{P}(\tilde{f})=0 \quad (\tilde{f}\prec z^{\tilde{\nu}}). \tag{3.6}$$

Transformations like (3.5) are called *refinements*. A refinement is said to be *admissible*, if the Newton degree of (3.6) does not vanish.

Now the process of computing starting terms and their corresponding refinements is generally infinite and even transfinite. *A priori*, the process therefore only generates an infinite number of necessary conditions for Puiseux series f to satisfy (3.4). In order to really *solve* (3.4), we have to prove that, after a finite number of steps of the Newton polygon method, and whatever starting terms we chose (when we have a choice), we obtain an asymptotic polynomial equation with a unique solution. In the next section, we will prove an implicit function theorem which guarantees the existence of such a unique solution for equations of Newton degree one. Such equations will be said to be *quasi-linear*.

Returning to our example equation (3.1), it can be checked that each of the refinements

$$\begin{aligned}
f &= -z^2+\tilde{f} && (\tilde{f}\prec z^2);\\
f &= -z+\tilde{f} && (\tilde{f}\prec z);\\
f &= -\mathrm{i}\,z^{-3/2}+\tilde{f} && (\tilde{f}\prec z^{-3/2});\\
f &= \mathrm{i}\,z^{-3/2}+\tilde{f} && (\tilde{f}\prec z^{-3/2})
\end{aligned}$$

leads to a quasi-linear equation in $\tilde{f}$. The case

$$f = 1 + \tilde{f} \quad (\tilde{f} \prec 1)$$

leads to an equation of Newton degree 2 (it will be shown later that the Newton degree of (3.6) coincides with the multiplicity of c as a root of N_{P,z^μ}). Therefore, the last case necessitates one more step of the Newton polygon method:

$$\begin{aligned} \tilde{f} &= -\mathrm{i}\sqrt{z} + \tilde{\tilde{f}} \quad && (\tilde{\tilde{f}} \prec z^{1/2}); \\ \tilde{f} &= \mathrm{i}\sqrt{z} + \tilde{\tilde{f}} && (\tilde{\tilde{f}} \prec z^{1/2}). \end{aligned}$$

For both refinements, it can be checked that the asymptotic equation in $\tilde{\tilde{f}}$ is quasi-linear. Hence, after a finite number of steps, we have obtained a complete description of the set of solutions to (3.1). The first terms of these solutions are as follows:

$$\begin{aligned} f_I &= -z^2 - 2\,z^3 - 4\,z^4 - 13\,z^5 - 50\,z^6 + O(z^7); \\ f_{II} &= -z + 3\,z^2 - 8\,z^3 + 46\,z^4 - 200\,z^5 + O(z^6); \\ f_{III} &= 1 - \mathrm{i}\,z^{1/2} + \tfrac{1}{2}\,z + \tfrac{5\,\mathrm{i}}{8}\,z^{3/2} - z^2 + O(z^{5/2}); \\ f_{IV} &= 1 + \mathrm{i}\,z^{1/2} + \tfrac{1}{2}\,z - \tfrac{5\,\mathrm{i}}{8}\,z^{3/2} - z^2 + O(z^{5/2}); \\ f_V &= -\mathrm{i}\,z^{-3/2} - 1 - \tfrac{1}{2}\,z - \mathrm{i}\,z^{3/2} - \tfrac{\mathrm{i}}{2}\,z^{5/2} + O(z^3); \\ f_{VI} &= \mathrm{i}\,z^{-3/2} - 1 - \tfrac{1}{2}\,z + \mathrm{i}\,z^{3/2} + \tfrac{\mathrm{i}}{2}\,z^{5/2} + O(z^3). \end{aligned}$$

3.1.3 Almost double roots

Usually the Newton degrees rapidly decreases during refinements and we are quickly left with only quasi-linear equations. However, in the presence of almost multiple roots, the Newton degree may remain bigger than two for quite a while. Consider for instance the equation

$$\left(f - \frac{1}{1-z}\right)^2 = \varepsilon^2 \qquad (3.7)$$

over $\mathbb{C}[[z;\varepsilon]]$, with $z \prec 1$ and $\varepsilon \prec 1$. This equation has Newton degree 2, and after n steps of the ordinary Newton polygon method, we obtain the equation

$$\left(\tilde{f} - \frac{z^n}{1-z}\right)^2 = \varepsilon^2 \quad (\tilde{f} \prec z^{n-1}),$$

which still has Newton degree 2. In order to enforce termination, an additional trick is applied: consider the first derivative

$$2\,f - \frac{2}{1-z} = 0$$

of the equation (3.7) w.r.t. f. This derived equation is quasi-linear, so it admits a unique solution

$$\varphi = \frac{1}{1-z}.$$

Now, instead of performing the usual refinement $f = 1 + \tilde{f}$ ($\tilde{f} \prec 1$) in the original equation (3.7), we perform refinement

$$f = \varphi + \tilde{f} \quad (\tilde{f} \prec 1).$$

This yields the equation

$$\tilde{f}^2 = \varepsilon^2 \quad (\tilde{f} \prec 1).$$

Applying one more step of the Newton polygon method yields the admissible refinements

$$\begin{aligned} \tilde{f} &= -\varepsilon + \tilde{\tilde{f}} && (\tilde{\tilde{f}} \prec \varepsilon); \\ \tilde{f} &= \varepsilon + \tilde{\tilde{f}} && (\tilde{\tilde{f}} \prec \varepsilon). \end{aligned}$$

In both cases, we obtain a quasi-linear equation in $\tilde{\tilde{f}}$:

$$\begin{aligned} -2\varepsilon \tilde{\tilde{f}} + \tilde{\tilde{f}}^2 &= 0 && (\tilde{\tilde{f}} \prec \varepsilon); \\ 2\varepsilon \tilde{\tilde{f}} + \tilde{\tilde{f}}^2 &= 0 && (\tilde{\tilde{f}} \prec \varepsilon). \end{aligned}$$

In section 3.3.2, we will show that this trick applies in general, and that the resulting method always yields a complete description of the solution set after a finite number of steps.

Remark 3.2. The idea of using repeated differentiation in order to handle almost multiple solutions is old [Smi75] and has been used in computer algebra before [Chi86, Gri91]. Our contribution has been to incorporate it directly into the Newton polygon process, as will be shown in more detail in section 3.3.2.

3.2 The implicit series theorem

In the previous section, we have stressed the particular importance of quasi-linear equations when solving asymptotic polynomial equations. In this section, we will prove an implicit series theorem for polynomial equations. In the next section, we will apply this theorem to show that quasi-linear equations admit unique solutions. The implicit series theorem admits several proofs (see the exercises). The proof we present here uses a powerful syntactic technique, which will be generalized in chapter 6.

Theorem 3.3. *Let C be a ring and $\mathfrak{M}$ a monomial monoid. Consider the polynomial equation*

$$P_n f^n + \cdots + P_0 = 0 \tag{3.8}$$

with coefficients P_0, ..., $P_n \in C[[\mathfrak{M}_{\preccurlyeq}]]$, such that $P_{0,1} = 0$ and $P_{1,1} \in C^*$. Then (3.8) admits a unique solution in $C[[\mathfrak{M}_{\prec}]]$.

Proof. Since $P_{1,1} \in C^*$, the series P_1 is invertible in $C[[\mathfrak{M}_{\preccurlyeq}]]$. Modulo division of (3.8) by P_1, we may therefore assume without loss of generality that $P_1 = 1$. Setting $Q_i = -P_i$ for all $i \neq 1$, we may then rewrite (3.8) as

$$f = Q_0 + Q_2 f^2 + \cdots + Q_n f^n. \tag{3.9}$$

Now consider the set $\mathscr{T}$ of trees with nodes of arities in $\{0, 2, \ldots, n\}$ and such that each node of arity i is labeled by a monomial in $\operatorname{supp} Q_i$. To each such tree

$$t = \begin{array}{c} \mathfrak{v} \\ \diagup \mid \diagdown \\ t_1 \ \cdots \ t_i \end{array} \in \mathscr{T}$$

we recursively associate a coefficient $c_t \in C$ and a monomial $\mathfrak{m}_t \in \mathfrak{M}$ by

$$\begin{aligned} c_t &= Q_{i,\mathfrak{v}}\, c_{t_1} \cdots c_{t_i}; \\ \mathfrak{m}_t &= \mathfrak{v}\, \mathfrak{m}_{t_1} \cdots \mathfrak{m}_{t_i}. \end{aligned}$$

Now we observe that each of these monomials $\mathfrak{m}_t$ is infinitesimal, with

$$\mathfrak{m}_t \in (\operatorname{supp} Q_0) \cdot (\operatorname{supp} Q_0 \cup \operatorname{supp} Q_2 \cup \cdots \cup \operatorname{supp} Q_n)^*. \tag{3.10}$$

Hence the mapping $t \mapsto \mathfrak{m}_t$ is strictly increasing, when $\mathscr{T}$ is given the embeddability ordering from section 1.4. From Kruskal's theorem, it follows that the family $(c_t\, \mathfrak{m}_t)_{t \in \mathscr{T}}$ is well-based and even grid-based, because of (3.10). We claim that $f = \sum_{t \in \mathscr{T}} c_t\, \mathfrak{m}_t$ is the unique solution to (3.9).

First of all, f is indeed a solution to (3.9), since

$$\begin{aligned} f &= \sum_{i \in \{0,2,\ldots,n\}} \sum_{\mathfrak{v} \in \operatorname{supp} Q_i} \sum_{t_1,\ldots,t_i \in \mathscr{T}} c_{\begin{array}{c} \scriptstyle\mathfrak{v} \\ \scriptstyle\diagup\mid\diagdown \\ \scriptstyle t_1 \cdots t_i \end{array}} \mathfrak{m}_{\begin{array}{c} \scriptstyle\mathfrak{v} \\ \scriptstyle\diagup\mid\diagdown \\ \scriptstyle t_1 \cdots t_i \end{array}} \\ &= \sum_{i \in \{0,2,\ldots,n\}} \sum_{\mathfrak{v} \in \operatorname{supp} Q_i} \sum_{t_1,\ldots,t_i \in \mathscr{T}} (Q_{i,\mathfrak{v}}\, \mathfrak{v})\, (c_{t_1}\, \mathfrak{m}_{t_1}) \cdots (c_{t_i}\, \mathfrak{m}_{t_i}) \\ &= \sum_{i \in \{0,2,\ldots,n\}} \left(\sum_{\mathfrak{v} \in \operatorname{supp} Q_i} Q_{i,\mathfrak{v}}\, \mathfrak{v} \right) \left(\prod_{j=1}^{i} \sum_{t_j \in \mathscr{T}} c_{t_j}\, \mathfrak{m}_{t_j} \right) \\ &= \sum_{i \in \{0,2,\ldots,n\}} Q_i f^i = Q_0 + Q_2 f^2 + \cdots + Q_n f^n. \end{aligned}$$

In order to see that f is the unique solution to (3.8), consider the polynomial $R(\delta) = P(f + \delta)$. Since $f \prec 1$, we have $R_i = P_i + o(1)$ for all i, whence in particular $R_1 = 1 + o(1)$. Furthermore, $P(f) = 0$ implies $R_0 = 0$. Now assume that $g \prec 1$ were another root of P. Then $\delta = g - f \prec 1$ would be a root of R, so that

$$\delta = (R_1 + R_2\, \delta + \cdots + R_{n-1}\, \delta^{n-1})^{-1} R(\delta) = 0, \tag{3.11}$$

since $R_1 + R_2\, \delta + \cdots + R_{n-1}\, \delta^{n-1} = 1 + o(1)$ is invertible. □

Exercise 3.1. Generalize theorem 3.3 to the case when (3.8) is replaced by

$$P_0 + P_1 f + P_2 f^2 + \cdots = 0,$$

where $(P_i)_{i \in \mathbb{N}} \in C[[\mathfrak{M}_{\preccurlyeq}]]$ is a grid-based family with $P_{0,1} = 0$ and $P_{1,1} \in C^*$.

Exercise 3.2. Give an alternative proof of theorem 3.3, using the fact that (3.9) admits a unique power series solution in $\mathbb{Z}[[Q_2\, Q_0, \ldots, Q_n\, Q_0^{n-1}]]\, Q_0$, when considered as an equation with coefficients in $\mathbb{Z}[[Q_0, Q_2, \ldots, Q_n]]$.

Exercise 3.3. Assuming that $\mathfrak{M}$ is totally ordered, give yet another alternative proof of theorem 3.3, by computing the terms of the unique solution by transfinite induction.

Exercise 3.4. Let $C\langle\langle z_1, \ldots, z_n\rangle\rangle$ denote the ring of non-commutative power series in $z_1, \ldots, z_n$ over C. Consider the equation

$$f(g(z_1, \ldots, z_n), z_1, \ldots, z_n) = 0 \tag{3.12}$$

with $f \in C\langle\langle y, z_1, \ldots, z_n\rangle\rangle$, $f_1 = 0$ and invertible f_y. Prove that (3.12) admits a unique infinitesimal solution $g \in C\langle\langle z_1, \ldots, z_n\rangle\rangle$.

3.3 The Newton polygon method

3.3.1 Newton polynomials and Newton degree

Let C be a constant field of characteristic zero and $\mathfrak{M}$ a totally ordered monomial group with $\mathbb{Q}$-powers. Consider the *asymptotic polynomial equation*

$$P_n\, f^n + \cdots + P_0 = 0 \quad (f \prec \mathfrak{v}), \tag{3.13}$$

with coefficients in $C[[\mathfrak{M}]]$ and $\mathfrak{v} \in \mathfrak{M}$. In order to capture ordinary polynomial equations, we will also allow $\mathfrak{v} = \top_{\mathfrak{M}}$, where $\top_{\mathfrak{M}}$ is a formal monomial with $\top_{\mathfrak{M}} \succ \mathfrak{M}$. A *starting monomial* of f relative to (3.13) is a monomial $\mathfrak{m} \prec \mathfrak{v}$ in $\mathfrak{M}$, such that there exist $0 \leqslant i < j \leqslant n$ and $\mathfrak{n} \in \mathfrak{M}$ with $P_i\, \mathfrak{m}^i \asymp P_j\, \mathfrak{m}^j \asymp \mathfrak{n}$ and $P_k\, \mathfrak{m}^k \preccurlyeq \mathfrak{n}$ for all other k. To such a starting monomial $\mathfrak{m}$ we associate the equation

$$N_{P,\mathfrak{m}}(c) = P_{n,\mathfrak{n}/\mathfrak{m}^d}\, c^n + \cdots + P_{0,\mathfrak{n}} = 0, \tag{3.14}$$

and $N_{P,\mathfrak{m}}$ is called the *Newton polynomial* associated to $\mathfrak{m}$. A *starting term* of f relative to (3.13) is a term $\tau = c\, \mathfrak{m}$, where $\mathfrak{m}$ is a starting monomial of f relative to (3.13) and $c \in C^{\neq}$ a non-zero root of $N_{P,\mathfrak{m}}$. In that case, the *multiplicity* of τ is defined to be the multiplicity of c as a root of $N_{P,\mathfrak{m}}$. Notice that there are only a finite number of starting terms relative to (3.13).

Proposition 3.4. *Let f be a non-zero solution to* (3.13). *Then τ_f is a starting term for* (3.13). □

The *Newton degree* d of (3.13) is defined to be the largest degree of the Newton polynomial associated to a monomial $\mathfrak{m} \prec \mathfrak{v}$. In particular, if there exists no starting monomial relative to (3.13), then the Newton degree equals the valuation of P in f. If $d = 1$, then we say that (3.13) is *quasi-linear*. The previous proposition implies that (3.13) does not admit any solution, if $d = 0$.

Lemma 3.5. *If* (3.13) *is quasi-linear, then it admits a unique solution in* $C[[\mathfrak{M}]]$.

Proof. If $P_0 = 0$, then our statement follows from proposition 3.4, since there are no starting monomials. Otherwise, our statement follows from theorem 3.3, after substitution of $f\mathfrak{n}$ for f in (3.13), where $\mathfrak{n}$ is chosen $\preccurlyeq$-maximal such that $\mathfrak{d}_{P_1} \succcurlyeq \mathfrak{d}_{P_i} \mathfrak{n}^{i-1}$ for all i, and after division of (3.13) by $\mathfrak{d}_{P_1}$. □

3.3.2 Decrease of the Newton degree during refinements

A *refinement* is a change of variables together with the imposition of an asymptotic constraint:

$$f = \varphi + \tilde{f} \quad (\tilde{f} \prec \tilde{\mathfrak{v}}), \tag{3.15}$$

where $\varphi \prec \mathfrak{v}$ and $\tilde{\mathfrak{v}} \preccurlyeq \mathfrak{v}$. Such a refinement transforms (3.13) into an asymptotic polynomial equation in $\tilde{f}$:

$$\tilde{P}_n \tilde{f}^n + \cdots + \tilde{P}_0 = 0 \quad (\tilde{f} \prec \tilde{\mathfrak{v}}), \tag{3.16}$$

where

$$\tilde{P}_i = \frac{1}{i!} P^{(i)}(\varphi) = \sum_{k=i}^{n} \binom{k}{i} P_k \varphi^{k-i}, \tag{3.17}$$

for each i. We say that the refinement (3.15) is *admissible* if the Newton degree of (3.16) is strictly positive.

Lemma 3.6. *Consider the refinement* (3.15) *with* $\tilde{\mathfrak{v}} = \mathfrak{d}_\varphi$. *Then*

a) *The Newton degree of* (3.16) *coincides with the multiplicity of* c *as a root of* $N_{P,\mathfrak{m}}$. *In particular,* (3.15) *is admissible if and only if* $c\,\mathfrak{m}$ *is a starting term for* (3.13).
b) *The Newton degree of* (3.16) *is bounded by the Newton degree of* (3.13).

Proof. Let d be maximal such that $P_d\, \mathfrak{m}^d \succcurlyeq P_i\, \mathfrak{m}^i$ for all i, and denote $\mathfrak{n} = \mathfrak{d}(P_d)\, \mathfrak{m}^d$. Then d is bounded by the Newton degree of (3.13) and

$$\begin{aligned}
\tilde{P}_i &= \frac{1}{i!} \sum_{k=i}^{n} \binom{k}{i} P_k \varphi^{k-i} \\
&= \frac{1}{i!} \sum_{k=i}^{n} \binom{k}{i} (P_{k,\mathfrak{n}\mathfrak{m}^{-k}} + o(1))\, \mathfrak{n}\, \mathfrak{m}^{-k} (c + o(1))^{k-i} \mathfrak{m}^{k-i} \\
&= \frac{1}{i!} N_{P,\mathfrak{m}}^{(i)}(c)\, \mathfrak{n}\, \mathfrak{m}^i + o(\mathfrak{n}\, \mathfrak{m}^i),
\end{aligned}$$

for all i. In particular, denoting the multiplicity of c as a root of $N_{P,\mathfrak{m}}$ by $\tilde{d}$, we have $\tilde{P}_{\tilde{d}} \asymp \mathfrak{n}\, \mathfrak{m}^{-\tilde{d}}$. Moreover, for all $i \geqslant \tilde{d}$, we have $\tilde{P}_i \preccurlyeq \mathfrak{n}\, \mathfrak{m}^{-i}$. Hence, for any $i > \tilde{d}$ and $\tilde{\mathfrak{m}} \prec \mathfrak{m}$, we have $\tilde{P}_i\, \tilde{\mathfrak{m}}^i \prec \tilde{P}_{\tilde{d}}\, \tilde{\mathfrak{m}}^{\tilde{d}}$. This shows that the Newton degree of (3.16) is at most $\tilde{d}$.

Let us now show that the Newton degree of (3.16) is precisely $\tilde{d}$. Choose $\tilde{\mathfrak{m}} \prec \mathfrak{m}$ large enough, so that

$$\tilde{\mathfrak{m}} \succcurlyeq \sqrt[\tilde{d}-i]{\frac{\mathfrak{d}_{\tilde{P}_i(\tilde{f})}}{\mathfrak{d}_{\tilde{P}_{\tilde{d}}(\tilde{f})}}}$$

for all $i < \tilde{d}$. Then $\deg N_{\tilde{P},\tilde{\mathfrak{m}}} = \tilde{d}$. □

If one step of the Newton polygon method does not suffice to decrease the Newton degree, then two steps do, when applying the trick from the next lemma:

Lemma 3.7. *Let d be the Newton degree of (3.13). If f admits a unique starting monomial $\mathfrak{m}$ and $N_{P,\mathfrak{m}}$ a unique root c of multiplicity d, then*

a) The equation

$$P^{(d-1)}(\varphi) = 0 \quad (\varphi \prec \mathfrak{v}) \tag{3.18}$$

is quasi-linear and its unique solution satisfies $\varphi = c\,\mathfrak{m} + o(\mathfrak{m})$.

b) The Newton degree of any refinement

$$\tilde{f} = \tilde{\varphi} + \tilde{\tilde{f}} \quad (\tilde{\tilde{f}} \prec \tilde{\tilde{\mathfrak{v}}})$$

relative to (3.16) with $\tilde{\tilde{\mathfrak{v}}} = \mathfrak{d}_{\tilde{\varphi}}$ is strictly inferior to d.

Proof. Notice first that $N_{P',\mathfrak{m}} = N'_{P,\mathfrak{m}}$ for all polynomials P and monomials $\mathfrak{m}$. Consequently, (3.18) is quasi-linear and c is a single root of $N_{P^{(d-1)},\mathfrak{m}}$. This proves ($a$).

As to (b), we first observe that $\tilde{P}_{d-1} = P^{(d-1)}(\varphi) = 0$. Given $\tilde{\mathfrak{m}} \prec \tilde{\mathfrak{v}}$, it follows that $N_{\tilde{P},\mathfrak{m},d-1} = 0$. In particular, there do not exist $\alpha \neq 0$, $\beta \neq 0$ with $N_{\tilde{P},\tilde{\mathfrak{m}}}(\tilde{c}) = \alpha\,(\tilde{c} - \beta)^d$. In other words, $N_{\tilde{P},\tilde{\mathfrak{m}}}$ does not admit roots of multiplicity d. We conclude by lemma 3.6. □

3.3.3 Resolution of asymptotic polynomial equations

Theorem 3.8. *Let C be an algebraically closed field of characteristic zero and $\mathfrak{M}$ a totally ordered monomial group with $\mathbb{Q}$-powers. Then $C[[\mathfrak{M}]]$ is algebraically closed.*

Proof. Consider the following algorithm:

Algorithm `polynomial_solve`
Input: An asymptotic polynomial equation (3.13).
Output: The set of solutions to (3.13).

1. Compute the starting terms $c_1\,\mathfrak{m}_1, \ldots, c_\nu\,\mathfrak{m}_\nu$ of f relative to (3.13).
2. If $\nu = 1$ and c_1 is a root of multiplicity d of $N_{P,\mathfrak{m}_1}$, then let φ be the unique solution to (3.18). Refine (3.15) and apply `polynomial_solve` to (3.16). Return the so obtained solutions to (3.13).

3. For each $1 \leqslant i \leqslant \nu$, refine

$$f = c_i\, \mathfrak{m}_i + \tilde{f} \quad (\tilde{f} \prec \mathfrak{m}_i)$$

and apply `polynomial_solve` to the new equation in $\tilde{f}$. Collect and return the so obtained solutions to (3.13), together with 0, if P is divisible by f.

The correctness of `polynomial_solve` is clear; its termination follows from lemmas 3.6(b) and 3.7(b). Since C is algebraically closed, all Newton polynomials considered in the algorithm split over C. Hence, `polynomial_solve` returns d solutions to (3.13) in $C[[\mathfrak{M}]]$, when counting with multiplicities. In particular, when taking $\mathfrak{v} = \top_{\mathfrak{M}} \succ \mathfrak{M}$, we obtain n solutions, so $C[[\mathfrak{M}]]$ is algebraically closed. □

Corollary 3.9. *Let C be a real closed field and $\mathfrak{M}$ a totally ordered monomial group with $\mathbb{Q}$-powers. Then $C[[\mathfrak{M}]]$ is real closed.*

Proof. By the theorem, a polynomial equation $P(n) = 0$ of degree n over $C[[\mathfrak{M}]]$ admits n solutions in $C[\mathrm{i}][[\mathfrak{M}]]$, when counting with multiplicities. Moreover, each root $\varphi \in C[\mathrm{i}][[\mathfrak{M}]] \setminus C[[\mathfrak{M}]]$ is imaginary, because

$$\mathrm{i} = \frac{\varphi - \operatorname{Re} \varphi}{\operatorname{Im} \varphi} \in C[[\mathfrak{M}]][\varphi]$$

for such φ. Therefore all real roots of P are in $C[[\mathfrak{M}]]$. □

Corollary 3.10. *The field $C[[z^{\mathbb{Q}}]]$ of Puiseux series over an algebraically resp. real closed field C is algebraically resp. real closed.* □

Exercise 3.5. Consider an asymptotic algebraic equation (3.13) of Newton degree d. Let $\tau_1, \ldots, \tau_k$ be the starting terms of (3.13), with multiplicities $\mu_1, \ldots, \mu_k$. Prove that

$$\mu_1 + \cdots + \mu_k \leqslant d.$$

Also show that $\mu_1 + \cdots + \mu_k = d$ if C is algebraically closed.

Exercise 3.6.

a) Show that the computation of all solutions to (3.13) can be represented by a finite tree, whose non-root nodes are labeled by refinements. Applied to (3.1), this would yields the following tree:

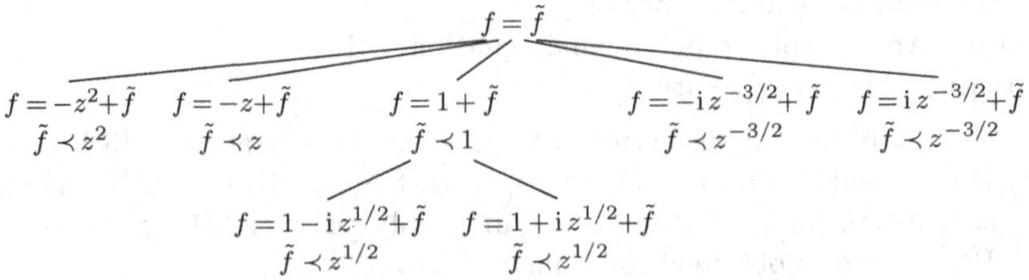

b) Show that the successors of each node may be ordered in a natural way, if C is a real field, and if we restrict our attention to real algebraic solutions. Prove that the natural ordering on the leaves, which is induced by this ordering, corresponds to the usual ordering of the solutions.

Exercise 3.7.

a) Generalize the results of this chapter to asymptotic equations of infinite degree in f, but of finite Newton degree.
b) Give an example of an asymptotic equation of infinite degree in f, with infinitely many solutions.

Exercise 3.8. Consider an asymptotic polynomial equation

$$P(f) = 0 \ (f \prec \mathfrak{v})$$

of Newton degree d, with $P \in C[[\mathfrak{M}]][F]$ and $\mathfrak{v} \in \mathfrak{M}$. Consider the monomial monoid $\mathfrak{U} = \mathfrak{M} \times F^{\mathbb{N}}$ with

$$\mathfrak{m} F^i \prec 1 \Leftrightarrow \mathfrak{m} \mathfrak{v}^i \prec 1 \vee (\mathfrak{m} \mathfrak{v}^i = 1 \wedge i > 0).$$

a) Show that there exists a unique invertible series $u \in C[[\mathfrak{U}]]$ such that $\tilde{P} = u P$ is a monoic polynomial in $C[[\mathfrak{M}]][F]$.
b) Show that $\deg \tilde{P} = d$.

3.4 Cartesian representations

In this section, we show that grid-based series may be represented by (finite sums of) multivariate Laurent series in which we substitute an infinitesimal monomial for each variable. Such representations are very useful for finer computations with grid-based series.

3.4.1 Cartesian representations

Let $C[[\mathfrak{M}]]$ be a grid-based algebra. A *Cartesian representation* for a series $f \in C[[\mathfrak{M}]]$ is a multivariate Laurent series $\check{f} \in C((\check{\mathfrak{z}}_1, \ldots, \check{\mathfrak{z}}_k))$, such that $f = \hat{\varphi}(\check{f})$ for some morphism of monomial monoids $\varphi: \check{\mathfrak{z}}_1^{\mathbb{Z}} \cdots \check{\mathfrak{z}}_k^{\mathbb{Z}} \to \mathfrak{M}$. Writing $\check{f} = \check{g}\, \check{\mathfrak{z}}_1^{\alpha_1} \cdots \check{\mathfrak{z}}_k^{\alpha_k}$, with $\check{g} \in C[[\check{\mathfrak{z}}_1, \ldots, \check{\mathfrak{z}}_k]]$, we may also interpret f as the product of a "series" $\hat{\varphi}(\check{g})$ in $\varphi(\check{\mathfrak{z}}_1), \ldots, \varphi(\check{\mathfrak{z}}_k)$ and the monomial $\mathfrak{m} = \varphi(\check{\mathfrak{z}}_1^{\alpha_1} \cdots \check{\mathfrak{z}}_k^{\alpha_k})$.

More generally, a *semi-Cartesian representation* for $f \in C[[\mathfrak{M}]]$ is an expression of the form

$$f = \hat{\varphi}(\check{g}_1)\, \mathfrak{m}_1 + \cdots + \hat{\varphi}(\check{g}_l)\, \mathfrak{m}_l,$$

where $g_1, \ldots, g_l \in C[[\check{\mathfrak{z}}_1, \ldots, \check{\mathfrak{z}}_k]]$, $\mathfrak{m}_1, \ldots, \mathfrak{m}_l \in \mathfrak{M}$ and $\varphi: \check{\mathfrak{z}}_1^{\mathbb{N}} \cdots \check{\mathfrak{z}}_k^{\mathbb{N}} \to \mathfrak{M}$ is a morphism of monomial monoids.

Proposition 3.11.

a) *Any grid-based series $f \in C[[\mathfrak{M}]]$ admits a semi-Cartesian representation.*
b) *If $\mathfrak{M}$ is a monomial group, which is generated by its infinitesimal elements, then each grid-based series $f \in C[[\mathfrak{M}]]$ admits a Cartesian representation.*

Proof.

a) Let $\mathfrak{m}_1, \ldots, \mathfrak{m}_k \in \mathfrak{M}^{\prec}$ and $\mathfrak{n}_1, \ldots, \mathfrak{n}_l \in \mathfrak{M}$ be such that

$$\operatorname{supp} f \subseteq \{\mathfrak{m}_1, \ldots, \mathfrak{m}_k\}^* \{\mathfrak{n}_1, \ldots, \mathfrak{n}_l\}.$$

For each $\mathfrak{v} \in \operatorname{supp} f$, let

$$n_{\mathfrak{v}} = \operatorname{card} \{(\alpha_1, \ldots, \alpha_k, i) \in \mathbb{N}^k \times \{1, \ldots, l\} : \mathfrak{v} = \mathfrak{m}_1^{\alpha_1} \cdots \mathfrak{m}_k^{\alpha_k} \mathfrak{n}_i\}.$$

Let

$$\breve{g}_i = \sum_{\alpha_1, \ldots, \alpha_k \in \mathbb{N}^k} \frac{f_{\mathfrak{m}_1^{\alpha_1} \cdots \mathfrak{m}_k^{\alpha_k} \mathfrak{n}_i}}{n_{\mathfrak{m}_1^{\alpha_1} \cdots \mathfrak{m}_k^{\alpha_k} \mathfrak{n}_i}} \breve{\mathfrak{z}}_1^{\alpha_1} \cdots \breve{\mathfrak{z}}_k^{\alpha_k}$$

for all $1 \leqslant i \leqslant l$ and let $\varphi: \breve{\mathfrak{z}}_1^{\mathbb{N}} \cdots \breve{\mathfrak{z}}_k^{\mathbb{N}} \to \mathfrak{M}, \breve{\mathfrak{z}}_1^{\alpha_1} \cdots \breve{\mathfrak{z}}_k^{\alpha_k} \mapsto \mathfrak{m}_1^{\alpha_1} \cdots \mathfrak{m}_k^{\alpha_k}$. Then

$$f = \hat{\varphi}(\breve{g}_1)\, \mathfrak{n}_1 + \cdots + \hat{\varphi}(\breve{g}_l)\, \mathfrak{n}_l.$$

b) For certain $\mathfrak{m}_{k+1}, \ldots, \mathfrak{m}_p \in \mathfrak{M}^{\prec}$ and $\beta_{i,j} \in \mathbb{Z}$, we may write

$$\mathfrak{n}_i = \mathfrak{m}_{k+1}^{\beta_{i,k+1}} \cdots \mathfrak{m}_p^{\beta_{i,p}},$$

for all $1 \leqslant i \leqslant l$. Let $\psi: \breve{\mathfrak{z}}_1^{\mathbb{Z}} \cdots \breve{\mathfrak{z}}_p^{\mathbb{Z}} \to \mathfrak{M}, \breve{\mathfrak{z}}_1^{\alpha_1} \cdots \breve{\mathfrak{z}}_p^{\alpha_p} \mapsto \mathfrak{m}_1^{\alpha_1} \cdots \mathfrak{m}_p^{\alpha_p}$ and

$$\breve{f} = \sum_{i=1}^{l} \breve{g}_i\, \breve{\mathfrak{z}}_{k+1}^{\beta_{i,k+1}} \cdots \breve{\mathfrak{z}}_p^{\beta_{i,p}}.$$

Then $f = \hat{\psi}(\breve{f})$. □

Cartesian or semi-Cartesian representations $f_1 = \hat{\varphi}_1(\breve{f}_1)$ and $f_2 = \hat{\varphi}_2(\breve{f}_2)$ are said to be *compatible*, if $\breve{f}_1$ and $\breve{f}_2$ belong to the same algebra $C((\breve{\mathfrak{z}}_1, \ldots, \breve{\mathfrak{z}}_k))$ of Laurent series, and if $\varphi_1 = \varphi_2$.

Proposition 3.12.

a) *Any $f_1, \ldots, f_n \in C[[\mathfrak{M}]]$ admit compatible semi-Cartesian representations.*

b) *If $\mathfrak{M}$ is a monomial group, which is generated by its infinitesimal elements, then any $f_1, \ldots, f_n \in C[[\mathfrak{M}]]$ admit compatible Cartesian representations.*

Proof. By the previous proposition, $f_1, \ldots, f_n$ admit semi-Cartesian representations $f_i = \hat{\varphi}_i(\breve{f}_i)$, where $\breve{f}_i \in C((\breve{\mathfrak{z}}_{i,1}, \ldots, \breve{\mathfrak{z}}_{i,k_i}))$ and $\varphi_i: \breve{\mathfrak{z}}_{i,1}^{\mathbb{N}} \cdots \breve{\mathfrak{z}}_{i,k_i}^{\mathbb{N}} \to \mathfrak{M}$ for each i. Now consider

$$\begin{aligned} \psi: \prod_{i=1}^{n} \prod_{j=1}^{k_i} \breve{\mathfrak{z}}_{i,j}^{\mathbb{N}} &\longrightarrow \mathfrak{M} \\ \prod_{i=1}^{n} \prod_{j=1}^{k_i} \breve{\mathfrak{z}}_{i,j}^{\alpha_{i,j}} &\longmapsto \prod_{i=1}^{n} \hat{\varphi}_i\left(\prod_{j=1}^{k_i} \breve{\mathfrak{z}}_{i,j}^{\alpha_{i,j}}\right). \end{aligned}$$

Then $f_i = \hat{\psi}(\breve{F}_i)$ for each i, where $\breve{F}_i$ is the image of $\breve{f}_i$ under the natural inclusion of $C((\breve{\mathfrak{z}}_{i,1}, \ldots, \breve{\mathfrak{z}}_{i,k_i}))$ into $C((\breve{\mathfrak{z}}_{1,1}, \ldots, \breve{\mathfrak{z}}_{1,k_1}, \ldots, \breve{\mathfrak{z}}_{n,1}, \ldots, \breve{\mathfrak{z}}_{n,k_n}))$. This proves (*a*); part (*b*) is proved in a similar way. □

3.4.2 Inserting new infinitesimal monomials

In proposition 3.12 we drastically increased the size of the Cartesian basis in order to obtain compatible Cartesian representations. The following lemma is often useful, if one wants to keep this size as low as possible.

Lemma 3.13. *Let $\mathfrak{z}_1, \ldots, \mathfrak{z}_k, \mathfrak{m}_1, \ldots, \mathfrak{m}_l$ be infinitesimal elements of a totally ordered monomial group $\mathfrak{M}$ with $\mathbb{Q}$-powers, such that $\mathfrak{m}_1, \ldots, \mathfrak{m}_l \in \mathfrak{z}_1^{\mathbb{Z}} \cdots \mathfrak{z}_k^{\mathbb{Z}}$. Then there exist infinitesimal $\mathfrak{z}_1', \ldots, \mathfrak{z}_k' \in \mathfrak{z}_1^{\mathbb{Q}} \cdots \mathfrak{z}_k^{\mathbb{Q}}$ with $\mathfrak{z}_1, \ldots, \mathfrak{z}_k, \mathfrak{m}_1, \ldots, \mathfrak{m}_l \in (\mathfrak{z}_1')^{\mathbb{N}} \cdots (\mathfrak{z}_k')^{\mathbb{N}}$.*

Proof. It suffices to prove the lemma for $l = 1$; the general case follows by induction over l. The case $l = 1$ is proved by induction over k. For $k = 0$, there is nothing to prove. So assume that $k \geqslant 1$ and let $\mathfrak{m}_1 = \mathfrak{z}_1^{\alpha_1} \cdots \mathfrak{z}_k^{\alpha_k}$ with $\alpha_1, \ldots, \alpha_k \in \mathbb{Z}$. Without loss of generality, we may assume that $\alpha_k > 0$, modulo a permutation of variables. Putting $\mathfrak{n} = \mathfrak{z}_1^{\alpha_1} \cdots \mathfrak{z}_{k-1}^{\alpha_{k-1}}$, we now distinguish the following three cases:

1. If $\mathfrak{n} \prec 1$, then there exist infinitesimal $\mathfrak{z}_1' \cdots \mathfrak{z}_{k-1}' \in \mathfrak{z}_1^{\mathbb{Z}} \cdots \mathfrak{z}_{k-1}^{\mathbb{Z}}$, such that $\mathfrak{z}_1, \ldots, \mathfrak{z}_{k-1}, \mathfrak{n} \in (\mathfrak{z}_1')^{\mathbb{N}} \cdots (\mathfrak{z}_{k-1}')^{\mathbb{N}}$, by the induction hypothesis. Taking $\mathfrak{z}_k' = \mathfrak{z}_k$, we now have $\mathfrak{z}_k, \mathfrak{m}_1 = \mathfrak{n}\, \mathfrak{z}_k^{\alpha_k} \in (\mathfrak{z}_1')^{\mathbb{N}} \cdots (\mathfrak{z}_k')^{\mathbb{N}}$, since $\alpha_k > 0$.
2. If $\mathfrak{n} = 1$, then $\mathfrak{m}_1 = \mathfrak{z}_k^{\alpha_k}$, and we may take $\mathfrak{z}_1' = \mathfrak{z}_1, \ldots, \mathfrak{z}_k' = \mathfrak{z}_k$.
3. If $\mathfrak{n} \succ 1$, then there exists infinitesimal $\mathfrak{z}_1' \cdots \mathfrak{z}_{k-1}' \in \mathfrak{z}_1^{\mathbb{Z}} \cdots \mathfrak{z}_{k-1}^{\mathbb{Z}}$, such that $\mathfrak{z}_1^{1/\alpha_k}, \ldots, \mathfrak{z}_{k-1}^{1/\alpha_k}, \mathfrak{n}^{-1/\alpha_k} \in (\mathfrak{z}_1')^{\mathbb{N}} \cdots (\mathfrak{z}_{k-1}')^{\mathbb{N}}$. Taking $\mathfrak{z}_k' = \mathfrak{z}_1^{\alpha_1/\alpha_k} \cdots \mathfrak{z}_{k-1}^{\alpha_{k-1}/\alpha_k} \mathfrak{z}_k$, we again have $\mathfrak{z}_k = \mathfrak{z}_k' \mathfrak{n}^{-1/\alpha_k}, \mathfrak{m}_1 = (\mathfrak{z}_k')^{\alpha_k} \in (\mathfrak{z}_1')^{\mathbb{N}} \cdots (\mathfrak{z}_k')^{\mathbb{N}}$. □

When doing computations on grid-based series in $C[[\mathfrak{M}]]$, one often works with respect to a *Cartesian basis* $\mathfrak{Z} = (\mathfrak{z}_1, \ldots, \mathfrak{z}_k)$ of infinitesimal elements in $\mathfrak{M}$. Each time one encounters a series $f \in C[[\mathfrak{M}]]$ which cannot be represented by a series in $C((\check{\mathfrak{z}}_1, \ldots, \check{\mathfrak{z}}_k))$, one has to replace $\mathfrak{Z}$ by a *wider* Cartesian basis $\mathfrak{Z}' = (\mathfrak{z}_1', \ldots, \mathfrak{z}_{k'}')$ with $\mathfrak{z}_1, \ldots, \mathfrak{z}_k \in (\mathfrak{z}_1')^{\mathbb{N}} \cdots (\mathfrak{z}_{k'}')^{\mathbb{N}}$. The corresponding mapping $C((\check{\mathfrak{z}}_1, \ldots, \check{\mathfrak{z}}_k)) \to C((\check{\mathfrak{z}}_1', \ldots, \check{\mathfrak{z}}_{k'}'))$ is called a *widening*. Lemma 3.13 enables us to keep the Cartesian basis reasonably small during the computation.

3.5 Local communities

Let C be a ring and $\mathfrak{M}$ a monomial group which is generated by its infinitesimal elements. Given a set $A_k \subseteq C[[\mathfrak{z}_1, \ldots, \mathfrak{z}_k]]$ for each $k \in \mathbb{N}$, we denote by $C[[\mathfrak{M}]]_A$ the set of all grid-based series $f \in C[[\mathfrak{M}]]$, which admit a Cartesian representation $\check{f} \in A_k\, \mathfrak{z}_1^{\mathbb{Z}} \cdots \mathfrak{z}_k^{\mathbb{Z}}$ for some $k \in \mathbb{N}$. In this section, we will show that if the A_k satisfy appropriate conditions, then many types of computations which can be carried out in $C[[\mathfrak{M}]]$ can also be carried out in $C[[\mathfrak{M}]]_A$.

3.5.1 Cartesian communities

Let C be a ring. A sequence $(A_k)_{k\in\mathbb{N}}$ with $A_k \subseteq C[[\mathfrak{z}_1, \ldots, \mathfrak{z}_k]]$ is said to be a *Cartesian community* over C, if the following conditions are satisfied:

CC1. $\mathfrak{z}_1 \in A_1$.

CC2. A_k is a C-algebra for each $k \in \mathbb{N}$.

CC3. The A_k are stable under strong monomial morphisms.

In **CC3**, a *strong monomial morphism* is strong C-algebra morphism which maps monomials to monomials. In our case, a monomial preserving strong morphism from $C[[\mathfrak{z}_1, \ldots, \mathfrak{z}_k]]$ into $C[[\mathfrak{z}_1, \ldots, \mathfrak{z}_{k'}]]$ is always of the form

$$\begin{aligned}\sigma: C[[\mathfrak{z}_1, \ldots, \mathfrak{z}_k]] &\longrightarrow C[[\mathfrak{z}_1, \ldots, \mathfrak{z}_{k'}]];\\ f(\mathfrak{z}_1, \ldots, \mathfrak{z}_k) &\longmapsto f(\mathfrak{z}_1^{\alpha_{1,1}} \cdots \mathfrak{z}_{k'}^{\alpha_{1,k'}}, \ldots, \mathfrak{z}_1^{\alpha_{k,1}} \cdots \mathfrak{z}_{k'}^{\alpha_{k,k'}}),\end{aligned}$$

where $\alpha_{i,j} \in \mathbb{N}$ and $\sum_j \alpha_{i,j} \neq 0$ for all i. In particular, **CA3** implies that the A_k are stable under widenings.

Proposition 3.14. *Let $(A_k)_{k\in\mathbb{N}}$ be a Cartesian community over C and let $\mathfrak{M}$ be a monomial group. Then $C[[\mathfrak{M}]]_A$ is a C-algebra.*

Proof. We clearly have $C \subseteq C[[\mathfrak{M}]]_A$. Let $\hat{f}, \hat{g} \in C[[\mathfrak{M}]]_A$. Mimicking the proof of proposition 3.12, we observe that f and g admit compatible Cartesian representations $f, g \in A_k\, \mathfrak{z}_1^{\mathbb{Z}} \cdots \mathfrak{z}_k^{\mathbb{Z}}$. Then $f+g$, $f-g$ and $f g$ are Cartesian representations of $\hat{f} + \hat{g}$, $\hat{f} - \hat{g}$ resp. $\hat{f}\hat{g}$. □

3.5.2 Local communities

A *local community* is a Cartesian community $(A_k)_{k\in\mathbb{N}}$, which satisfies the following additional conditions:

LC1. For each $f \in A_k$ with $[\mathfrak{z}_k^0] f = 0$, we have $f/\mathfrak{z}_k \in A_k$.

LC2. Given $g \in A_k$ and $f_1, \ldots, f_k \in A_l^{\prec}$, we have $g \circ (f_1, \ldots, f_k) \in A_l$.

LC3. Given $f \in A_{k+1}$ with $[\mathfrak{z}_1^0 \cdots \mathfrak{z}_{k+1}^0] f = 0$ and $[\mathfrak{z}_1^0 \cdots \mathfrak{z}_k^0\, \mathfrak{z}_{k+1}^1] f \in C^*$, the unique series $\varphi \in C[[\mathfrak{z}_1, \ldots, \mathfrak{z}_k]]$ with $f \circ (\mathfrak{z}_1, \ldots, \mathfrak{z}_k, \varphi) = 0$ belongs to A_k.

In **LC1** and **LC3**, the notation $[\mathfrak{z}_1^{\alpha_1} \cdots \mathfrak{z}_p^{\alpha_p}] f$ stands for the coefficient of $\mathfrak{z}_1^{\alpha_1} \cdots \mathfrak{z}_p^{\alpha_p}$ in f. The condition **LC3** should be considered as an implicit function theorem for the local community. Notice that A_k is stable under $\partial/\partial \mathfrak{z}_i$ for all $\{i \in 1, \ldots, k\}$, since

$$\frac{\partial f}{\partial \mathfrak{z}_i} = \frac{f \circ (\mathfrak{z}_1, \ldots, \mathfrak{z}_i + \mathfrak{z}_{k+1}, \ldots, \mathfrak{z}_k) - f}{\mathfrak{z}_{k+1}} \circ (\mathfrak{z}_1, \ldots, \mathfrak{z}_k, 0). \tag{3.19}$$

Remark 3.15. In [vdH97], the conditions **LC2** and **LC3** were replaced by a single, equivalent condition: given $f \in A_{k+1}$ as in **LC3**, we required that $\operatorname{im} \varphi \subseteq A_k$, for the unique strong C-algebra morphism $\varphi: C[[\mathfrak{z}_1, \ldots, \mathfrak{z}_{k+1}]] \to C[[\mathfrak{z}_1, \ldots, \mathfrak{z}_k]]$, such that $\varphi_{|C[[\mathfrak{z}_1, \ldots, \mathfrak{z}_k]]} = \mathrm{Id}_{C[[\mathfrak{z}_1, \ldots, \mathfrak{z}_k]]}$ and $\varphi(f) = 0$. We also explicitly requested the stability under differentiation, although (3.19) shows that this is superfluous.

Example 3.16. Let C be a subfield of $\mathbb{C}$ and let $A_k = C\{\{\mathfrak{z}_1, \ldots, \mathfrak{z}_k\}\}$ be the set of convergent power series in k variables over C, for each $k \in \mathbb{N}$. Then the A_k form a local community. If $\mathfrak{M}$ is a monomial group, then $C\{\!\{\mathfrak{M}\}\!\} = C[\![\mathfrak{M}]\!]_A$ will also be called the set of *convergent grid-based series* in $\mathfrak{M}$ over C.

Example 3.17. For each $k \in \mathbb{N}$, let A_k be the set of power series in $C[[\mathfrak{z}_1, \ldots, \mathfrak{z}_k]]$, which satisfy an algebraic equation over $C[\mathfrak{z}_1, \ldots, \mathfrak{z}_k]$. Then the A_k form a local community.

3.5.3 Faithful Cartesian representations

In this and the next section, $A = (A_k)_{k \in \mathbb{N}}$ is a local community. A Cartesian representation $f \in C((\mathfrak{z}_1, \ldots, \mathfrak{z}_k))$ is said to be *faithful*, if for each dominant monomial $\mathfrak{d}$ of f, there exists a dominant monomial $\hat{\mathfrak{d}}'$ of $\hat{f}$, with $\hat{\mathfrak{d}} \preccurlyeq \hat{\mathfrak{d}}'$.

Proposition 3.18. *Let $(A_i)_{i \in \mathbb{N}}$ be a local community and $f \in A_k$. Then*

a) For each $1 \leqslant i \leqslant k$ and $\alpha \in \mathbb{Z}$, we have $[\mathfrak{z}_k^\alpha] f \in A_{k-1}$.
b) For each initial segment $\mathfrak{I} \subseteq \mathfrak{z}_1^{\mathbb{Z}} \cdots \mathfrak{z}_k^{\mathbb{Z}}$, we have

$$f_{\mathfrak{I}} = \sum_{\mathfrak{m} \in \mathfrak{I}} f_{\mathfrak{m}} \mathfrak{m} \in A_k.$$

Proof. For each α, let $f_\alpha = [\mathfrak{z}_k^\alpha] f$. We will prove (a) by a weak induction over α. If $\alpha = 0$, then $[\mathfrak{z}_k^0] f = f \circ (\mathfrak{z}_1, \ldots, \mathfrak{z}_{k-1}, 0) \in A_{k-1}$. If $\alpha > 0$, then

$$[\mathfrak{z}_k^\alpha] f = \frac{f - ([\mathfrak{z}_k^0] f) \mathfrak{z}_k^0 - \cdots - ([\mathfrak{z}_k^{\alpha-1}] f) \mathfrak{z}_k^{\alpha-1}}{\mathfrak{z}_k^\alpha}.$$

By the weak induction hypothesis and **LC1**, we thus have $[\mathfrak{z}_k^\alpha] f \in A_k$.

In order to prove (b), let $\mathfrak{D} = \{\mathfrak{d}_1, \ldots, \mathfrak{d}_l\}$ be the finite anti-chain of maximal elements of $\mathfrak{I}$, so that $\mathfrak{I} = \operatorname{in}(\mathfrak{d}_1, \ldots, \mathfrak{d}_l)$. Let n be the number of variables which effectively occur in $\mathfrak{D}$, i.e. the number of $i \in \{1, \ldots, k\}$, such that $\mathfrak{d}_j = \mathfrak{z}_1^{\alpha_1} \cdots \mathfrak{z}_k^{\alpha_k}$ with $\alpha_i \neq 0$ for some j. We prove (b) by weak induction over n. If $n = 0$, then either $l = 0$ and $f_{\mathfrak{I}} = 0$, or $l = 1$, $\mathfrak{d}_1 = \{1\}$ and $f_{\mathfrak{I}} = f$.

Assume now that $n > 0$ and order the variables $\mathfrak{z}_1, \ldots, \mathfrak{z}_k$ in such a way that $\mathfrak{z}_k$ effectively occurs in one of the $\mathfrak{d}_i$. For each $\alpha \in \mathbb{N}$, let

$$\begin{aligned} \mathfrak{I}_\alpha &= \{\mathfrak{m} \in \mathfrak{z}_1^{\mathbb{N}} \cdots \mathfrak{z}_{k-1}^{\mathbb{N}} : \mathfrak{m} \mathfrak{z}_k^\alpha \in \mathfrak{I}\}; \\ \mathfrak{D}_\alpha &= \{\mathfrak{m} \in \mathfrak{z}_1^{\mathbb{N}} \cdots \mathfrak{z}_{k-1}^{\mathbb{N}} : \mathfrak{m} \mathfrak{z}_k^\alpha \in \mathfrak{D}\}. \end{aligned}$$

We observe that

$$\mathfrak{I}_\alpha = \operatorname{in}(\mathfrak{D}_0 \amalg \cdots \amalg \mathfrak{D}_\alpha) \cap \mathfrak{z}_1^{\mathbb{N}} \cdots \mathfrak{z}_{k-1}^{\mathbb{N}}.$$

In particular, if ν is maximal with $\mathfrak{D}_\nu \neq \varnothing$, then $\mathfrak{I}_\alpha = \mathfrak{I}_\nu$ for all $\alpha \geqslant \nu$ and

$$\mathfrak{I} = \mathfrak{I}_0 \amalg \cdots \amalg \mathfrak{I}_{\nu-1} \mathfrak{z}_k^{\nu-1} \amalg \mathfrak{I}_\nu \mathfrak{z}_k^{\nu+\mathbb{N}},$$

so that

$$f_{\mathfrak{I}} = f_{0,\mathfrak{I}_0}\mathfrak{z}_k^0 + \cdots + f_{\nu-1,\mathfrak{I}_{\nu-1}}\mathfrak{z}_k^{\nu-1} + \left(\frac{f - f_0\mathfrak{z}_k^0 - \cdots - f_{k-1}\mathfrak{z}_k^{\nu-1}}{\mathfrak{z}_k^\nu}\right)_{\mathfrak{I}_\nu \mathfrak{z}_k^{\mathbb{N}}} \mathfrak{z}_k^\nu.$$

Moreover, for each α, at most $n-1$ variables effectively occur in the set $\mathfrak{D}_0 \amalg \cdots \amalg \mathfrak{D}_\alpha$ of dominant monomials of $\mathfrak{I}_\alpha$. Therefore $f_{\mathfrak{I}} \in A_k$, by the induction hypothesis. □

Proposition 3.19. *Given a Cartesian representation*

$$f \in A_k \mathfrak{z}_1^{\mathbb{Z}} \cdots \mathfrak{z}_k^{\mathbb{Z}}$$

of a series $\hat{f} \in C[[\mathfrak{M}]]$, *its truncation*

$$\tilde{f} = f_{\{\mathfrak{m} \in \mathfrak{z}_1^{\mathbb{N}} \cdots \mathfrak{z}_k^{\mathbb{N}} : \exists \hat{\mathfrak{n}} \in \operatorname{supp} \hat{f},\ \hat{\mathfrak{m}} \preccurlyeq \hat{\mathfrak{n}}\}} \in A_k \mathfrak{z}_1^{\mathbb{Z}} \cdots \mathfrak{z}_k^{\mathbb{Z}}$$

is a faithful Cartesian representation of the same series $\hat{f}$. □

3.5.4 Applications of faithful Cartesian representations

Proposition 3.20. *Let* $\hat{f} \in C[[\mathfrak{M}]]_A$ *be series, which is either*

a) infinitesimal,
b) bounded, or
c) regular.

Then $\hat{f}$ *admits a Cartesian representation in* $A_k \mathfrak{z}_1^{\mathbb{Z}} \cdots \mathfrak{z}_k^{\mathbb{Z}}$ *for some* $k \in \mathbb{N}$, *which is also infinitesimal, bounded resp. regular.*

Proof. Assume that $\hat{f}$ is infinitesimal and let $f \in A_k \mathfrak{z}_1^{\mathbb{Z}} \cdots \mathfrak{z}_k^{\mathbb{Z}}$ be a faithful Cartesian representation of $\hat{f}$, with dominant monomials $\mathfrak{d}_1, \ldots, \mathfrak{d}_l \prec 1$. For each $i \in \{1, \ldots, l\}$, let

$$f_i = f_{\operatorname{in}(\mathfrak{d}_1, \ldots, \mathfrak{d}_i)} - f_{\operatorname{in}(\mathfrak{d}_1, \ldots, \mathfrak{d}_{i-1})} \in A_k \mathfrak{z}_1^{\mathbb{Z}} \cdots \mathfrak{z}_k^{\mathbb{Z}},$$

with $\mathfrak{d}_{f_i} = \mathfrak{d}_i$. Then $f = f_1 + \cdots + f_l$ and

$$\tilde{f} = \sum_{i=1}^{l} f_i \frac{\mathfrak{d}_1}{\mathfrak{d}_i} \mathfrak{z}_{k+i}$$

is an infinitesimal Cartesian representation of $\hat{f}$ in A_{k+l}, when setting $\hat{\mathfrak{z}}_{k+i} = \hat{\mathfrak{d}}_i / \hat{\mathfrak{d}}_1$ for each $i \in \{1, \ldots, l\}$. This proves (a).

If $\hat{f}$ is bounded, then let $g \in A_k$ be an infinitesimal Cartesian representation of $\hat{g} = \hat{f} - \hat{f}_{\{1\}}$. Now $f = g + \hat{f}_{\{1\}} \mathfrak{z}_1^0 \cdots \mathfrak{z}_k^0 \in A_k$ is a bounded Cartesian representation of $\hat{f}$. This proves (b).

Assume finally that $\hat{f} \neq 0$ is regular, with dominant monomial $\hat{\mathfrak{d}}$. Let $g \in A_k$ be a bounded Cartesian representation of $\hat{g} = \hat{f}/\hat{\mathfrak{d}}$. Since $\hat{g}_0 \neq 0$, the series g is necessarily regular. Now take a Cartesian monomial $\mathfrak{d}$ which represents $\hat{\mathfrak{d}}$ (e.g. among the dominant monomials of a faithful Cartesian representation of $\hat{\mathfrak{d}}$). Then $f = g\,\mathfrak{d}$ is a regular Cartesian representation of $\hat{f}$. □

3.5.5 The Newton polygon method revisited

Theorem 3.21. *Let $(A_k)_{k\in\mathbb{N}}$ be a local community over a ring C and let $\mathfrak{M}$ be a monomial monoid. Consider the polynomial equation*

$$\hat{P}_n \hat{f}^n + \cdots + \hat{P}_0 = 0 \tag{3.20}$$

with coefficients $\hat{P}_0, \ldots, \hat{P}_n \in C[\![\mathfrak{M}_{\preccurlyeq}]\!]_A$, such that $(\hat{P}_0)_1 = 0$ and $(\hat{P}_1)_1 \in C^*$. Then (3.20) admits a unique solution in $C[\![\mathfrak{M}_{\prec}]\!]_A$.

Proof. By proposition 3.20, there exist bounded Cartesian representations $P_0, \ldots, P_n \in A_k$ for certain $\hat{\mathfrak{z}}_1, \ldots, \hat{\mathfrak{z}}_k \in \mathfrak{M}$. Now consider the series

$$P = P_0 + P_1\,\mathfrak{z}_{k+1} + \cdots + P_n\,\mathfrak{z}_{k+1}^n \in A_{k+1}.$$

We have $[\mathfrak{z}_1^0 \cdots \mathfrak{z}_{k+1}^0]P = 0$ and $[\mathfrak{z}_1^0 \cdots \mathfrak{z}_k^0 \mathfrak{z}_{k+1}^1]P \in C^*$, so there exists a $f \in A_k$ with

$$P \circ (\mathfrak{z}_1, \ldots, \mathfrak{z}_k, f) = P_0 + P_1 f + \cdots + P_n f^n = 0,$$

by **LC3**. We conclude that $\hat{f} \in C[\![\mathfrak{M}]\!]_A$ satisfies $\hat{P}_n \hat{f}^n + \cdots + \hat{P}_0 = 0$. The uniqueness of $\hat{f}$ follows from theorem 3.3. □

Theorem 3.22. *Let $(A_k)_{k\in\mathbb{N}}$ be a local community over a (real) algebraically closed field C and $\mathfrak{M}$ a totally ordered monomial group with $\mathbb{Q}$-powers. Then $C[\![\mathfrak{M}]\!]_A$ is a (real) algebraically closed field.*

Proof. The proof is analogous to the proof of theorem 3.8. In the present case, theorem 3.21 ensures that $\varphi \in C[\![\mathfrak{M}]\!]_A$ in step 2 of `polynomial_solve`. □

Exercise 3.9. Let C be a ring, $\mathfrak{M}$ a monomial monoid and $(A_k)_{k\in\mathbb{N}}$ a local community. We define $C[\![\mathfrak{M}]\!]_A$ to be the set of series f in $C[\![\mathfrak{M}]\!]$, which admit a semi-Cartesian representation

$$f = \hat{\varphi}(\check{f}_1)\,\mathfrak{m}_1 + \cdots + \hat{\varphi}(\check{f}_p)\,\mathfrak{m}_p$$

with $\check{f}_1, \ldots, \check{f}_p \in A_k$ for some $k \in \mathbb{N}$, $\varphi: \check{\mathfrak{z}}_1^{\mathbb{N}} \cdots \check{\mathfrak{z}}_k^{\mathbb{N}} \to \mathfrak{M}$ and $\mathfrak{m}_1, \ldots, \mathfrak{m}_p \in \mathfrak{M}$. Which results from this section generalize to this more general setting?

Exercise 3.10. Let C be a field. A series f in $C[[\mathfrak{z}_1, \ldots, \mathfrak{z}_k]]$ is said to be *differentially algebraic*, if the field generated by its partial derivatives $\partial^{i_1+\cdots+i_k} f / (\partial\mathfrak{z}_1)^{i_1} \cdots (\partial\mathfrak{z}_k)^{i_k}$ has finite transcendence degree over C. Prove that the collection of such series forms a local community over C.

Exercise 3.11. Assume that C is an effective field, i.e. all field operations can be performed by algorithm. In what follows, we will measure the complexity of algorithms in terms of the number of such field operations.

a) A series $f \in C[[\mathfrak{z}_1, \dots, \mathfrak{z}_k]]$ is said to be *effective*, if there is an algorithm which takes $\alpha_1, \dots, \alpha_k \in \mathbb{N}$ on input, and which outputs $f_{\alpha_1, \dots, \alpha_k}$. Show that the collection of effective series form a local community.
b) An effective series $f \in C[[\mathfrak{z}_1, \dots, \mathfrak{z}_k]]$ is said to be of *polynomial time complexity*, if there is an algorithm, which takes $n \in \mathbb{N}$ on input and which computes $f_{\alpha_1, \dots, \alpha_n}$ for all $\alpha_1, \dots, \alpha_n$ with $\alpha_1 + \dots + \alpha_n \leqslant k$ in time $\binom{n+k}{n}^{O(1)}$. Show that the collection of such series forms a local community. What about even better time complexities?

Exercise 3.12. Let $(A_k)_{k \in \mathbb{N}}$ be a local community and let

$$f \in A_k \, \mathfrak{z}_1^{\mathbb{Z}} \cdots \mathfrak{z}_k^{\mathbb{Z}}$$

be a Cartesian representation of an infinitesimal, bounded or regular grid-based series $\hat{f}$ in $C[[\mathfrak{M}]]$. Show that, modulo widenings, there exists an infinitesimal, bounded resp. regular Cartesian representation of $\hat{f}$, with respect to a Cartesian basis with at most k elements.

Exercise 3.13. Let $(A_k)_{k \in \mathbb{N}}$ be a local community over a field C.

a) If $f \in C[[\mathfrak{M}]]_{A, \prec}$ and $g \in A_1$, then show that $g \circ f \in C[[\mathfrak{M}]]_A$.
b) If $\mathfrak{M}$ is totally ordered, then prove that $C[[\mathfrak{M}]]_A$ is a field.

Exercise 3.14. Let $(A_k)_{k \in \mathbb{N}}$ be a local community over a field C and let $\mathfrak{M}$ be a totally ordered monomial group. Prove that $f_{\succ}, f_{\asymp}, f_{\prec} \in C[[\mathfrak{M}]]_A$ for any $f \in C[[\mathfrak{M}]]_A$, and

$$C[[\mathfrak{M}]]_A = C[[\mathfrak{M}]]_{A, \succ} \oplus C \oplus C[[\mathfrak{M}]]_{A, \prec}.$$

Exercise 3.15. Let $(A_k)_{k \in \mathbb{N}}$ be a Cartesian community. Given monomial groups $\mathfrak{M}$ and $\mathfrak{N}$, let $\mathscr{A}(C[[\mathfrak{M}]], C[[\mathfrak{N}]])$ be the set of strong C-algebra morphisms from $C[[\mathfrak{M}]]$ into $C[[\mathfrak{N}]]$ and $\mathscr{A}(C[[\mathfrak{M}]], C[[\mathfrak{N}]])_A$ the set of $\varphi \in \mathscr{A}(C[[\mathfrak{M}]], C[[\mathfrak{N}]])$, such that $\varphi(\mathfrak{m}) \in C[[\mathfrak{N}]]_A$ for all $\mathfrak{m} \in \mathfrak{M}$.

a) Given $\varphi \in \mathscr{A}(C[[\mathfrak{M}]], C[[\mathfrak{N}]])_A$ and $\psi \in \mathscr{A}(C[[\mathfrak{N}]], C[[\mathfrak{V}]])_A$, where $\mathfrak{V}$ is a third monomial group, prove that $\psi \circ \varphi \in \mathscr{A}(C[[\mathfrak{M}]], C[[\mathfrak{V}]])_A$.
b) Given $\varphi \in \mathscr{A}(C[[\mathfrak{M}]], C[[\mathfrak{N}]])_A$ and $\psi \in \mathscr{A}(C[[\mathfrak{N}]], C[[\mathfrak{M}]])$ such that $\psi \circ \varphi = \mathrm{Id}_{C[[\mathfrak{M}]]}$, prove that $\psi \in \mathscr{A}(C[[\mathfrak{N}]], C[[\mathfrak{M}]])_A$.

Exercise 3.16. Let C be a subfield of $\mathbb{C}$ and let $\mathfrak{M}$ and $\mathfrak{N}$ be monomial groups with $\mathfrak{M} \subseteq \mathfrak{N}$. Prove that $C\{\{\mathfrak{M}\}\} = C[[\mathfrak{M}]] \cap C\{\{\mathfrak{N}\}\}$. Does this property generalize to other local communities?

Exercise 3.17. Let $(A_k)_{k \in \mathbb{N}}$ be the local community from example 3.17 and let $\mathfrak{M}$ be a totally ordered monomial group. Prove that $C[[\mathfrak{M}]]_A$ is isomorphic to the algebraic closure of $C[\mathfrak{M}]$.

Exercise 3.18. Does theorem 3.22 still hold if we remove condition **LC2** in the definition of local communities?

Exercise 3.19. Consider the resolution of $P(f)=0$ $(f \prec \mathfrak{v})$, with $P \in C[[\mathfrak{M}]]_A$ and $\mathfrak{v} \in \mathfrak{M}$.

a) Given a starting term $c\,\mathfrak{m}$ of multiplicity d, let $\mathfrak{n}$ be minimal for $\preccurlyeq$ such that $P_i\,\mathfrak{m}^i \preccurlyeq \mathfrak{n}$ for all i. Show that there exist Cartesian coordinates $\mathfrak{z}_1, \ldots, \mathfrak{z}_k$ with $\mathfrak{m}, \mathfrak{n} \in \mathfrak{z}_1^{\mathbb{Z}} \cdots \mathfrak{z}_k^{\mathbb{Z}}$, in which $P_i\,\mathfrak{m}^i/\mathfrak{n}$ admits a bounded Cartesian representations u_i for all $0 \leqslant i \leqslant n = \deg P$.
b) Consider a bounded Cartesian representation $\varphi \in A_k$ with $\varphi \sim c$ and let $\tilde{u}_i = \sum_{k=i}^{n} \binom{k}{i} u_k\,\varphi^{k-i}$. Given $\mathfrak{w} \in \mathfrak{z}_1^{\mathbb{Q}^{\geqslant}} \cdots \mathfrak{z}_k^{\mathbb{Q}^{\geqslant}}$, let

$$Q_{\mathfrak{w}} = \sum_{i=0}^{n} \tilde{u}_{i,\mathfrak{w}^{d-i}}\,F^i.$$

Show that $Q = \sum_{\mathfrak{w}} Q_{\mathfrak{w}}\,\mathfrak{w}$ is a series in $C[F][[\mathfrak{z}_1^{1/d!}, \ldots, \mathfrak{z}_k^{1/d!}]]_A$.
c) For each $\mu \in \{0, \ldots, d\}$, let $\mathfrak{I}_\mu$ be initial segment generated by the $\mathfrak{w}$ such that $\operatorname{val} Q_{\mathfrak{w}} < \mu$, and $\mathfrak{F}_\mu$ its complement. We say that $\varphi_{\mathfrak{F}_\mu}$ is the part of multiplicity $\geqslant \mu$ of φ as a zero of $u_0 + \cdots + u_n\,F^n$. Show that $\varphi_{\mathfrak{F}_\mu} \in A_k$ can be determined effectively for all μ.
d) In `polynomial_solve`, show that refinements of the type

$$f = \hat{\varphi}\,\mathfrak{m} + \tilde{f} \quad (\tilde{f} \prec \mathfrak{m}),$$

where $\varphi \in C_k$ is the unique solution to $\partial^{d-1}(u_0 + \cdots + u_n\,F^n)/\partial F^{d-1}$, may be replaced by refinements

$$f = \widehat{\varphi_{\mathfrak{F}_{d-1}}}\,\mathfrak{m} + \tilde{f} \quad (\tilde{f} \prec \mathfrak{m}).$$

4

Transseries

Let C be a totally ordered exp-log field. This means that C is a totally ordered field with an exponential function and a partial logarithmic function which satisfies similar properties as those defined on the real numbers. Axioms for exp-log fields will be given in section 4.1. For the moment, the reader may assume that $C = \mathbb{R}$.

The aim of this chapter is the construction of the totally ordered exp-log field $C[\![[x]]\!]$ of grid-based transseries in x over C. This means that $C[\![[x]]\!]$ is a field of grid-based series with an additional exponential structure. Furthermore, $C[\![[x]]\!]$ contains x as an infinitely large monomial. Actually, the field $C[\![[x]]\!]$ carries much more structure: in the next chapter, we will show how to differentiate, integrate, compose and invert transseries. From corollary 3.9, it also follows that $C[\![[x]]\!]$ is real closed. In chapter 9, this theorem will be generalized to algebraic differential equations.

As to the construction of $C[\![[x]]\!]$, let us first consider the field $C[\![x]\!] = C[\![x^C]\!]$. Given an infinitesimal series f, we may naturally define

$$\begin{aligned} \exp f &= 1 + f + \tfrac{1}{2} f^2 + \cdots \\ \log(1 + f) &= f - \tfrac{1}{2} f^2 + \cdots \end{aligned}$$

Using the exp-log structure of C, these definitions may be extended to $C[\![x]\!]_{\preccurlyeq} = C \oplus C[\![x]\!]_{\prec}$ for exp and to $C^{>} + C[\![x]\!]_{\prec}$ for log. However, nor the logarithm of x, nor the exponential of any infinitely large series f are defined. Consequently, we have to add new monomials to x^C in order to obtain a field of grid-based series which is stable under exponentiation and logarithm.

Now it is easy to construct a field of grid-based series $\mathbb{L}$ which is stable under logarithm (in the sense that $\log f$ is defined for all strictly positive f). Indeed, taking $\mathbb{L} = C[\![\ldots; \log\log x; \log x; x]\!]$, we set

$$\log x^{\alpha_0} \cdots \log_n^{\alpha_n} x = \alpha_0 \log x + \cdots + \alpha_n \log_{n+1} x$$

for monomials $\log x^{\alpha_0} \cdots \log_n^{\alpha_n} x$ (here $\log_n = \log \circ \overset{n\times}{\cdots} \circ \log$ stands for the n-th iterated logarithm). For general $f \in \mathbb{L}^>$, we define

$$\log f = \log \mathfrak{d}_f + \log c_f + \log(1 + \delta_f),$$

where $\log(1 + \delta_f) = \delta_f - \frac{1}{2}\delta_f^2 + \cdots$ as above.

In order to construct a field $\mathbb{T} = C[[\mathfrak{M}]]$ of grid-based series with an exponentiation, we first have to decide what monomial group $\mathfrak{M}$ to take. The idea is to always take exponentials of purely infinite series for the monomials in $\mathfrak{M}$. For instance, e^{x^2+x} is a monomial. On the other hand, $\mathrm{e}^{x^2+x+x^{-1}}$ is not a monomial and we may expand it in terms of e^{x^2+x} using

$$\mathrm{e}^{x^2+x+x^{-1}} = \mathrm{e}^{x^2+x} + x^{-1}\mathrm{e}^{x^2+x} + \frac{1}{2}x^{-2}\mathrm{e}^{x^2+x} + \cdots.$$

More generally, as soon as each purely infinite series in $\mathbb{T}$ admits an exponential, then $\mathbb{T}$ is closed under exponentiation: for all $f \in \mathbb{T}$ we take

$$\exp f = \exp f_{\succ} \exp f_{\asymp} \exp f_{\prec},$$

with $\exp f_{\prec} = 1 + f_{\prec} + \frac{1}{2}(f_{\prec})^2 + \cdots$ as above.

In section 4.2, we first study abstract fields of transseries. These are totally ordered fields of grid-based series, with logarithmic functions that satisfy some natural compatibility conditions with the serial structures. Most of these conditions were briefly motivated above. In section 4.3 we construct the field $C[[[x]]]$ of transseries in x. We start with the construction of the field $\mathbb{L}$ of logarithmic transseries. Next, we close this field under exponentiation by repeatedly inserting exponentials of purely infinite series as new monomials. In section 4.4, we prove the incomplete transbasis theorem, which provides a convenient way to represent and compute with grid-based transseries.

4.1 Totally ordered exp-log fields

A *partial exponential ring* is a ring R with a partial *exponential function* $\exp: R \rightarrow R$, such that

E1. $\exp 0 = 1$.

E2. $\exp y = \exp(y - x) \exp x$, for all $x, y \in \operatorname{dom} \exp$.

The second condition stipulates in particular that $y - x \in \operatorname{dom} \exp$, whenever $x, y \in \operatorname{dom} \exp$. If $\operatorname{dom} \exp = R$, then we say that R is an *exponential ring*. If $\exp$ is an exponential function, then we will also write e^x for $\exp x$ and $\exp_n$ for the n-th iterate of exp (i.e. $\exp_0 = \mathrm{Id}$ and $\exp_{n+1} = \exp \circ \exp_n$ for all $b \in \mathbb{N}$).

The field $\mathbb{R}$ of real numbers is a classical example of an exponential field. Moreover, the real numbers carry an ordering and it is natural to search for axioms which model the compatibility of the exponential function with this ordering. Unfortunately, an explicit set of axioms which imply all relations satisfied by the exponential function on $\mathbb{R}$ has not been found yet. Nevertheless, Dahn and Wolter have proposed a good candidate set of axioms [DW84].

We will now propose a similar system of axioms in the partial context. For each $n \in \mathbb{N}$, we denote the Taylor expansion of $\exp x$ at order n by

$$E_n(x) = 1 + x + \cdots + \frac{1}{(n-1)!} x^{n-1}.$$

We also denote

$$N_n = \begin{cases} 1, & \text{if } n = 0 \\ (n-1)!, & \text{otherwise} \end{cases}$$

so that $N_n E_n \in \mathbb{Z}[x]$. An *ordered partial exponential ring* is a partially ordered ring R, with a partial exponential function $\exp: R \rightarrow R$, which satisfies **E1**, **E2** and

E3. $N_{2n} \exp x \geqslant N_{2n} E_{2n}(x)$, for all $x \in \operatorname{dom} \exp$ and $n \in \mathbb{N}$.

If $\operatorname{dom} \exp = R$, then we say that R is an *ordered exponential ring*.

Proposition 4.1. *Let R be an ordered ring in which $x \neq 0 \Rightarrow x^2 > 0$. Given a partial exponential function on R which satisfies* **E1**, **E1** *and* **E1**, *we have*

$$N_{2n} \exp x = N_{2n} E_{2n}(x) \Rightarrow x = 0,$$

for all $n \in \mathbb{N}$ and $x \in \operatorname{dom} \exp$.

Proof. Assume that $\exp x = E_{2n}(x)$. We cannot have $x \leqslant -2n$, since otherwise

$$N_{2n} E_{2n}(x) = \sum_{k=0}^{n-1} \frac{N_{2n}}{(2k+1)!} (2k+1+x) x^{2k} < 0.$$

If $-2n < x$, then

$$\begin{aligned} 0 &\geqslant N_{2n+2} (E_{2n+2}(x) - \exp x) \\ &= N_{2n+2} (E_{2n+2}(x) - E_{2n}(x)) = x^{2n} (2n+1+x), \end{aligned}$$

whence $x = 0$. □

Proposition 4.2. $\mathbb{R}$ *is a totally ordered exponential field.*

Proof. Let $n \in \mathbb{N}$. For $x \geqslant -2n$, we have

$$\exp x - E_{2n}(x) = \sum_{k=n}^{\infty} \frac{x^{2k}}{(2k)!} \left(1 + \frac{x}{2k+1}\right) \geqslant 0.$$

For $x < -2n$, we have shown above that $E_{2n}(x) < 0 \leqslant \exp x$. □

Proposition 4.3. *Let R be an ordered partial exponential ring. Then*

a) $\exp$ *is injective.*
b) $x < y \Leftrightarrow \exp x < \exp y$, *for all* $x, y \in \operatorname{dom} \exp$.
c) If $\operatorname{dom} \exp$ *is a* $\mathbb{Q}$*-module, then*

$$\forall n \in \mathbb{N}, \exists x_n \in R, \forall x \in \operatorname{dom} \exp, \quad x > x_0 \Rightarrow \exp x > x^n.$$

Proof. Assume that $\exp x = \exp y$, for some $x, y \in R$. Then

$$\exp(y-x) = \exp y \exp(-x) = \exp x \exp(-x) = 1$$

and similarly $\exp(x-y) = 1$. Hence,

$$\begin{aligned} 1 &= \exp(y-x) \geqslant 1+y-x \\ 1 &= \exp(x-y) \geqslant 1+x-y, \end{aligned}$$

so that both $y \leqslant x$ and $x \leqslant y$. This proves that $x = y$, whence exp is injective.

Assume now that $x < y$ for some $x, y \in \operatorname{dom}\exp$. Then

$$\exp(y-x) \geqslant 1+y-x \geqslant 1.$$

Consequently,

$$\exp y = \exp(y-x)\exp x \geqslant \exp x$$

and $\exp y > \exp x$, by the injectivity of exp. Inversely, assume that $\exp x < \exp y$ for some $x, y \in \operatorname{dom}\exp$. Then

$$1+x-y \leqslant \exp(x-y) = \exp x \exp(-y) \leqslant \exp y \exp(-y) = 1,$$

whence $x \leqslant y$. We again conclude that $x < y$, since $\exp y \neq \exp x$. This proves (*b*).

If $n = 0$, then (*c*) follows from (*b*). If $n > 0$, then $\exp(x/2n) \geqslant (x/2n)+1$ implies

$$\exp x \geqslant \left(\frac{x}{2n}+1\right)^{2n} > \left(\frac{x}{2n}\right)^{2n} > x^n,$$

for all $x > (2n)^2$. □

Instead of axiomatizing partial exponential functions on a ring, it is also possible to axiomatize partial logarithmic functions. The natural counterparts of **E1**, **E2** and **E3** are

L1. $\log 1 = 0$.
L2. $\log y = \log \frac{y}{x} + \log x$, for all $x, y \in \operatorname{dom}\log$.
L3. $N_{2n}\, x \geqslant N_{2n}\, E_{2n}(\log x)$, for all $x \in \operatorname{dom}\log$ and $n \in \mathbb{N}$.

Notice that the second condition assumes the existence of a partial inversion $x \mapsto \frac{1}{x}$, whose domain contains dom log. The n-th iterate of log will be denoted by $\log_n$.

In a similar fashion, we define a partial logarithmic ring to be a ring R with a partial logarithmic function which satisfies **L1** and **L2**. An ordered partial logarithmic ring is an ordered ring R with a partial logarithmic function which satisfies **L1**, **L2** and **L3**. In the case when $\operatorname{dom}\log = R^{>}$ for such a ring, then we say that R is an ordered logarithmic ring.

Proposition 4.4.

a) *Let R be a partial exponential ring, such that* exp *is injective. Then the partial inverse* log *of* exp *satisfies* **L1** *and* **L2**.
b) *If R is an ordered partial exponential ring, then* exp *is injective, and its partial inverse* log *satisfies* **L1**, **L2** *and* **L3**.

c) *Let* R *be a partial logarithmic ring, such that* $\log$ *is injective. Then the partial inverse* $\exp$ *of* $\log$ *satisfies* **E1** *and* **E2**.

d) *If* R *is an ordered partial logarithmic ring, then* $\log$ *is injective, and its partial inverse* $\exp$ *satisfies* **E1**, **E2** *and* **E3**.

Proof. Let R be a partial exponential ring, such that exp is injective. Then we clearly have **L1**. Now assume that $x = \exp x' \in \operatorname{dom} \log = \operatorname{im} \exp$. Then $(\exp x') \cdot (\exp(-x')) = 1$, whence $\exp(-x') = 1/x \in \operatorname{dom} \log$. Furthermore, if $y = \exp y' \in \operatorname{dom} \log = \operatorname{im} \exp$, then $\exp y' = \exp(y' - x') \exp x'$, so that $\exp(y' - x') = y/x$. Consequently, $y/x \in \operatorname{im} \exp = \operatorname{dom} \log$ and $\log y - \log x = y' - x' = \log(y/x)$. This proves **L2** and (a). As to (b), if R is an ordered partial exp-log ring, then exp is injective by proposition 4.3(a). The property **L3** directly follows from **E3**.

Assume now that R is a partial logarithmic ring, such that log is injective. We clearly have **E1**. Given $x = \log x'$ and $y = \log y'$ in $\operatorname{dom} \exp = \operatorname{im} \log$, we have $\log y = \log(y/x) + \log x$ and in particular $\log(y/x) \in \operatorname{dom} \exp$. It follows that $\exp y' / \exp x' = y/x = \exp(\log y - \log x) = \exp(y' - x')$. This proves **E2** and (c).

Assume finally that R is an ordered partial logarithmic ring. Let $x, y \in \operatorname{dom} \log$ be such that $\log x = \log y$. Then

$$x/y \geqslant 1 + \log(x/y) = 1 + \log x - \log y = 1.$$

Hence $x \geqslant y$, since $y \in \operatorname{dom} \log \Rightarrow y \geqslant 0$. Similarly, $y \geqslant x$ and $x = y$, which proves that log is injective. The property **E3** directly follows from **L3**. □

If (a) and (c) (resp. (b) and (d)) are satisfied in the above proposition, then we say that R is a *partial exp-log ring* (resp. an *ordered partial exp-log ring*). An *ordered exp-log ring* is an ordered partial exp-log ring R, such that $\operatorname{dom} \exp = R$ and $\operatorname{im} \exp = R^{>}$. An ordered (partial) exponential, logarithmic resp. exp-log ring, which is also an ordered field is called an *ordered (partial) exponential, logarithmic resp. exp-log field*. In a partial exp-log ring, we extend the notations $\exp_n$ and $\log_n$ to the case when $n \in \mathbb{Z}$, by setting $\exp_n = \log_{-n}$ and $\log_n = \exp_{-n}$, if $n < 0$.

Assume now that R is a ring with C-powers, for some subring $C \subseteq R$. An exponential resp. logarithmic function is said to be compatible with the C-powers structure on R if

E4. $\exp(\lambda f) = (\exp f)^{\lambda}$, for all $f \in \operatorname{dom} \exp$ and $\lambda \in C$; resp.

L4. $\log f^{\lambda} = \lambda \log f$, for all $f \in \operatorname{dom} \exp$ and $\lambda \in C$.

Here we understand that $\operatorname{im} \exp \subseteq R^{\times}$ in **E4** and $\operatorname{dom} \log \subseteq R^{\times}$ in **L4**. Notice that **E4** and **L4** are equivalent, if exp and log are partial inverses. Notice also that any totally ordered exp-log field C naturally has C-powers: set $\lambda^{\mu} = \exp(\mu \log \lambda)$ for all $\lambda \in C^{>}$ and $\mu \in C$.

Exercise 4.1. Let R be an exponential ring. Show that for all $x \in R$, we have $\exp x = 0 \Rightarrow 1 = 0$.

Exercise 4.2. Show that the only exponential function on the totally ordered field of real numbers $\mathbb{R}$ is the usual exponential function.

Exercise 4.3. Let R be a totally ordered exponential field. Show that the exponential function on R is continuous. That is, for all x and $\varepsilon > 0$ in R, there exists a $\delta > 0$, such that $|\exp x' - \exp x| < \varepsilon$, for all $x' \in R$ with $|x' - x| < \delta$. Show also that the exponential function is equal to its own derivative.

Exercise 4.4. Let R be an ordered partial exponential ring. Given $x \in \operatorname{dom} \exp$ and $n \in \mathbb{N}$, prove that

a) $\exp x > E_{2n+1}(x)$, if $x > 0$.
b) $\exp x < E_{2n+1}(x)$, if $x < 0$.

4.2 Fields of grid-based transseries

Let C be a totally ordered exp-log field, $\mathfrak{T}$ a totally ordered monomial group with C-powers. Assume that we have a partial logarithmic function on the totally ordered field $\mathbb{T} = C[[\mathfrak{T}]]$ with C-powers. We say that $\mathbb{T}$ is a *field of grid-based transseries* (or a field of transseries) if

T1. $\operatorname{dom} \log = \mathbb{T}^{>}$.
T2. $\log \mathfrak{m} \in \mathbb{T}_{\succ}$, for all $\mathfrak{m} \in \mathfrak{T}$.
T3. $\log (1 + \varepsilon) = l \circ \varepsilon$, for all $\varepsilon \in \mathbb{T}_{\prec}$, where $l = \sum_{k=1}^{\infty} \frac{(-1)^{k+1}}{k} z^k \in C[[z]]$.

Intuitively speaking, the above conditions express a strong compatibility between the logarithmic and the serial structure of $\mathbb{T}$.

Example 4.5. Assume that $\mathbb{T}$ is a field of transseries, such that $x \in \mathfrak{T}_{\succ}$, and such that $\mathbb{T}$ is stable under exponentiation. Then $x^5\, \mathrm{e}^{x^3 + x^{\mathrm{e}} + x^2 + x^{\mathrm{e}-1}}$ is a monomial in $\mathfrak{T}$. The series $\mathrm{e}^{x^2/(1-x^{-1})}$ is not a monomial, since $x^2/(1-x^{-1})$. We have

$$\begin{aligned}\exp\left(\frac{x^2}{1-x^{-1}}\right) &= \mathrm{e}^{x^2+x} \exp\left(\frac{1}{1-x^{-1}}\right)\\ &= \mathrm{e} \cdot \mathrm{e}^{x^2+x} + \mathrm{e} \cdot \frac{\mathrm{e}^{x^2+x}}{x} + \frac{3\,\mathrm{e}}{2} \cdot \frac{\mathrm{e}^{x^2+x}}{x^2} + \cdots.\end{aligned}$$

On the other hand, $\mathrm{e}^{\mathrm{e}^x/(1-x^{-1})}$ is a monomial, since

$$\mathrm{e}^x/(1-x^{-1}) = \mathrm{e}^x + \mathrm{e}^x/x + \cdots \in \mathbb{T}_{\succ}.$$

Proposition 4.6. *Let $\mathbb{T}$ be a field of transseries. Then*

a) Given $f \in \mathbb{T}^{>}$, the canonical decomposition of $\log f$ *is given by*

$$\begin{array}{ccccccc}\log f & = & (\log f)_{\succ} & + & (\log f)_{\asymp} & + & (\log f)_{\prec}\\ & & \| & & \| & & \|\\ & & \log \mathfrak{d}_f & & \log c_f & & \log (1+\delta_f)\end{array}$$

b) Given $f, g \in \mathbb{T}^{>}$, we have

$$\begin{aligned} f \preccurlyeq g &\Leftrightarrow (\log f)_{\succ} \leqslant (\log g)_{\succ};\\ f \asymp g &\Leftrightarrow (\log f)_{\succ} = (\log g)_{\succ};\\ f \prec g &\Leftrightarrow (\log f)_{\succ} < (\log g)_{\succ};\\ f \sim g &\Leftrightarrow (\log f)_{\succcurlyeq} = (\log g)_{\succcurlyeq}.\end{aligned}$$

c) *Given* $f, g \in \mathbb{T}^{>,\succ}$, *we have*

$$\begin{aligned} f \preccurlyeq\!\!\preccurlyeq g &\Leftrightarrow \log f \preccurlyeq \log g; \\ f \prec\!\!\prec g &\Leftrightarrow \log f \prec \log g; \\ f \asymp\!\!\asymp g &\Leftrightarrow \log f \asymp \log g; \\ f \approx g &\Leftrightarrow \log f \sim \log g \end{aligned}$$

d) *For all* $f \in \mathbb{T}^{>,\succ}$, *we have* $\log f \in \mathbb{T}^{>}$, $\log f \prec f$ *and* $\log f \prec\!\!\prec f$.

Proof.

a) Follows from **L1**, **L2**, **T2** and **T3**.
b) We have

$$\begin{aligned} f \preccurlyeq g &\Leftrightarrow \mathfrak{d}_f \preccurlyeq \mathfrak{d}_g \\ &\Leftrightarrow \mathfrak{d}_f \leqslant \mathfrak{d}_g \\ &\Leftrightarrow (\log f)_{\succ} = \log \mathfrak{d}_f \leqslant \log \mathfrak{d}_g = (\log g)_{\succ}. \end{aligned}$$

The other relations are proved in a similar way.
c) We have

$$\begin{aligned} f \preccurlyeq\!\!\preccurlyeq g &\Leftrightarrow \exists \mu \in C, f \preccurlyeq g^{\mu} \\ &\Leftrightarrow \exists \mu \in C, (\log f)_{\succ} \leqslant (\mu \log g)_{\succ} = \mu\, (\log g)_{\succ} \\ &\Leftrightarrow \log f \preccurlyeq \log g. \end{aligned}$$

The other relations are proved similarly.
d) Let $f \in \mathbb{T}^{>,\succ}$. Then $(\log f)_{\succ} > 0$, by (b), whence $g = \log f \in \mathbb{T}^{>,\succ}$. Furthermore,

$$\log(g/3) + 1 < (g/3) \Rightarrow 3\,(\log g - \log 3) < g.$$

Consequently, $2 \log g < g$, since $\log g \in \mathbb{T}^{>,\succ} \Rightarrow \log g > 3 \log 3$. It follows that

$$C + \log g < 2 \log g < g.$$

Since exp is total on C, we infer that $C^{>} \log f < f$. Therefore, $\log f \prec f$. Finally, $\log_2 f \prec \log f$ and (c) imply that $\log f \prec\!\!\prec f$ by (c). □

The following lemma, which is somehow the inverse of proposition 4.6(a) and (d), will be useful for the construction of fields of transseries.

Lemma 4.7. *Let* log *be a partial function on* $\mathbb{T}$, *which satisfies* **T1**, **T2** *and*

a) $\log(\mathfrak{m}^{\lambda} \mathfrak{n}) = \lambda \log \mathfrak{m} + \log \mathfrak{n}$, *for all* $\mathfrak{m}, \mathfrak{n} \in \mathfrak{T}$ *and* $\lambda \in C$.
b) $\log f = \log \mathfrak{d}_f + \log c_f + l \circ \delta_f$, *for all* $f \in \mathbb{T}^{>}$.
c) $0 < \log \mathfrak{m} \prec \mathfrak{m}$, *for all* $\mathfrak{m} \in \mathfrak{T}_{\succ}$.

Then log *is a logarithmic function, which is compatible with the ordering and* C*-powers on* $\mathbb{T}$. *Hence,* $\mathbb{T}$ *is a field of grid-based transseries.*

Proof. We clearly have **L1**. Given $f, g \in \mathbb{T}^{>}$, we also have

$$\begin{aligned}\log(f/g) &= \log \mathfrak{d}_{f/g} + \log c_{f/g} + l \circ \delta_{f/g} \\ &= \log(\mathfrak{d}_f/\mathfrak{d}_g) + \log(c_f/c_g) + l \circ \left(\tfrac{\delta_f - \delta_g}{1+\delta_g}\right) \\ &= \log \mathfrak{d}_f - \log \mathfrak{d}_g + \log c_f - \log c_g + l \circ \delta_f - l \circ \delta_g \\ &= \log f - \log g.\end{aligned}$$

Here

$$l \circ \left(\frac{\delta_f - \delta_g}{1+\delta_g}\right) = l \circ \delta_f - l \circ \delta_g$$

by proposition 2.18 and the fact that $l(\frac{z_1 - z_2}{1+z_2}) = l(z_1) - l(z_2)$ in $C[[z_1, z_2]]$. This proves **L2**.

Let us now show that

$$f > E_{2n}(\log f) = 1 + \log f + \cdots + \frac{1}{(2n-1)!} (\log f)^{2n-1},$$

for all $f \in \mathbb{T}^{>} \setminus \{1\}$ and $n \in \mathbb{N}$. Assume first that $f \asymp 1$. If $f \sim 1$, then we have

$$f - E_{2n}(\log f) = \left(\frac{z^{2n}}{(2n)!} + \frac{z^{2n+1}}{(2n+1)!} + \cdots\right) \circ l \circ \delta_f \sim \frac{\delta_f^{2n}}{(2n)!} > 0.$$

Otherwise, $c_f > E_{2n}(\log c_f)$ and

$$f - E_{2n}(\log f) \sim c_f - \log c_f - 1 > 0.$$

If $f \prec 1$, then $\log f = -\log(1/f) \in \mathbb{T}^{<,\succ}$. Consequently,

$$f - E_{2n}(\log f) \sim -\frac{1}{(2n-1)!} (\log f)^{2n-1} > 0.$$

If $f \succ 1$, let us show that $(\log f)^k \prec f$, for all $k \in \mathbb{N}$, which clearly implies that $f > E_{2n}(\log f)$. We first observe that $\log \varphi \in \mathbb{T}^{>,\succ}$ for all $\varphi \in \mathbb{T}^{>,\succ}$, since $\log \mathfrak{d}_\varphi \in \mathbb{T}^{>}_{\succ} \subseteq \mathbb{T}^{>,\succ}$ and $\log \varphi = \log \mathfrak{d}_\varphi + O(1)$. Furthermore, $\log \varphi \sim \log \mathfrak{d}_\varphi \prec \mathfrak{d}_\varphi \asymp \varphi$, for all $\varphi \in \mathbb{T}^{>,\succ}$. Taking $\varphi = \mathfrak{d}_{\log f} = \mathfrak{d}_{\log \mathfrak{d}_f}$, we get $\log \mathfrak{d}_{\log f} \prec \mathfrak{d}_{\log \mathfrak{d}_f} \Rightarrow k \log \mathfrak{d}_{\log f} < \log \mathfrak{d}_f \Rightarrow \mathfrak{d}_{\log f}^k \prec \mathfrak{d}_f \Rightarrow (\log f)^k \prec f$. This proves **L3**.

Let us finally show that $\log f^\lambda = \lambda \log f$ for any $f \in \mathbb{T}^{>}$ and $\lambda \in C$. Denoting $\pi_\lambda = (1+z)^\lambda \in C[[z]]$, we have

$$\begin{aligned}\log f^\lambda &= \log(c_f^\lambda \mathfrak{d}_f^\lambda \pi_\lambda \circ \delta_f) \\ &= \lambda \log \mathfrak{d}_f + \lambda \log c_f + l \circ (\pi_\lambda \circ \delta_f - 1) \\ &= \lambda \log \mathfrak{d}_f + \lambda \log c_f + \lambda\, l \circ \delta_f \\ &= \lambda \log f.\end{aligned}$$

Indeed, proposition 2.18 implies that $l \circ (\pi_\lambda \circ \delta_f - 1) = \lambda\, l \circ \delta_f$, since $l(\pi_\lambda(z) - 1) = \lambda\, l(z)$ is a formal identity in $C[[z]]$. □

Exercise 4.5. Let $\mathbb{T}$ be a field of transseries.

a) Show that $\exp f = e \circ f$ for all $f \in \mathbb{T}_{\prec}$, where $e = \sum_{k=0}^{\infty} \frac{1}{k!} z^k \in C[[z]]$.
b) For each $f \in \operatorname{dom} \exp$, show that

$$\begin{array}{ccccccc} \exp f & = & (\exp f_{\succ}) & \cdot & (\exp f_{\asymp}) & \cdot & (\exp f_{\prec}) \\ & & \| & & \| & & \| \\ & & \mathfrak{d}_{\exp f} & & c_{\exp f} & & (1+\delta_{\exp f}) \end{array}$$

c) For each $f \in \operatorname{dom} \exp \cap \mathbb{T}^{>,\succ}$, show that $\exp f \in \mathbb{T}^{>,\succ}$, $f \prec \exp f$ and $f \prec\!\!\!\prec \exp f$.

Exercise 4.6. Let $e(z) = \sum_{k=0}^{\infty} \frac{1}{k!} z^k$, $l(z) = \sum_{k=1}^{\infty} \frac{(-1)^{k+1}}{k} z^k$ and $\pi_\lambda = \sum_{k=0}^{\infty} \frac{\lambda \cdots (\lambda - k + 1)}{k!} z^k$ be as above. Prove the following formal identities:

a) $e(z_1 + z_2) = e(z_1)\, e(z_2)$;
b) $l(\frac{z_1 - z_2}{1 + z_2}) = l(z_1) - l(z_2)$;
c) $e(l(z)) = 1 + z$;
d) $l(\pi_\lambda(z) - 1) = \lambda\, l(z)$.

Hint: prove that the left and right hands sides satisfy the same (partial) differential equations and the same initial conditions.

Exercise 4.7. Let $\mathbb{T} = C[\![\mathfrak{T}]\!]$ be a field of transseries and consider a flat subset $\mathfrak{T}^\flat$ of $\mathfrak{T}$ (i.e. $\forall \mathfrak{m} \in \mathfrak{T}, \forall \mathfrak{n} \in \mathfrak{T}^\flat \colon \mathfrak{m} \preceq\!\!\!\prec \mathfrak{n} \Rightarrow \mathfrak{m} \in \mathfrak{T}^\flat$).

a) Show that there exists an initial segment $\mathfrak{I}$ of $\mathfrak{T}_{\succ}$ such that

$$\mathfrak{T}^\flat = \{e^f \colon f \in \mathbb{T}_{\succ}, \mathfrak{d}_f \in \mathfrak{I}\}.$$

b) Show that $\mathbb{T} = C[\![\mathfrak{T}^\flat \dot{\times} \mathfrak{T}^\sharp]\!]$, where

$$\mathfrak{T}^\sharp = \{e^f \colon f \in \mathbb{T}_{\succ}, \operatorname{supp} f \cap \mathfrak{I} = \varnothing\}.$$

We call $\mathfrak{T}^\sharp$ the *steep complement* of $\mathfrak{T}^\flat$.

4.3 The field of grid-based transseries in x

Let C be a fixed totally ordered exp-log field, such as $\mathbb{R}$, and x a formal infinitely large variable. In this section, we will construct the field $C[\![\![x]\!]\!]$ of grid-based transseries in x over C. We proceed as follows:

- We first construct the field
$$\mathbb{L} = C[\![\mathfrak{L}]\!] = C[\![\ldots; \log\log x; \log x; x]\!]$$
of logarithmic transseries in x.
- Given a field of transseries $\mathbb{T} = C[\![\mathfrak{T}]\!]$, we next show how to construct its exponential extension $\mathbb{T}_{\exp} = C[\![\mathfrak{T}_{\exp}]\!]$: this is the smallest field of transseries with $\mathfrak{T}_{\exp} \supseteq \mathfrak{T}$ and such that $\exp f$ is defined in $\mathbb{T}_{\exp}$ for all $f \in \mathbb{T}$.

- We finally consider the sequence

$$\mathbb{L} \subseteq \mathbb{L}_{\text{exp}} \subseteq \mathbb{L}_{\text{exp},\text{exp}} \subseteq \cdots$$

of successive exponential extensions of $\mathbb{L}$. Their union

$$C[\![[x]\!]] = \mathbb{L} \cup \mathbb{L}_{\text{exp}} \cup \cdots = C[\![\mathfrak{L} \cup \mathfrak{L}_{\text{exp}} \cup \cdots]\!]$$

is the desired field of grid-based transseries in x over C.

4.3.1 Logarithmic transseries in x

Consider the field $\mathbb{L} = C[\![\mathfrak{L}]\!]$, where

$$\mathfrak{L} = \cdots \dot{\times} \log_2^C x \dot{\times} \log^C x \dot{\times} x^C.$$

Given a monomial $\mathfrak{m} = x^{\alpha_0} \cdots \log_k^{\alpha_k} x \in \mathfrak{L}$, we define $\log \mathfrak{m}$ by

$$\log (x^{\alpha_0} \cdots \log_k^{\alpha_k} x) = \alpha_0 \log x + \cdots + \alpha_k \log_{k+1} x.$$

We extend this definition to $\mathbb{L}^>$, by setting

$$\log f = \log \mathfrak{d}_f + \log c_f + l \circ \delta_f$$

for each $f \in \mathbb{L}^>$. Here we recall that $l = \sum_{k=1}^{\infty} \frac{(-1)^{k+1}}{k} z^k \in C[[z]]$.

Proposition 4.8. *$\mathbb{L}$ is a field of transseries.*

Proof. Clearly, $\log (\mathfrak{m}^{\lambda} \mathfrak{n}) = \lambda \log \mathfrak{m} + \log \mathfrak{n}$, for all $\mathfrak{m}, \mathfrak{n} \in \mathfrak{L}$ and $\lambda \in C$. Now let $\mathfrak{m} \in \mathfrak{L}_{\succ}$. Then $\mathfrak{m} = \log_i^{\alpha_i} x \cdots \log_k^{\alpha_k} x$, for certain $\alpha_i, \ldots, \alpha_k \in C$ with $\alpha_i > 0$. Hence, $0 < \log \mathfrak{m} \prec \mathfrak{m}$, since $\log \mathfrak{m} \sim \alpha_i \log_{i+1} x$ and $0 < \alpha_i \log_{i+1} x \prec \log_i^{\alpha_i} x \cdots \log_k^{\alpha_k} x = \mathfrak{m}$. Now the proposition follows from lemma 4.7. □

4.3.2 Exponential extensions

Let $\mathbb{T} = C[\![\mathfrak{T}]\!]$ be a field of transseries and let

$$\mathfrak{T}_{\text{exp}} = \exp \mathfrak{T}_{\succ}$$

be the monomial group of formal exponentials $\exp f$ with $f \in \mathfrak{T}_{\succ}$, which is isomorphic to the totally ordered C-module $\mathfrak{T}_{\succ}$: we define $(\exp f)^{\lambda} (\exp g) = \exp(\lambda f + g)$ and $\exp f \succcurlyeq \exp g \Leftrightarrow f \geqslant g$ for all $f, g \in \mathbb{T}_{\succ}$ and $\lambda \in C$.

Now the mapping $\nu: \mathfrak{T} \rightarrow \mathfrak{T}_{\text{exp}}, \mathfrak{m} \mapsto \exp(\log \mathfrak{m})$ is an injective morphism of monomial groups, since $\mathfrak{m} \preccurlyeq \mathfrak{n} \Leftrightarrow \log \mathfrak{m} \leqslant \log \mathfrak{n} \Leftrightarrow \nu(\mathfrak{m}) \preccurlyeq \nu(\mathfrak{n})$ for all $\mathfrak{m}, \mathfrak{n} \in \mathfrak{T}$. Therefore, we may identify $\mathfrak{T}$ with its image in $\mathfrak{T}_{\text{exp}}$ and $\mathbb{T}$ with the image of the strongly linear extension $\hat{\nu}$ of ν in $\mathbb{T}_{\text{exp}} = C[\![\mathfrak{T}_{\text{exp}}]\!]$. We extend the logarithm on $\mathbb{T}$ to $\mathbb{T}_{\text{exp}}$ by setting $\log \mathfrak{m} = f \in \mathbb{T}_{\succ}$ for monomials $\mathfrak{m} = \exp f \in \mathfrak{T}_{\text{exp}}$, and $\log f = \log \mathfrak{d}_f + \log c_f + l \circ \delta_f$ for general $f \in (\mathbb{T}_{\text{exp}})^>$.

Proposition 4.9. *$\mathbb{R}_{\text{exp}}$ is a field of transseries.*

Proof. By construction, $\log(\mathfrak{m}^\lambda \mathfrak{n}) = \lambda \log \mathfrak{m} + \log \mathfrak{n}$, for all $\mathfrak{m}, \mathfrak{n} \in \mathfrak{T}_{\exp}$ and $\lambda \in C$. Given $\mathfrak{m} \in \mathfrak{T}_{\exp,\succ}$, we have $\log \mathfrak{m} \in \mathbb{T}^{>}_{\succ} \subseteq \mathbb{T}^{>}$. Consequently, $\log \mathfrak{m}$ and $\log\log \mathfrak{m}$ are both in $\mathbb{T}$, and proposition 4.6(d) implies that $\log\log \mathfrak{m} \prec \log \mathfrak{m}$. Hence, $(\log\log \mathfrak{m})_\succ < (\log \mathfrak{m})_\succ$ and $\log f \asymp \exp((\log\log \mathfrak{m})_\succ) \prec \exp((\log f)_\succ) = f$. We conclude by lemma 4.7. □

4.3.3 Increasing unions

Proposition 4.10. *Let I be a totally ordered set and let $(\mathbb{T}_i)_{i \in I}$ be a family of fields of transseries of the form $\mathbb{T}_i = C[[\mathfrak{T}_i]]$, such that $\mathfrak{T}_i \subseteq \mathfrak{T}_j$ and $\mathbb{T}_i \subseteq \mathbb{T}_j$, whenever $i \leqslant j$. Then $\mathbb{T} = C[[\bigcup_{i \in I} \mathfrak{T}_i]] = \bigcup_{i \in I} \mathbb{T}_i$ is a field of transseries.*

Proof. Clearly $\bigcup_{i \in I} \mathbb{T}_i \subseteq C[[\bigcup_{i \in I} \mathfrak{T}_i]]$. Inversely, assume that

$$f \in C[[\bigcup_{i \in I} \mathfrak{T}_i]].$$

Since f is grid-based, there exist $\mathfrak{m}_1, \ldots, \mathfrak{m}_n, \mathfrak{n} \in \bigcup_{i \in I} \mathfrak{T}_i$, such that

$$\operatorname{supp} f \subseteq \{\mathfrak{m}_1, \ldots, \mathfrak{m}_n\}^* \mathfrak{n}.$$

For sufficiently large $i \in I$, we have $\mathfrak{m}_1, \ldots, \mathfrak{m}_n, \mathfrak{n} \in \mathfrak{T}_i$, since I is totally ordered. Hence, $\operatorname{supp} f \subseteq \mathfrak{T}_i$ and $f \in \mathbb{T}_i$. This proves that $C[[\bigcup_{i \in I} \mathfrak{T}_i]] \subseteq \bigcup_{i \in I} \mathbb{T}_i$. Similarly, one verifies that $\mathbb{T}$ is a field of transseries, using the fact that given $f_1, \ldots, f_n \in \mathbb{T}$, we actually have $f_1, \ldots, f_n \in \mathbb{T}_i$ for some $i \in I$. □

4.3.4 General transseries in x

Let $(\mathbb{L}_n)_{n \in \mathbb{N}}$ be the sequence defined by $\mathbb{L}_0 = \mathbb{L}$ and $\mathbb{L}_{n+1} = \mathbb{L}_{n,\exp}$ for all n. By propositions 4.8, 4.9 and 4.10,

$$C[[[x]]] = \mathbb{L}_0 \cup \mathbb{L}_1 \cup \mathbb{L}_2 \cup \cdots$$

is a field of transseries. We call it the *field of grid-based transseries in x* over C. The *exponential height* of a transseries in $C[[[x]]]$ is the smallest index n, such that $f \in \mathbb{L}_n$. A transseries of exponential height 0 is called a *logarithmic transseries*.

Intuitively speaking, we have constructed $C[[[x]]]$ by closing $C[[x]]$ first under logarithm and next under exponentiation. It is also possible to construct $C[[[x]]]$ the other way around: let $\mathbb{E}_n$ be the smallest subfield of $C[[[x]]]$, which contains $\log_n x$ and which is stable under grid-based summation and exponentiation. We have $C[[[x]]] = \mathbb{E}_0 \cup \mathbb{E}_1 \cup \mathbb{E}_2 \cup \cdots$ of $C[[[x]]]$. The *logarithmic depth* of a transseries in $C[[[x]]]$ is the smallest number $n \in \mathbb{N}$, such that $f \in \mathbb{E}_n$.

We will write $C_p^q[[[x]]]$ for the field of transseries of exponential height $\leqslant p$ and logarithmic depth $\leqslant q$. We will also write $C_p[[[x]]] = \mathbb{L}_p = \bigcup_{q \in \mathbb{N}} C_p^q[[[x]]]$ and $C^q[[[x]]] = \mathbb{E}_q = \bigcup_{p \in \mathbb{N}} C_p^q[[[x]]]$.

Example 4.11. The divergent transseries

$$1+\log x\,\mathrm{e}^{-x}+2!\log^2 x\,\mathrm{e}^{-2x}+3!\log^3 x\,\mathrm{e}^{-3x}+\cdots \qquad (4.1)$$

is an example of a transseries of exponential height and logarithmic depth 1. The transseries $\mathrm{e}^{x^2/(1-x^{-1})}$ and $\mathrm{e}^{\mathrm{e}^x/(1-x^{-1})}$ from example 4.5 have exponential height 1 resp. 2 and logarithmic depth 0.

For the purpose of differential calculus, it is convenient to introduce slight variations of the notions of exponential height and logarithmic depth. The *level* of a transseries is the smallest number $n\in\mathbb{Z}$ for which $f\in\mathbb{E}_{-n}$. The field $\mathbb{E}=\mathbb{E}_{-1}$ of transseries of level $\geqslant 1$ is called the field of *exponential transseries*. The *depth* of a transseries is the smallest number $n\in\mathbb{N}$ with $f\in\mathbb{E}_{n-1}$.

Example 4.12. The transseries (4.1) has level -1 and depth 2. Both transseries $\mathrm{e}^{x^2/(1-x^{-1})}$ and $\mathrm{e}^{\mathrm{e}^x/(1-x^{-1})}$ have level 0 and depth 1. The transseries $\exp\exp(x+\mathrm{e}^{-\mathrm{e}^x})$ has level 2 and depth 0.

4.3.5 Upward and downward shifting

In this section, we define the right compositions of transseries in x with $\exp x$ and $\log x$. Given $f\in C[\![x]\!]$, we will also denote $f\circ\exp x$ and $f\circ\log x$ by $f\uparrow$ resp. $f\downarrow$ and call them the *upward* and *downward shifts* of f. The mappings $\uparrow,\downarrow: C[\![x]\!]\to C[\![x]\!]$ are strong difference operators and will be constructed by induction over the exponential height.

For monomials $\mathfrak{m}=x^{\alpha_0}\log^{\alpha_1}x\cdots\log_n^{\alpha_n}x\in\mathfrak{L}$, we define

$$\begin{aligned}(x^{\alpha_0}\log^{\alpha_1}x\cdots\log_n^{\alpha_n}x)\uparrow &= \exp^{\alpha_0}x\,x^{\alpha_1}\cdots\log_{n-1}^{\alpha_n}x;\\ (x^{\alpha_0}\log^{\alpha_1}x\cdots\log_n^{\alpha_n}x)\downarrow &= \log^{\alpha_0}x\,\log_2^{\alpha_1}x\cdots\log_{n+1}^{\alpha_n}x.\end{aligned}$$

Extending these definitions by strong linearity, we obtain mappings

$$\begin{aligned}\uparrow: C^0[\![x]\!] &\to C^1[\![x]\!]\\ \downarrow: C^0[\![x]\!] &\to C^0[\![x]\!].\end{aligned}$$

Now assume that we have further extended these mappings into mappings

$$\begin{aligned}\uparrow: C^p[\![x]\!] &\to C^{p+1}[\![x]\!]\\ \downarrow: C^p[\![x]\!] &\to C^p[\![x]\!].\end{aligned}$$

Then we define

$$\begin{aligned}(\exp f)\uparrow &= \exp(f\uparrow);\\ (\exp f)\downarrow &= \exp(f\downarrow),\end{aligned}$$

for monomials $\mathfrak{m}=\exp f\in\exp C^p[\![x]\!]_{\succ}$. Extending these definitions by strong linearity, we obtain mappings

$$\begin{aligned}\uparrow: C^{p+1}[\![x]\!] &\to C^{p+2}[\![x]\!]\\ \downarrow: C^{p+1}[\![x]\!] &\to C^{p+1}[\![x]\!].\end{aligned}$$

By induction over p, we have thus defined $\uparrow$ and $\downarrow$ on $C[\![x]\!]$. Notice that $\uparrow$ and $\downarrow$ are mutually inverse, since $f\uparrow\downarrow = f$ for all $f \in C^p[\![x]\!]$ and $p \in \mathbb{N}$, by induction over p.

There is another way of interpreting right compositions of transseries in x with $\exp x$ and $\log x$ as formal substitutions $x \mapsto \exp x$ and $x \mapsto \log x$, considered as mappings from $C[\![x]\!]$ into $C[\![\exp x]\!]$ resp. $C[\![\log x]\!]$. Postulating that these mappings coincide with the upward and downward shiftings amounts to natural isomorphisms between $C[\![x]\!]$ and $C[\![\exp x]\!]$ resp. $C[\![\log x]\!]$.

Exercise 4.8. Let $\mathbb{T}$ be any non-trivial field of grid-based transseries. Prove that there exists a strongly linear ring homomorphism $\varphi: \mathbb{L} \rightarrow \mathbb{T}$.

Exercise 4.9. For all $p, q \in \mathbb{N}$, prove that

a) $C_p^q[\![\log x]\!] \subseteq C_p^{q+1}[\![x]\!]$;
b) $C_p^{q+1}[\![x]\!] \subseteq C_{p+1}^q[\![\log x]\!]$;
c) $\mathbb{E}_p = C^0[\![\log_p x]\!]$;
d) $C_{p+1}^q[\![x]\!] = C[\![\log_p^C x \times \exp C_p^q[\![x]\!]_{\succ}]\!]$.

Exercise 4.10. Given $f \in C[\![x]\!]^{>,\succ}$, we call $\operatorname{con} f = \log \circ f \circ \exp$ the *contraction* and $\operatorname{dil} f = \exp \circ f \circ \log$ the *dilatation* of f. Determine $\operatorname{dil}(x+1)$, $\operatorname{dil}\operatorname{dil}(x+1)$ and $\operatorname{dil}\operatorname{dil}\operatorname{dil}(x+1)$. Prove that for any $f \in C[\![x]\!]^{>,\succ}$, we have $\operatorname{con}_k f \sim \exp_l x$ for some $l \in \mathbb{Z}$ and all sufficiently large $k \in \mathbb{N}$. Here con_k denotes the k-th iterate of con.

Exercise 4.11. A field of *well-based transseries* is a field of well-based series of the form $\mathbb{T} = C[[\mathfrak{T}]]$, which satisfies **T1**, **T2**, **T3** and

T4. Let $(\mathfrak{m}_i)_{i \in \mathbb{N}}$ be a sequence of monomials in $\mathfrak{T}$, such that $\mathfrak{m}_{i+1} \in \operatorname{supp} \log \mathfrak{m}_i$, for each $i \in \mathbb{N}$. Then there exists an index i_0, such that for all $i \geqslant i_0$ and all $\mathfrak{n} \in \operatorname{supp} \log \mathfrak{m}_i$, we have $\mathfrak{n} \succcurlyeq \mathfrak{m}_{i+1}$ and $(\log \mathfrak{m}_i)_{\mathfrak{m}_{i+1}} = \pm 1$.

Show that the results from sections 4.3.1, 4.3.2 and 4.3.3 generalize to the well-based context.

Exercise 4.12. Define a transfinite sequence $(C^\alpha[[[x]]])_\alpha = (C[[[\mathfrak{T}_\alpha]]])_\alpha$ of fields of well-based transseries as follows: we take $\mathfrak{T}_0 = \mathfrak{L}$, $\mathfrak{T}_{\alpha+1} = (\mathfrak{T}_\alpha)_{\exp}$ for each ordinal α and $\mathfrak{T}_\lambda = \bigcup_{\alpha<\lambda} \mathfrak{T}_\alpha$, for each limit ordinal λ.

a) Prove that $C^\alpha[[[x]]] \subsetneq C^\beta[[[x]]]$ for all ordinals $\alpha < \beta$. Hint: one may consider the transfinite sequence of transseries $(f_\alpha)_{\alpha>0}$ defined by

$$f_\alpha = x^2 - \sum_{0<\beta<\alpha} e^{f_\beta \circ \log}.$$

b) If we restrict the supports of well-based transseries to be countable, then prove that the transfinite sequence $(C^\alpha[[[x]]])_\alpha$ stabilizes. Hint: find a suitable representation of transseries by labeled trees.

Exercise 4.13.

a) Prove that **T1**, **T2** and **T3** do not imply **T4**.
b) A transseries $f \in \mathbb{T}^{>,\succ}$ is said to be *log-confluent*, if there exists an index i_0, such that for all $i \geqslant i_0$, we have $\partial_{\log_{i+1} f} = \log \partial_{\log_i f}$. Prove that **T4** implies the log-confluence of all transseries in $\mathbb{T}^{>,\succ}$.
c) Prove that **T1**, **T2**, **T3** and the log-confluence of all transseries in $\mathbb{T}^{>,\succ}$ do not imply **T4**.

Exercise 4.14.

a) Prove that there exists a field of well-based transseries $\mathbb{T}$ in the sense of exercise 4.11, which contains the transseries

$$f = \mathrm{e}^{x^2 + \mathrm{e}^{\log_2^2 x + \mathrm{e}^{\log_4^2 x + \cdots}}}$$

b) Prove that the functional equation

$$g(x) = \mathrm{e}^{x^2 + g(\log_2 x) + \log x}$$

admits a solution in $\mathbb{T}$.

4.4 The incomplete transbasis theorem

A *transbasis* is a finite basis $\mathfrak{B} = (\mathfrak{b}_1, \ldots, \mathfrak{b}_n)$ of an asymptotic scale, such that

TB1. $\mathfrak{b}_1, \ldots, \mathfrak{b}_n \succ 1$ and $\mathfrak{b}_1 \prec\!\!\!\prec \cdots \prec\!\!\!\prec \mathfrak{b}_n$.
TB2. $\mathfrak{b}_1 = \exp_l x$, for some $l \in \mathbb{Z}$.
TB3. $\log \mathfrak{b}_i \in C[[\mathfrak{b}_1; \ldots; \mathfrak{b}_{i-1}]]_\succ$ for all $1 < i \leqslant n$.

The integer l in **TB2** is called the *level* of the transbasis $\mathfrak{B}$. We say that $\mathfrak{B}$ is a transbasis for $f \in \mathbb{T}$ (or that f can be expanded w.r.t. $\mathfrak{B}$), if $f \in C[[\mathfrak{b}_1; \ldots; \mathfrak{b}_n]]$.

Remark 4.13. Although the axiom **TB3** is well-suited to the purpose of this book, there are several variants which are more efficient from a computational point of view: see exercise 4.15.

Example 4.14. The tuple $(x, \mathrm{e}^{\sqrt{x}}, \mathrm{e}^{x\sqrt{x}})$ is a transbasis for $\mathrm{e}^{(x+1)^{3/2}}$ and so is $(x, \mathrm{e}^{(x+1)\sqrt{x}})$. Neither $(x, \mathrm{e}^x, \mathrm{e}^{\mathrm{e}^x + x^{-1}})$ nor $(x, \mathrm{e}^x, \mathrm{e}^{x\mathrm{e}^{\mathrm{e}^x}}, \mathrm{e}^{x\mathrm{e}^{\mathrm{e}^x} + \mathrm{e}^x})$ is a transbasis.

Theorem 4.15. *Let $\mathfrak{B}$ be a transbasis and $f \in C[[x]]$ a transseries. Then f can be expanded w.r.t. a super-transbasis $\hat{\mathfrak{B}}$ of $\mathfrak{B}$. Moreover, $\hat{\mathfrak{B}}$ may be chosen so as to fulfill the following requirements:*

a) The level of $\hat{\mathfrak{B}}$ is the minimum of the levels of $\mathfrak{B}$ and f.
b) If $\mathfrak{B}$ and f belong to a flat subring of $C[[x]]$ of the form $C[[x]]^\flat = C[[\mathfrak{T}^\flat]]$, then so does $\hat{\mathfrak{B}}$.

Proof. Let l be the level of $\mathfrak{B} = (\mathfrak{b}_1, \ldots, \mathfrak{b}_n)$. Without loss of generality, we may assume that $f \in C^0[[\exp_l x]]$. Indeed, there exists an l' with $f \in C^0[[\exp_{l'} x]]$; if $l' < l$, then we insert $\exp_{l'} x, \ldots, \exp_{l-1} x$ into $\mathfrak{B}$. We will now prove the theorem by induction over the minimal p, such that $f \in C_p^0[[\exp_l x]]$. If $p = 0$, then we clearly have nothing to prove. So assume that $p > 0$.

Let us consider the case when $f = \mathrm{e}^g$, with $g \in C[\![\mathfrak{b}_1; \ldots; \mathfrak{b}_n]\!]$. We distinguish three cases:

g is bounded. We may take $\hat{\mathfrak{B}} = \mathfrak{B}$.

$g \not\asymp \log \mathfrak{b}_i$ for each i. $\hat{\mathfrak{B}} = (\mathfrak{b}_1, \ldots, \mathfrak{b}_i, \mathrm{e}^{|g_\succ|}, \mathfrak{b}_{i+1}, \ldots, \mathfrak{b}_n)$ is again a transbasis for some $i \in \{1, \ldots, n\}$ and

$$f = \mathrm{e}^{g_\succ}\, \mathrm{e}^{g_\asymp}\, \mathrm{e}^{g_\prec} = \mathrm{e}^{\pm |g_\succ|}\, \mathrm{e}^{g_\asymp} \left(1 + g^- + \frac{1}{2} (g^-)^2 + \cdots\right)$$

can be expanded w.r.t. $\hat{\mathfrak{B}}$. Moreover, $\hat{\mathfrak{B}}$ satisfies the extra requirements (a) and (b). Indeed, $\hat{\mathfrak{B}}$ has level l and

$$\mathrm{e}^g \in C[\![x]\!]^\flat \Rightarrow \mathrm{e}^{|g_\succ|} \in C[\![x]\!]^\flat,$$

since $\mathrm{e}^{|g_\succ|} \succcurlyeq\!\!\!\!\!\preccurlyeq \mathrm{e}^g$.

$g \asymp \log \mathfrak{b}_i$ for some i. We rewrite $g = \lambda_i \log \mathfrak{b}_i + \tilde{g}$, with $\tilde{g} \prec g$. If $\tilde{g}$ is again equivalent to some $\log \mathfrak{b}_{\tilde{\imath}}$, then $\tilde{\imath} < i$, and we may rewrite $\tilde{g} = \lambda_{\tilde{\imath}} \log \mathfrak{b}_{\tilde{\imath}} + \tilde{\tilde{g}}$, with $\tilde{\tilde{g}} \prec \tilde{g}$. Repeating this procedure, we end up with an expression of the form

$$g = \lambda_{i_1} \log \mathfrak{b}_{i_1} + \cdots + \lambda_{i_k} \log \mathfrak{b}_{i_k} + h,$$

with $i_1 > \cdots > i_k$ and where h is either bounded or infinitely large with $h \not\asymp \log \mathfrak{b}_j$, for all j. By what precedes, e^h and $f = \mathrm{e}^g = \mathfrak{b}_{i_1}^{\lambda_{i_1}} \cdots \mathfrak{b}_{i_k}^{\lambda_{i_k}} \mathrm{e}^h$ may be expanded w.r.t. a super-transbasis $\hat{\mathfrak{B}}$ of $\mathfrak{B}$ which satisfies the additional requirements (a) and (b).

This proves the theorem in the case when $f = \mathrm{e}^g$, with $g \in C[\![\mathfrak{b}_1; \ldots; \mathfrak{b}_n]\!]$.

Assume now that f is a general grid-based transseries in $C_p^0[\![\exp_l x]\!]$. Then $\operatorname{supp} f$ is contained in a set of the form $\mathrm{e}^{g_0 + g_1 \mathbb{N} + \cdots + g_k \mathbb{N}}$, where $g_0, \ldots, g_k \in C_{p-1}^0[\![\exp_l x]\!]_\succ$ and $\mathrm{e}^{g_1}, \ldots, \mathrm{e}^{g_k} \prec 1$. Moreover, if $f \in C[\![x]\!]^\flat$, then we may choose $g_0, g_1, \ldots, g_k \in C[\![x]\!]^\flat$. Indeed, setting

$$\tilde{g}_i = \sum_{\mathfrak{m} \in \mathfrak{T}_\succ, \mathrm{e}^{\mathfrak{m}} \in \mathfrak{T}^\flat} g_{i,\mathfrak{m}}\, \mathfrak{m} \in C[\![x]\!]^\flat$$

for all i, we have

$$\mathrm{e}^{g_0 + g_1 \mathbb{N} + \cdots + g_k \mathbb{N}} \cap \mathfrak{T}^\flat \subseteq \mathrm{e}^{\tilde{g}_0 + \tilde{g}_1 \mathbb{N} + \cdots + \tilde{g}_k \mathbb{N}}.$$

Using the induction hypothesis, and modulo an extension of $\mathfrak{B}$, we may therefore assume without loss of generality that $g_0, \ldots, g_k \in C[\![\mathfrak{b}_1; \ldots; \mathfrak{b}_n]\!]$. By what precedes, it follows that there exists a super-transbasis $\hat{\mathfrak{B}}$ of $\mathfrak{B}$ for $\mathrm{e}^{g_0}, \ldots, \mathrm{e}^{g_k}$ which satisfies the requirements (a) and (b). By strong linearity, we conclude that $\hat{\mathfrak{B}}$ is the required transbasis for f. □

Exercise 4.15. Consider the following alternatives for **TB3**:

TB3-a. $\log \mathfrak{b}_i \in C[\![\mathfrak{b}_1; \ldots; \mathfrak{b}_n]\!]_\succ$, for all $1 < i \leqslant n$;

TB3-b. $\log \mathfrak{b}_i \in C[\![\mathfrak{b}_1; \ldots; \mathfrak{b}_{i^*}]\!]$, for all $1 < i \leqslant n$, where i^* is such that $\mathfrak{d}_{\log \mathfrak{b}_i} \succcurlyeq\!\!\!\!\!\preccurlyeq \mathfrak{b}_{i^*}$;

TB3-c. $\log \mathfrak{b}_i \in C[\![\mathfrak{b}_1; \ldots; \mathfrak{b}_{i-1}]\!]$ for all $1 < i \leqslant n$;
TB3-d. $\log \mathfrak{b}_i \in C[\![\mathfrak{b}_1; \ldots; \mathfrak{b}_n]\!]$ for all $1 < i \leqslant n$.

We respectively say that $\mathfrak{B}$ is a heavy, normal, light or sloppy transbasis.

a) Show that **TB3-a** $\Rightarrow$ **TB3-b** $\Rightarrow$ **TB3-c** $\Rightarrow$ **TB3-d**.
b) Show that theorem 4.15 holds for any of the above types of transbases.

Exercise 4.16. Find heavy, normal, light and sloppy normal transbases with respect to which the following "exp-log transseries" can be expanded:

a) $\mathrm{e}^{\mathrm{e}^{\mathrm{e}^{x} + \mathrm{e}^{-\mathrm{e}^{x}}}}$;
b) $\mathrm{e}^{\mathrm{e}^{\mathrm{e}^{x} + \mathrm{e}^{-\mathrm{e}^{\mathrm{e}^{x}}}}}$;
c) $\mathrm{e}^{\frac{x^{1000}}{1 - x^{-1}}}$;
d) $\mathrm{e}^{\frac{x^{1000}}{1 - \mathrm{e}^{-x}}} + \mathrm{e}^{\frac{x^{1000}}{1 - \mathrm{e}^{-\mathrm{e}^{x}}}}$;
e) $\log \log (x\, \mathrm{e}^{x \mathrm{e}^{x}} + 1) - \exp \exp (\log \log x + \frac{1}{x})$.

More precisely, an exp-log transseries (resp. function) is a transseries (resp. function) built up from x and constants in C, using the field operations $+$, $-$, $\times$, $/$, exponentiation and logarithm.

Exercise 4.17. Let $\mathfrak{B} = (\mathfrak{b}_1, \ldots, \mathfrak{b}_n)$ be a transbasis. Prove that there exists a unique transbasis $\tilde{\mathfrak{B}} = (\tilde{\mathfrak{b}}_1, \ldots, \tilde{\mathfrak{b}}_n)$, such that

i. $\tilde{\mathfrak{B}}^C = \mathfrak{B}^C$
ii. $c_{\log \tilde{\mathfrak{b}}_i} = 1$ for all $1 \leqslant i \leqslant n$.
iii. $(\log \tilde{\mathfrak{b}}_j)_{\mathfrak{d}_{\log \tilde{\mathfrak{b}}_i}} = 0$ for all $1 \leqslant i < j \leqslant n$.

Exercise 4.18. Let A be a local community.

a) If f and $\mathfrak{B}$ belong to $C[\![[x]]\!]_A$ in theorem 4.15, then show that $\tilde{\mathfrak{B}}$ may be chosen to belong to $C[\![[x]]\!]_A$ as well.
b) Show that (a) remains valid if **LC3** is replaced by the weaker axiom that for all $f \in A_{k+1}$ we have $f(z_1, \ldots, z_n, 0) \in A_k$.
c) Given a transbasis $\mathfrak{B} \subseteq C[\![[x]]\!]_A$, show that $C[\![\mathfrak{b}_1; \ldots; \mathfrak{b}_n]\!]_A \subseteq C[\![[x]]\!]_A$ and that the coefficients of recursive expansions of $f \in C[\![\mathfrak{b}_1; \ldots; \mathfrak{b}_n]\!]_A$ are again in $C[\![\mathfrak{b}_1; \ldots; \mathfrak{b}_n]\!]_A$.
d) Given $f \in C[\![[x]]\!]_A$, show that $f_{\succ}, f_{\prec} \in C[\![[x]]\!]_A$.

4.5 Convergent transseries

Assume now that $C = \mathbb{R}$ and let us define the exp-log subfield $C\{\{\{x\}\}\}$ of $C[\![[x]]\!]$ convergent transseries in x. The field $C_p\{\{\{x\}\}\}$ of convergent transseries of exponentiality $\leqslant p$ is defined by induction over p by taking $C_0\{\{\{x\}\}\} = C\{\{\mathfrak{L}\}\}$ and $C_{p+1}\{\{\{x\}\}\} = C\{\{\exp C_p\{\{\{x\}\}\}_{\succ}\}\}$. Here we notice that $\log \mathfrak{L} \subseteq C_0\{\{\{x\}\}\}_{\succ}$, so that $C_0\{\{\{x\}\}\} \subseteq C_1\{\{\{x\}\}\} \subseteq \cdots$, by induction. Now we define $C\{\{\{x\}\}\} = \bigcup_{l \in \mathbb{N}} C_l\{\{\{x\}\}\}$. By exercises 3.13 and 3.14, the set $C\{\{\{x\}\}\}$ is an exp-log subfield of $C[\![[x]]\!]$.

Let $\mathscr{G}$ be the ring of germs at infinity of real analytic functions at infinity. We claim that there exists a natural embedding $C\{\{x\}\} \hookrightarrow \mathscr{G}$, which preserves the ordered exp-log field structure. Our claim relies on the following lemma:

Lemma 4.16. *Let $\mathfrak{M}$ be a totally ordered monomial group and $\varphi: \mathfrak{M} \hookrightarrow \mathscr{G}^>$ an injection, which preserves multiplication and $\prec$. Then for each $f \in C\{\mathfrak{M}\}$,*

$$\hat{\varphi}(f) = \sum_{\mathfrak{m} \in \operatorname{supp} f} f_{\mathfrak{m}} \varphi(\mathfrak{m})$$

is a well-defined function in $\mathscr{G}$ and the mapping $\hat{\varphi}: C\{\mathfrak{M}\} \hookrightarrow \mathscr{G}$ is an injective morphism of totally ordered fields.

Proof. Let $f = \psi(\check{f})$ be a regular convergent Cartesian representation for f, with $\check{f} \in C((z_1, \ldots, z_k))$. Let $U = (0, \varepsilon)^k$ be such that $\check{f}$ is real analytic on U. Consider the mapping

$$\xi: x \mapsto (\varphi(\psi(z_1))(x), \ldots, \varphi(\psi(z_k))(x)).$$

Since φ preserves $\prec$, we have $\xi(x) \in U$, for sufficiently large x. Hence, $\hat{\varphi}(f)(x) = \check{f} \circ \xi(x)$ is defined and real analytic for all sufficiently large x.

Assume now that $f > 0$ and write $\check{f} = \check{g} z_1^{\alpha_1} \cdots z_k^{\alpha_k}$, where $\check{g}$ is a convergent series in $z_1, \ldots, z_k$ with $\check{g}(0, \ldots, 0) > 0$. Then

$$\check{g}(z_1, \ldots, z_k) > \frac{1}{2} \check{g}(0, \ldots, 0) > 0$$

for $(z_1, \ldots, z_k) \in U$, when choosing ε sufficiently small. Hence,

$$\hat{\varphi}(f)(x) = \check{g} \circ \xi(x) \, \varphi(\psi(z_1^{\alpha_1} \cdots z_k^{\alpha_k}))(x) > 0,$$

for all sufficiently large x, i.e. $\hat{\varphi}(f) > 0$. Consequently, $\hat{\varphi}$ is an injective, increasing mapping and it is clearly a ring homomorphism. □

Let us now construct embeddings $\hat{\varphi}_p: C_p\{\{x\}\} \hookrightarrow \mathscr{G}$, by induction over p. For $p = 0$, the elements in $\mathfrak{L}$ may naturally be interpreted as germs at infinity, which yields a natural embedding $\hat{\varphi}_0: C_0\{\{x\}\} \hookrightarrow \mathscr{G}$ by lemma 4.16. Assume that we have constructed the embedding $\hat{\varphi}_p$ and consider the mapping

$$\begin{aligned} \varphi_{p+1}: \exp C_p\{\{x\}\}_{\succ} &\hookrightarrow \mathscr{G} \\ \exp f &\mapsto \exp \hat{\varphi}_p(f). \end{aligned}$$

Clearly, φ_{p+1} is an injective multiplicative mapping. Given $f, g \in C_p\{\{x\}\}_{\succ}$, we also have

$$\begin{aligned} \exp f \prec \exp g &\Leftrightarrow f < g \\ &\Rightarrow g - f \in \mathbb{T}^{>,\succ} \\ &\Rightarrow 0 < \hat{\varphi}_p(g) - \hat{\varphi}_p(f) \succ 1 \\ &\Rightarrow \exp \hat{\varphi}_p(g) / \exp \hat{\varphi}_p(f) \succ 1 \\ &\Leftrightarrow \varphi_{p+1}(\exp f) \prec \varphi_{p+1}(\exp g). \end{aligned}$$

Applying lemma 4.16 on φ_{p+1}, we obtain the desired embedding $\hat{\varphi}_{p+1}$: $C_{p+1}\{\{\{x\}\}\} \hookrightarrow \mathscr{G}$. Using induction over p, we also observe that $\hat{\varphi}_{p+1}$ coincides with $\hat{\varphi}_p$ on $C_p\{\{\{x\}\}\}$ for each p. Therefore, we have a natural embedding of $C\{\{\{x\}\}\}$ into $\mathscr{G}$, which coincides with $\hat{\varphi}_p$ on each $C_p\{\{\{x\}\}\}$.

Remark 4.17. In the case of well-based transseries, the notion of convergence is more complicated. In general, sums like $\mathrm{e}^{-x} + \mathrm{e}^{-\exp x} + \mathrm{e}^{-\exp_2 x} + \cdots$ only yield *quasi-analytic* functions and for a more detailed study we refer to [É92, É93]. For natural definitions of convergence like in exercise 4.21, it can be hard to show that convergence is preserved under simple operations, like differentiation.

Exercise 4.19.

a) Given $f \in \mathbb{C}[[\mathfrak{M}]]$, let

$$\bar{f} = \sum_{\mathfrak{m} \in \mathfrak{M}} |f_{\mathfrak{m}}| \, \mathfrak{m}.$$

We say that $\mathcal{F} \in \mathscr{F}(\mathbb{C}\{\{\mathfrak{M}\}\})$ is summable in $\mathbb{C}\{\{\mathfrak{M}\}\}$, if $\overline{\mathcal{F}}$ is grid-based and $\sum \overline{\mathcal{F}} \in \mathbb{C}\{\{\mathfrak{M}\}\}$. Show that this defines a strong ring structure on $C\{\{\mathfrak{M}\}\}$.

b) Let $\mathcal{F}$ be a family of elements in $\mathscr{G}$. Define $f = \sum \mathcal{F}$ by $f(x) = \sum_{f \in \mathcal{F}} f(x)$, whenever there exists a neighbourhood U of infinity, such that f is defined on U for each $f \in \mathcal{F}$ and such that $\sum \mathcal{F}$ is normally convergent on each compact subset of U. Show that this defines a strong ring structure on $\mathscr{G}$.

c) Reformulate lemma 4.16 as a principle of "convergent extension by strong linearity".

Exercise 4.20. Prove that

$$\int \mathrm{e}^{x^2} = \frac{1}{2\,x}\,\mathrm{e}^{x^2} + \frac{1}{4\,x^3}\,\mathrm{e}^{x^2} + \frac{3}{8\,x^5}\,\mathrm{e}^{x^2} + \cdots \notin C\{\{\{x\}\}\}.$$

Exercise 4.21. Let $\mathbb{T} = C[[[x]]]$ be the field of well-based transseries of finite exponential and logarithmic depths. Given $\sigma \in \mathbb{R}$, let $\mathscr{C}$ be the set of infinitely differentiable real germs at infinity and $\mathscr{C}^\sigma$ the set of infinitely differentiable real functions on $(\sigma, \rightarrow)$.

a) Construct the smallest subset $\mathbb{T}^{\mathrm{cv},\sigma}$ of $\mathbb{T}$, together with a mapping $\varphi^\sigma \colon \mathbb{T}^{\mathrm{cv},\sigma} \rightarrow \mathscr{C}^\sigma$, such that

   CT1. If $\sigma \geqslant \exp_l 0$, then $\log_l x \in \mathbb{T}^{\mathrm{cv},\sigma}$ and $\varphi(\log_l x) = \log_l$.

   CT2. If $f \in \mathbb{T}$ is such that $\log \mathfrak{m} \in \mathbb{T}^{\mathrm{cv},\sigma}$ for all $\mathfrak{m} \in \operatorname{supp} f$ and $\sum_{\mathfrak{m}} |f_{\mathfrak{m}}\, \varphi(\mathfrak{m})|$ is convergent on $(\sigma, \rightarrow)$, then $f \in \mathbb{T}^{\mathrm{cv},\sigma}$ and $\varphi^\sigma(f) = \sum_{\mathfrak{m}} f_{\mathfrak{m}}\, \varphi(\mathfrak{m})$.

   Show that $\mathbb{T}^{\mathrm{cv},\sigma}$ is a ring.

b) Show that $\mathbb{T}^{\mathrm{cv},\sigma} \subseteq \mathbb{T}^{\mathrm{cv},\tau}$ for $\tau \geqslant \sigma$. Denoting $\mathbb{T}^{\mathrm{cv}} = \bigcup_{\sigma \in \mathbb{R}} \mathbb{T}^{\mathrm{cv},\sigma}$, show that there exists a mapping $\varphi \colon \mathbb{T}^{\mathrm{cv}} \rightarrow \mathscr{C}$, such that $\varphi(f)$ is the germ associated to $\varphi^\sigma(f)$ for every σ with $f \in \mathbb{T}^{\mathrm{cv},\sigma}$. Show also that $\mathbb{T}^{\mathrm{cv}}$ is a field.

5

Operations on transseries

One of the major features of the field $\mathbb{T} = C[\![x]\!]$ of grid-based transseries in x is its stability under the usual operations from calculus: differentiation, integration, composition and inversion.

What is more, besides the classical properties from calculus, these operations satisfy interesting additional properties, which express their compatibility with infinite summation, the ordering, and the asymptotic relations $\preccurlyeq$, $\prec$, etc. Therefore, the field of transseries occurs as a natural model of "ordered or asymptotic differential algebra", in addition to the more classical Hardy fields. It actually suggests the development of a whole new branch of model theory, which integrates the infinitary summation operators. Also, not much is known on the model theory of compositions.

In section 5.1, we start by defining the differentiation w.r.t. x as the unique strongly linear C-differentiation with $x' = 1$ and $(\mathrm{e}^f)' = f' \mathrm{e}^f$ for all f. This differentiation satisfies

$$\begin{aligned} f \prec g \wedge g \not\asymp 1 &\Rightarrow f' \prec g' \\ f > 0 \wedge f \succ 1 &\Rightarrow f' > 0 \end{aligned}$$

In section 5.2, we show that the differentiation has a unique right inverse $\int$ with the property that $(\int f)_{\asymp} = 0$ for all $f \in \mathbb{T}$; for this reason, we call $\int f$ the "distinguished integral" of f. Moreover, the distinguished integration is strongly linear and we will see in the exercises that one often has $(\int f)(\int g) = \int f \int g + \int g \int f$.

In section 5.3, we proceed with the definition of a composition on $\mathbb{T}$. More precisely, given $g \in \mathbb{T}^{>,\succ}$, we will show that there exists a unique strongly linear C-difference operator $\circ_g$ with $\circ_g(x) = g$ and $\circ_g(\mathrm{e}^f) = \mathrm{e}^{\circ_g(f)}$ for all f. This difference operator satisfies

$$\begin{aligned} f \succ 1 &\Rightarrow \circ_g(f) \succ 1 \\ f \geqslant 0 &\Rightarrow \circ_g(f) \geqslant 0 \end{aligned}$$

Moreover, the composition defined by $f \circ g = \circ_g(f)$ is associative and compatible with the differentiation: $(f \circ g)' = g'(f' \circ g)$ for all $f \in \mathbb{T}$ and $g \in \mathbb{T}^{>,\succ}$. Finally, the Taylor series expansion $f \circ (x + \delta) = f + f'\delta + \frac{1}{2} f''\delta^2 + \cdots$ holds under mild hypotheses on f and δ.

In section 5.4, we finally show that each $g \in \mathbb{T}^{>,\succ}$ admits a unique functional inverse g^{inv} with $g \circ g^{\text{inv}} = g^{\text{inv}} \circ g = x$. We conclude this chapter with Écalle's "Translagrange theorem" [É03], which generalizes Lagrange's classical inversion formula.

5.1 Differentiation

Let R be a strong totally ordered partial exp-log C-algebra. A *strong derivation* on R is a mapping $\partial: R \to R; f \mapsto f' = \partial f$, which satisfies

D1. $\partial c = 0$, for all $c \in C$.
D2. ∂ is strongly linear.
D3. $\partial(fg) = (\partial f)\, g + f \partial g$, for all $f, g \in R$.

We say that ∂ is an *exp-log derivation*, if we also have

D4. $\partial(\exp f) = (\partial f) \exp f$, for all $f \in \operatorname{dom} \exp \subseteq R$.

We say that ∂ is (strictly) *asymptotic* resp. *positive*, if

D5. $f \prec g \Rightarrow \partial f \prec \partial g$, for all $f, g \in R$ with $g \not\asymp 1$.
D6. $f \succ 1 \Rightarrow (f > 0 \Rightarrow \partial f > 0)$, for all $f \in R$.

In this section, we will show that there exists a unique strong exp-log derivation ∂ on $\mathbb{T}$, such that $\partial x = 1$. This derivation turns out to be asymptotic and positive. In what follows, given a derivation ∂ on a field, we will denote by $f^\dagger = f'/f$ the *logarithmic derivative* of $f \neq 0$.

Lemma 5.1. *Let* $\mathbb{T} = C[[\mathfrak{T}]]$ *be an arbitrary field of transseries and let* $\partial: \mathfrak{T} \to \mathbb{T}$ *be a mapping, which satisfies* $\partial(\mathfrak{m}\,\mathfrak{n}) = (\partial \mathfrak{m})\, \mathfrak{n} + \mathfrak{m}\, \partial \mathfrak{n}$ *for all* $\mathfrak{m}, \mathfrak{n} \in \mathfrak{T}$. *Then*

a) ∂ *is a grid-based mapping, which extends uniquely to a strong derivation on* $\mathbb{T}$.
b) If $\partial(\log \mathfrak{m}) = \partial \mathfrak{m}/\mathfrak{m}$ *for all* $\mathfrak{m} \in \mathfrak{T}$, *then* ∂ *is an exp-log derivation on* $\mathbb{T}$.

Proof. Let $\mathfrak{G}$ be a grid-based subset of $\mathbb{T}$, so that

$$\mathfrak{G} \subseteq \{\mathfrak{m}_1, \ldots, \mathfrak{m}_n\}^* \,\mathfrak{n}$$

for certain monomials $\mathfrak{m}_1 \prec 1, \ldots, \mathfrak{m}_n \prec 1$ and $\mathfrak{n}$ in $\mathfrak{T}$. For any $\mathfrak{m}_1^{\alpha_1} \cdots \mathfrak{m}_n^{\alpha_n} \mathfrak{n} \in \mathfrak{T}$, we have

$$(\mathfrak{m}_1^{\alpha_1} \cdots \mathfrak{m}_n^{\alpha_n} \mathfrak{n})' = \left(\alpha_1 \mathfrak{m}_1^\dagger + \cdots + \alpha_n \mathfrak{m}_n^\dagger + \mathfrak{n}^\dagger\right) \mathfrak{m}_1^{\alpha_1} \cdots \mathfrak{m}_n^{\alpha_n} \mathfrak{n}.$$

Hence $\operatorname{supp} \mathfrak{v}' \subseteq (\operatorname{supp} \mathfrak{m}_1^\dagger \cup \cdots \cup \operatorname{supp} \mathfrak{m}_n^\dagger \cup \operatorname{supp} \mathfrak{n}^\dagger)\mathfrak{v}$ for all $\mathfrak{v} \in \mathfrak{G}$, and ∂ a grid-based mapping. The strongly linear extension of ∂ is indeed a derivation, since $(f, g) \mapsto (fg)'$ and $(f, g) \mapsto f' g + f g'$ are both strongly bilinear mappings from $\mathbb{T}^2$ into $\mathbb{T}$, which coincide on $\mathfrak{T}^2$ (a proof which does not use strong bilinearity can be given in a similar way as for proposition 2.16). This proves (a).

As to (b), assume that $(\log \mathfrak{m})' = \mathfrak{m}^\dagger$ for all $\mathfrak{m} \in \mathfrak{T}$. Obviously, in order to prove that ∂ is a strong exp-log derivation, it suffices to prove that $(\log f)' = f^\dagger$ for all $f \in \mathbb{T}^>$. Now each $f \in \mathbb{T}^>$ may be decomposed as $f = c\mathfrak{m}(1+\varepsilon)$, with $c \in C^>$, $\mathfrak{m} \in \mathfrak{T}$ and $\varepsilon \prec 1$. For each $k \in \mathbb{N}^>$, we have $(\frac{(-1)^{k-1}}{k} \varepsilon^k)' = (-1)^{k-1} \varepsilon^{k-1} \varepsilon'$. Hence,

$$(\log(1+\varepsilon))' = \varepsilon'/(1+\varepsilon) = (1+\varepsilon)^\dagger,$$

by strong linearity. We conclude that

$$\begin{aligned} (\log f)' &= (\log c)' + (\log \mathfrak{m})' + (\log(1+\varepsilon))' \\ &= \mathfrak{m}^\dagger + (1+\varepsilon)^\dagger = (c\mathfrak{m}(1+\varepsilon))^\dagger. \end{aligned}$$ □

Proposition 5.2. *There exists a unique strong exp-log derivation ∂ on $\mathbb{T}$ with $\partial x = 1$.*

Proof. We will show by induction over $p \in \mathbb{N}$ that there exists a unique strong exp-log derivation ∂ on $C_p[\![x]\!] = C[\![\mathfrak{T}_p]\!]$ with $\partial x = 1$. Since this mapping ∂ is required to be strongly linear, it is determined uniquely by its restriction to $\mathfrak{T}_p$. Furthermore, ∂ will be a strong exp-log derivation, if its restriction to $\mathfrak{T}_p$ satisfies the requirements of lemma 5.1.

For $p = 0$, the derivative of a monomial $\mathfrak{m} = x^{\alpha_0} \cdots \log_q^{\alpha_q} x \in \mathfrak{T}_0$ must be given by

$$(x^{\alpha_0} \cdots \log_q^{\alpha_q} x)' = \left(\frac{\alpha_0}{x} + \cdots + \frac{\alpha_q}{x \cdots \log_q x}\right) x^{\alpha_0} \cdots \log_q^{\alpha_q} x$$

in view of axioms **D3** and **D4** and the requirements of lemma 5.1 are easily checked.

If $p > 0$, then the induction hypothesis states that there exists a unique strong exp-log derivation ∂ on $C_{p-1}[\![x]\!]$ with $\partial x = 1$. In view of **D4**, any strong exp-log derivation on $C_p[\![x]\!]$ should therefore satisfy

$$(\mathrm{e}^f)' = f' \mathrm{e}^f,$$

for all $\mathrm{e}^f \in \mathfrak{T}_p = \exp C_{p-1}[\![x]\!]_\succ$. On the other hand, when defining $(\mathrm{e}^f)'$ in this way, we have

$$(\mathrm{e}^f \mathrm{e}^g)' = (f' + g')\mathrm{e}^{f+g} = (f' \mathrm{e}^f)\mathrm{e}^g + \mathrm{e}^f(g' \mathrm{e}^g) = (\mathrm{e}^f)' \mathrm{e}^g + \mathrm{e}^f (\mathrm{e}^g)'$$

for all $\mathrm{e}^f, \mathrm{e}^g \in \mathfrak{T}_p$. Hence, there exists a unique strong derivation ∂ with $\partial x = 1$ on $C_p[\![x]\!]$, by lemma 5.1. Moreover, ∂ is a strong exp-log derivation, since

$$(\log \mathrm{e}^f)' = f' = (f' \mathrm{e}^f)/\mathrm{e}^f = (\mathrm{e}^f)'/\mathrm{e}^f$$

for all monomials $\mathrm{e}^f \in \mathfrak{T}_p$. □

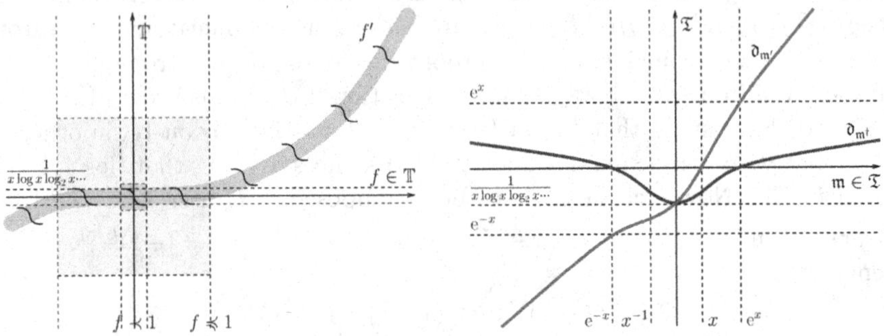

Fig. 5.1. We will often adopt a geometric point of view for which the derivative ∂ is a function on the "transline" $\mathbb{T}$. Due to the highly non-archimedean character of $\mathbb{T}$, it is difficult to sketch the behaviour of this function. An attempt has been made in the left figure above. The two squares correspond to the regions where both coordinates are infinitesimal resp. bounded. Notice that ∂ is locally decreasing everywhere (the small curves), although its restriction to $\mathbb{T}_{\succ}$ is increasing (the fat curve). At the right hand side, we also sketched the behaviour of the functions $\mathfrak{m} \mapsto \partial_{\mathfrak{m}'}$ and $\mathfrak{m} \mapsto \partial_{\mathfrak{m}^\dagger}$ for transmonomials (using logarithmic coordinates).

Proposition 5.3. *For all $f \in \mathbb{T}$, we have*

$$\begin{aligned} f{\uparrow}' &= \mathrm{e}^x\,(f'{\uparrow}); \\ f{\downarrow}' &= \tfrac{1}{x}\,(f'{\downarrow}). \end{aligned}$$

Proof. The mappings $d_1 \colon f \mapsto (\mathrm{e}^{-x}\,(f{\uparrow}')){\downarrow}$ and $d_2 \colon f \mapsto (x\,(f{\downarrow}')){\uparrow}$ are both strong exp-log derivations with $d_1\,x = d_2\,x = 1$. We conclude by proposition 5.2. □

Proposition 5.4. *Let $\mathfrak{B} = (\mathfrak{b}_1, \ldots, \mathfrak{b}_n)$ be a transbasis.*

a) If $\mathfrak{b}_1 = x$ or $\mathfrak{b}_1 = \exp x$, then $C[[\mathfrak{b}_1; \ldots; \mathfrak{b}_n]]$ is stable under ∂.
b) If $\mathfrak{b}_1 = \log_l x$ and $\log_{l-1} x, \ldots, x \in \mathfrak{B}$, then $C[[\mathfrak{b}_1; \ldots; \mathfrak{b}_n]]$ is stable under ∂.

Proof. Let us prove (a) by induction over n. Clearly, $C[[x]]$ and $C[[\exp x]]$ are stable under differentiation. So assume that $n > 1$ and that $C[[\mathfrak{b}_1; \ldots; \mathfrak{b}_{n-1}]]$ is stable under differentiation. Then $\mathfrak{b}_n' = (\log \mathfrak{b}_n)'\,\mathfrak{b}_n \in C[[\mathfrak{b}_1; \ldots; \mathfrak{b}_n]]$. Hence

$$(\mathfrak{b}_1^{\alpha_1} \cdots \mathfrak{b}_n^{\alpha_n})' = \left(\alpha_1\,\mathfrak{b}_1^\dagger + \cdots + \alpha_n\,\mathfrak{b}_n^\dagger\right) \mathfrak{b}_1^{\alpha_1} \cdots \mathfrak{b}_n^{\alpha_n} \in C[[\mathfrak{b}_1; \ldots; \mathfrak{b}_n]],$$

for all monomials $\mathfrak{b}_1^{\alpha_1} \cdots \mathfrak{b}_n^{\alpha_n} \in \mathfrak{B}^C$. Consequently, $C[[\mathfrak{b}_1; \ldots; \mathfrak{b}_n]]$ is stable under differentiation, by strong linearity.

As to (*b*), we first observe that $(\mathfrak{b}_1 \circ \exp_l, \ldots, \mathfrak{b}_n \circ \exp_l)$ is also a transbasis, so $C[\![\mathfrak{b}_1 \circ \exp_l; \ldots; \mathfrak{b}_n \circ \exp_l]\!]$ is stable under differentiation. Given $f \in C[\![\mathfrak{b}_1; \ldots; \mathfrak{b}_n]\!]$, we now have

$$\begin{aligned} f' &= (f \circ \exp_l \circ \log_l)' \\ &= \frac{1}{x \log x \cdots \log_{l-1} x}((f \circ \exp_l)' \circ \log_l) \in C[\![\mathfrak{b}_1; \ldots; \mathfrak{b}_n]\!]. \end{aligned}$$

□

Proposition 5.5. *The derivation ∂ on $\mathbb{T}$ is asymptotic and positive.*

Proof. Let $\mathfrak{B} = (\mathfrak{b}_1, \ldots, \mathfrak{b}_n)$ be a transbasis with $\mathfrak{b}_1 = \mathrm{e}^x$. We will first prove by induction over n, that ∂ is asymptotic and positive on $C[\![\mathfrak{b}_1; \ldots; \mathfrak{b}_n]\!]$, and $f' \succ 1$, for all $f \succ 1$ in $C[\![\mathfrak{b}_1; \ldots; \mathfrak{b}_n]\!]$. This is easy in the case when $n = 1$. So assume that $n > 1$.

Given a monomial $\mathfrak{m} = \mathfrak{b}_1^{\alpha_1} \cdots \mathfrak{b}_n^{\alpha_n}$, we first observe that

$$\begin{aligned} \mathfrak{m}^\dagger &= \alpha_1 \mathfrak{b}_1^\dagger + \cdots + \alpha_n \mathfrak{b}_n^\dagger \\ &= \alpha_1 + \alpha_2 (\log \mathfrak{b}_2)' + \cdots + \alpha_n (\log \mathfrak{b}_n)' \end{aligned}$$

belongs to $C[\![\mathfrak{b}_1; \ldots; \mathfrak{b}_{n-1}]\!]$. Moreover,

$$\mathfrak{b}_i^\dagger = (\log \mathfrak{b}_i)' \prec (\log \mathfrak{b}_n)' = \mathfrak{b}_n^\dagger,$$

for all $1 < i < n$, by the induction hypothesis. Actually, the induction hypothesis also implies that $\mathfrak{b}_1^\dagger = 1 \prec \mathfrak{b}_n^\dagger$, since $\log \mathfrak{b}_n \succ 1$. Consequently, $\mathfrak{m}^\dagger \asymp \mathfrak{b}_n^\dagger$, if $\alpha_n \neq 0$.

Secondly, let $\mathfrak{m} = \mathfrak{b}_1^{\alpha_1} \cdots \mathfrak{b}_n^{\alpha_n}$ and $\mathfrak{n} = \mathfrak{b}_1^{\beta_1} \cdots \mathfrak{b}_n^{\beta_n}$ be monomials with $\mathfrak{m} \prec \mathfrak{n} \neq 1$. If $\alpha_n = \beta_n = 0$, then $\mathfrak{m}' \prec \mathfrak{n}'$ by the induction hypothesis. If $\alpha_n < \beta_n$, then

$$\begin{aligned} \mathfrak{m}' &\in C[\![\mathfrak{b}_1; \ldots; \mathfrak{b}_{n-1}]\!]\, \mathfrak{b}_n^{\alpha_n} \\ \mathfrak{n}' &\in C[\![\mathfrak{b}_1; \ldots; \mathfrak{b}_{n-1}]\!]\, \mathfrak{b}_n^{\beta_n}, \end{aligned}$$

whence $\mathfrak{m}' \prec \mathfrak{n}'$. If $\alpha_n = \beta_n \neq 0$, then

$$\mathfrak{m}' \asymp \mathfrak{b}_n^\dagger \mathfrak{m} \prec \mathfrak{b}_n^\dagger \mathfrak{n} \asymp \mathfrak{n}'.$$

Hence $\mathfrak{m}' \prec \mathfrak{n}'$ in all cases. Given $f \in C[\![\mathfrak{b}_1; \ldots; \mathfrak{b}_n]\!]$ with $f \neq 0$ and $f \not\asymp 1$, we thus get $\mathfrak{m}' \prec \mathfrak{d}_f'$, for all $\mathfrak{m} \in \operatorname{supp} f \setminus \{\mathfrak{d}_f\}$, whence $f' \sim c_f \mathfrak{d}_f'$, by strong linearity.

Let us now prove that the induction hypothesis is satisfied at order n. Given $f, g \in C[\![\mathfrak{b}_1; \ldots; \mathfrak{b}_n]\!]$, with $1 \not\asymp f \prec g \not\asymp 1$, we have

$$f' \sim c_f \mathfrak{d}_f' \prec c_g \mathfrak{d}_g' \sim g'.$$

If $f \asymp 1$, we still have $f' \prec g'$, since $f' = f_{\not\asymp}'$ and $f_{\not\asymp} \prec f \prec g$. Now let $f \in C[\![\mathfrak{b}_1; \ldots; \mathfrak{b}_n]\!]^{>,\succ}$. By the induction hypothesis, we have $\mathfrak{d}_f^\dagger > 0$, since $\log f \in C[\![\mathfrak{b}_1; \ldots; \mathfrak{b}_{n-1}]\!]^{>,\succ}$. We conclude that

$$f' \sim c_f \mathfrak{d}_f' = c_f \mathfrak{d}_f^\dagger \mathfrak{d}_f > 0.$$

At this point, we have proved that ∂ is asymptotic and positive on $C[[\mathfrak{b}_1; ...; \mathfrak{b}_n]]$. By theorem 4.15($a$), this also proves that ∂ asymptotic and positive on $C_0[[[\exp x]]]$. Now let $f, g \in C_l[[[\exp x]]]$ be such that $f \prec g \not\asymp 1$. Then

$$\begin{aligned} f = (f \circ \exp_l \circ \log_l)' &= \\ \frac{(f \circ \exp_l)'}{x \cdots \log_{l-1} x} &\prec \frac{(g \circ \exp_l)'}{x \cdots \log_{l-1} x} \\ &= (g \circ \exp_l \circ \log_l)' = g'. \end{aligned}$$

Similarly, if $f \in C_l[[[\exp x]]]$ is such that $f \succ 1$ and $f > 0$, then

$$f' = \frac{(f \circ \exp_l)'}{x \cdots \log_l x} > 0. \qquad \square$$

Remark 5.6. A transbasis $\mathfrak{B} = (\mathfrak{b}_1, ..., \mathfrak{b}_n)$ of level 1 will also be called a *plane transbasis*. The two facts that $C[[\mathfrak{b}_1; ...; \mathfrak{b}_i]]$ is stable under differentiation for each i and $\mathfrak{m}^\dagger \prec\!\!\prec \mathfrak{m}$ for all $\mathfrak{m} = \mathfrak{b}_1^{\alpha_1} \cdots \mathfrak{b}_n^{\alpha_n} \neq 1$, make plane transbases particularly useful for differential calculus.

By theorem 4.15(a), we notice that any exponential transseries can be expanded with respect to a plane transbases. Computations which involve more general transseries can usually be reduced to the exponential case using the technique of upward and downward shifting.

Exercise 5.1. For all $f, g \in \mathbb{T}$, prove that

$$\begin{aligned} f \preccurlyeq g \wedge f \prec 1 \wedge g \prec 1 &\Rightarrow f^\dagger \succcurlyeq g^\dagger; \\ f \prec 1 \wedge g \not\asymp 1 &\Rightarrow f' \prec g^\dagger. \end{aligned}$$

Exercise 5.2. For all $f, g \in \mathbb{T}^{\neq}$ with $f \not\asymp 1$ and $g \not\asymp 1$, show that

$$\begin{aligned} f \preceq\!\!\preccurlyeq g &\Leftrightarrow f^\dagger \preccurlyeq g^\dagger; \\ f \prec\!\!\prec g &\Leftrightarrow f^\dagger \prec g^\dagger; \\ f \asymp\!\!\asymp g &\Leftrightarrow f^\dagger \asymp g^\dagger; \\ f \approx g &\Leftrightarrow f^\dagger \sim g^\dagger. \end{aligned}$$

Exercise 5.3. Let $f \in \mathbb{T}$. Prove that

a) $f \succ 1 \Leftrightarrow f' \succ \dfrac{1}{x \log x \log\log x \cdots}$.
b) $f' > 0 \Leftrightarrow ((f \succ 1 \wedge f > 0) \vee (f \preccurlyeq 1 \wedge f_{\not\asymp} < 0))$.
c) $f' > 0 \Leftrightarrow ((\forall \lambda \in C, f > \lambda) \vee (\exists \lambda \in C, \forall \mu \in C, \mu < \lambda \Rightarrow \mu < f < \lambda))$.

In the case of (a), notice that we may for instance interpret $f' \succ \frac{1}{x \log x \log\log x \cdots}$ as a relation in a field of well-based transseries in x.

Exercise 5.4. Consider a derivation ∂ on a totally ordered C-algebra R, which is also a field. We say that ∂ is asymptotic resp. positive, but not necessarily strictly, if

D5′. $f \preccurlyeq g \Rightarrow \partial f \preccurlyeq \partial g$, for all $f, g \in R$ with $g \not\asymp 1$.
D6′. $f \succ 1 \Rightarrow (f \geqslant 0 \Rightarrow \partial f \geqslant 0)$, for all $f \in R$.

If d is an asymptotic derivation, prove that fd is again an asymptotic derivation for any $f \in R$. Given positive derivations $d_1, \ldots, d_n$, prove that $f_1 d_1 + \cdots + f_n d_n$ is again a positive derivation. Prove that neither the set of asymptotic, nor the set of positive derivations necessarily form a module.

Exercise 5.5. Let $\mathbb{T} = C[\![x_1]\!] \cdots [\![x_n]\!]$. Characterize

a) The strong C-module of all strong exp-log derivations on $\mathbb{T}$.
b) The set of all (not necessarily strictly) asymptotic strong exp-log derivations on $\mathbb{T}$.
c) The set of all (not necessarily strictly) positive strong exp-log derivations on $\mathbb{T}$.

Exercise 5.6. Let $\mathfrak{T}^\flat \ni x$ be a flat subset of the set $\mathfrak{T}$ of transmonomials and let $\mathfrak{T}^\sharp$ be its steep complement (see exercise 4.7).

a) Show that $\mathbb{T}^\flat = C[\![\mathfrak{T}^\flat]\!]$ is stable under differentiation.
b) Considering $\mathbb{T}$ as a strong $\mathbb{T}^\flat$-algebra, show that there exists a unique strongly $\mathbb{T}^\flat$-linear mapping $\partial^\sharp : \mathbb{T} \to \mathbb{T}$ with $\partial^\sharp \mathfrak{m}^\sharp = (\mathfrak{m}^\sharp)'$ for all $\mathfrak{m}^\sharp \in \mathfrak{T}^\sharp$.
c) Show that

$$\left(\sum_{\mathfrak{m}^\sharp \in \mathfrak{T}^\sharp} f_{\mathfrak{m}^\sharp} \mathfrak{m}^\sharp\right)' = \sum_{\mathfrak{m}^\sharp \in \mathfrak{T}^\sharp} f'_{\mathfrak{m}^\sharp} \mathfrak{m}^\sharp + \sum_{\mathfrak{m}^\sharp \in \mathfrak{T}^\sharp} f_{\mathfrak{m}^\sharp} \partial^\sharp \mathfrak{m}^\sharp$$

for all $f \in \mathbb{T}$.

Exercise 5.7. Let f be a convergent transseries. Prove that f' is convergent and that the germ at infinity associated to f' coincides with the derivative of the germ at infinity associated to f. In other words, $C\{\!\{x\}\!\}$ is a Hardy field.

Exercise 5.8. Construct a strong exp-log derivation on the field $C[[[x]]]$ of well-based transseries of finite exponential and logarithmic depths. Show that there exists a unique such derivation ∂ with $\partial x = 1$, and show that ∂ is asymptotic and positive. Hint: see [vdH97].

5.2 Integration

In this section, we show that each transseries $f \in \mathbb{T}$ admits an integral in $\mathbb{T}$. Since the derivative of a transseries vanishes if and only if it is a constant, we infer that f admits a unique, *distinguished integral* $\int f$, whose constant term $(\int f)_\asymp$ vanishes. The distinguished property immediately implies that mapping $\int : f \mapsto \int f$ is linear. We will show that $\int$ is actually strongly linear.

Proposition 5.7. *There exists a unique right inverse $\int : \mathbb{T} \to \mathbb{T}$ of ∂, such that the constant term $(\int f)_\asymp$ of $\int f$ vanishes for all $f \in \mathbb{T}$. This right inverse is strongly linear.*

Proof. We will first consider the case when $f \in \mathbb{E}$ is exponential. Let $\mathfrak{B} = (\mathfrak{b}_1, ..., \mathfrak{b}_n)$ be a plane transbasis for f. Consider the double sum

$$\int f = f_{\asymp} x + \sum_{\mathfrak{m} \in \operatorname{supp} f \setminus \{1\}} \sum_{k \geqslant 0} f_{\mathfrak{m}} F_{\mathfrak{m},k} \mathfrak{m}, \tag{5.1}$$

where

$$\begin{aligned} F_{\mathfrak{m},0} &= \frac{1}{\mathfrak{m}^{\dagger}}; \\ F_{\mathfrak{m},k} &= -\frac{1}{\mathfrak{m}^{\dagger}} F'_{\mathfrak{m},k-1} \quad \text{for } k \geqslant 1. \end{aligned}$$

We will show that the family $(f_{\mathfrak{m}} F_{\mathfrak{m},k} \mathfrak{m})_{\mathfrak{m} \in \operatorname{supp} f \setminus \{1\}, k \geqslant 0}$ is grid-based, so that (5.1) defines an integral of f.

Let us first study the $F_{\mathfrak{m},k}$ for a monomial $\mathfrak{m} = \mathfrak{b}_1^{\alpha_1} \cdots \mathfrak{b}_i^{\alpha_i}$ with $\alpha_i \neq 0$. We observe that $\mathfrak{m}' = (\alpha_1 \mathfrak{b}_1^{\dagger} + \cdots + \alpha_i \mathfrak{b}_i^{\dagger}) \mathfrak{m} \asymp \mathfrak{d}_{\mathfrak{b}_i^{\dagger}} \mathfrak{m}$. Setting

$$\begin{aligned} \mathfrak{d}_i &= \mathfrak{d}_{\mathfrak{b}_i^{\dagger}}; \\ \mathfrak{D}_i &= ((\operatorname{supp} \mathfrak{b}_1^{\dagger} \cup \cdots \cup \operatorname{supp} \mathfrak{b}_i^{\dagger}) \mathfrak{d}_i^{-1})^* \mathfrak{d}_i, \\ \mathfrak{D}_{<i} &= \mathfrak{D}_1 \cup \cdots \cup \mathfrak{D}_{i-1} \end{aligned}$$

we thus have $\operatorname{supp} \mathfrak{m}' \subseteq \mathfrak{D}_i \, \mathfrak{m}$ and $\operatorname{supp} F_{\mathfrak{m},0} \subseteq \mathfrak{D}_i / \mathfrak{d}_i^2$. Moreover, for any $\mathfrak{v} \in \operatorname{supp} F_{\mathfrak{m},k}$, we have $\operatorname{supp} \mathfrak{v}' \subseteq \mathfrak{D}_{<i} \mathfrak{v}$. Now define families $\mathcal{T}_{\mathfrak{m},k}$ by

$$\begin{aligned} \mathcal{T}_{\mathfrak{m},0} &= \operatorname{term}\left(\frac{1}{\mathfrak{m}^{\dagger}}\right) \\ \mathcal{T}_{\mathfrak{m},k} &= -\mathcal{T}_{\mathfrak{m},0} \mathcal{T}'_{\mathfrak{m},k-1} \end{aligned}$$

where

$$\mathcal{T}'_{\mathfrak{m},k-1} = ((\mathfrak{v}')_{\mathfrak{v}\mathfrak{w}} \, \mathfrak{v} \, \mathfrak{w})_{\mathfrak{v} \in \mathcal{T}_{k-1}, \mathfrak{w} \in \mathfrak{D}_{<i}}.$$

Then $F_{\mathfrak{m},k} = \sum \mathcal{T}_{\mathfrak{m},k}$ for all $k \in \mathbb{N}$. Setting $\mathcal{T}_{\mathfrak{m}} = \bigcup_{k \in \mathbb{N}} \mathcal{T}_{\mathfrak{m},k}$, we have

$$\begin{aligned} \operatorname{mon} \mathcal{T}_{\mathfrak{m}} &\precsim ((\operatorname{mon} \mathfrak{D}_{<i}) (\operatorname{mon} \mathcal{T}_{\mathfrak{m},0}))^* (\operatorname{mon} \mathcal{T}_{\mathfrak{m},0}) \\ \operatorname{mon} \mathcal{T}_{\mathfrak{m},0} &\precsim \operatorname{mon} \mathfrak{D}_i / \mathfrak{d}_i^2, \end{aligned}$$

whence $\mathcal{T}_{\mathfrak{m}}$ is grid-based by proposition 2.14(c) and (2.7). We conclude that $\int \mathfrak{m} = \sum_{k \geqslant 0} F_{\mathfrak{m},k} \mathfrak{m}$ is well-defined, and

$$\begin{aligned} \int \mathfrak{m} &= \sum_{k \geqslant 0} (F'_{\mathfrak{m},k} + \mathfrak{m}^{\dagger} F_{\mathfrak{m},k}) \, \mathfrak{m} \\ &= \sum_{k \geqslant 0} \mathfrak{m}^{\dagger} (F_{\mathfrak{m},k} - F_{\mathfrak{m},k+1}) \, \mathfrak{m} = \mathfrak{m}^{\dagger} F_{\mathfrak{m},0} \, \mathfrak{m} = \mathfrak{m}. \end{aligned}$$

Let us now show that the mapping $\int : \mathfrak{B}^C \to \mathbb{T}$ is grid-based. Given a grid-based subset $\mathfrak{G}$ of $\mathfrak{B}^C$, we may decompose

$$\mathfrak{G} \setminus \{1\} = \mathfrak{G}_1 \amalg \cdots \amalg \mathfrak{G}_n,$$

where the $\mathfrak{G}_i$ $(i = 1, ..., n)$ are given by

$$\mathfrak{G}_i = \{\mathfrak{m} \in \mathfrak{G} : \mathfrak{m} \asymp\!\!\!\!\!\!\!\; \succ \mathfrak{b}_i\}.$$

By what precedes, $\coprod_{\mathfrak{m} \in \mathfrak{S}_i} \mathcal{T}_{\mathfrak{m}}$ is grid-based for each i. Hence, $\int$ is a grid-based mapping which extends uniquely to $\mathbb{T}$ by strong linearity. Furthermore, given $\mathfrak{m} = \mathfrak{b}_1^{\alpha_1} \cdots \mathfrak{b}_i^{\alpha_i}$ with $\alpha_i \neq 0$, we have $\mathfrak{D}_i \subseteq C[\![\mathfrak{b}_1; \ldots; \mathfrak{b}_{i-1}]\!]$, so that

$$\operatorname{supp} f_{\mathfrak{m}} F_{\mathfrak{m},k} \mathfrak{m} \subseteq C[\![\mathfrak{b}_1; \ldots; \mathfrak{b}_{i-1}]\!] \, \mathfrak{b}_i^{\alpha_i} \not\ni \{1\}.$$

This implies that $\int$ is a distinguished, strongly linear integral on $C^0[\![\exp x]\!]$.

Assume now that we have defined a distinguished, strongly linear integral $\int$ on $C^p[\![\exp x]\!]$. We claim that we may extend $\int$ to $C^{p+1}[\![\exp x]\!]$ by

$$\int f = \left(\int \mathrm{e}^x f\uparrow\right)\downarrow. \tag{5.2}$$

Indeed, (5.2) defines a distinguished integral, since

$$\left(\int \mathrm{e}^x f\uparrow\right)\downarrow' = \frac{1}{x}\left(\left(\mathrm{e}^x f\uparrow\right)\downarrow\right) = f$$

and

$$\left(\int \mathrm{e}^x f\uparrow\right)\downarrow_{\asymp} = \left(\int \mathrm{e}^x f\downarrow\right)_{\asymp} = 0,$$

for all $f \in C^{p+1}[\![\exp x]\!]$. Its distinguished property implies that it extends the previous integral on $C^p[\![\exp x]\!]$. Its strong linearity follows from the fact that we may see $\int$ as the composition of four strongly linear operations. Our proposition now follows by induction over p. □

Proposition 5.8. *Let $\mathfrak{B} = (\mathfrak{b}_1, \ldots, \mathfrak{b}_n)$ be a transbasis.*

a) If $\mathfrak{b}_1 = x$ or $\mathfrak{b}_1 = \exp x$, then $C[\![\mathfrak{b}_1; \ldots; \mathfrak{b}_i]\!][\log \mathfrak{b}_1]$ is stable under $\int$ for all i.

b) If $\mathfrak{b}_1 = \log_l x$ and $\log_{l-1} x, \ldots, x \in \mathfrak{B}$, then $C[\![\mathfrak{b}_1; \ldots; \mathfrak{b}_n]\!][\log \mathfrak{b}_1]$ is stable under $\int$.

Proof. We will consider the case when $\mathfrak{b}_1 = \mathrm{e}^x$ and $i = n$. The other cases follow by upward shifting. Now given

$$f = f_d x^d + \cdots + f_0$$

with $f_0, \ldots, f_d \in C[\![\mathfrak{b}_1; \ldots; \mathfrak{b}_n]\!]$, we claim that

$$\int f = F := g_{d+1} x^{d+1} + \cdots + g_0,$$

where $g_0, \ldots, g_{d+1} \in C[\![\mathfrak{b}_1; \ldots; \mathfrak{b}_n]\!]$ are given by

$$\begin{aligned}
g_{d+1} &= f_{d,\asymp}/(d+1); \\
g_d &= f_{d-1,\asymp}/d + \int (f_d - (d+1)\, g_{d+1})_{\not\asymp}; \\
g_{d-1} &= f_{d-2,\asymp}/(d-1) + \int (f_{d-1} - d\, g_d)_{\not\asymp}; \\
&\;\;\vdots \\
g_0 &= \int (f_0 - g_1)_{\not\asymp}.
\end{aligned}$$

Indeed, it is easily checked that $F' = f$. Furthermore,

$$F_{\asymp} = g_{0,\asymp} = \left(\int (f_0 - g_1)_{\not\asymp}\right)_{\asymp} = 0,$$

whence $F = \int f$, by the distinguished property of integration. □

Exercise 5.9. Let $\mathfrak{m} \neq 1$ be a transmonomial. Show that there exists a unique transmonomial $\mathfrak{n} \asymp \mathfrak{m}$, so that $\mathfrak{n}'$ is a transmonomial.

Exercise 5.10. Let $f, g \in \mathbb{T}$.

a) If $\int f \prec 1$ and $\int g \prec 1$, then show that

$$\left(\int f\right)\left(\int g\right) = \int f \int g + \int g \int f. \tag{5.3}$$

b) Give a necessary and sufficient condition for (5.3) to hold.
c) Prove that there does not exist a strong integration on $C((\mathrm{e}^x))$ so that (5.3) holds for all $f, g \in C((\mathrm{e}^x))$.

Exercise 5.11. Show that $\int \mathrm{e}^{x^2}$ is divergent. Deduce that $\int \mathrm{e}^{x^2}$ is not an exp-log function.

Exercise 5.12. Let $\varphi: H \hookrightarrow \mathbb{T}$ an embedding of a Hardy field into $\mathbb{T} = \mathbb{R}[[[x]]]$. The embedding φ is assumed to preserve the differential ring structure and the ordering. Given $f \in H$, show that φ can be extended into an embedding $\hat{\varphi}: H(\int f) \hookrightarrow \mathbb{T}$.

5.3 Functional composition

Let R and S be strong totally ordered partial exp-log C-algebras. A *strong difference operator* of R into S is an injection $\delta: R \to S$, which satisfies

Δ1. $\delta c = c$, for all $c \in C$.
Δ2. δ is strongly linear.
Δ3. $\delta(f g) = \delta(f)\, \delta(g)$, for all $f, g \in R$.

If $S = R$, then we say that δ is a strong difference operator on R. We say that δ is an *exp-log difference operator*, if we also have

Δ4. $\delta(\exp f) = \exp \delta(f)$, for all $f \in R \cap \operatorname{dom} \exp$.

We say that δ is *asymptotic* resp. *increasing*, if

Δ5. $f \prec 1 \Rightarrow \delta(f) \prec 1$, for all $f \in R$.
Δ6. $f \geqslant 0 \Rightarrow \delta(f) \geqslant 0$, for all $f \in R$.

In this section, we will show that for each $g \in \mathbb{T}^{>,\succ}$, there exists a unique strong exp-log difference operator $\circ_g$ on $\mathbb{T}$, such that $\circ_g(x) = g$. This allows us to define a composition on $\mathbb{T}$ by

$$\begin{aligned} \circ: \mathbb{T} \times \mathbb{T}^{>,\succ} &\to \mathbb{T} \\ (f, g) &\mapsto \circ_g(f). \end{aligned}$$

We will show that this composition is associative, that it satisfies the chain rule, and that we can perform Taylor series expansion under certain conditions.

Lemma 5.9. *Let* $\mathbb{T}=C[[\mathfrak{T}]]\subseteq\hat{\mathbb{T}}=C[[\hat{\mathfrak{T}}]]$ *be arbitrary fields of transseries and let* $\delta\colon\mathfrak{T}\rightarrow\hat{\mathbb{T}}$ *be a mapping, which satisfies* $\delta(\mathfrak{m}\ \mathfrak{n})=\delta(\mathfrak{m})\ \delta(\mathfrak{n})$ *and* $1\prec\mathfrak{m}\Rightarrow\delta(\mathfrak{m})\in\hat{\mathbb{T}}^{>,\succ}$ *for all* $\mathfrak{m},\mathfrak{n}\in\mathfrak{T}$. *Then*

a) δ *is a grid-based mapping, which extends uniquely to a strong, asymptotic and increasing difference operator from* $\mathbb{T}$ *into* $\hat{\mathbb{T}}$.

b) If $\delta(\log\mathfrak{m})=\log\delta(\mathfrak{m})$ *for all* $\mathfrak{m}\in\mathfrak{T}$, *then the extension of* δ *to* $\mathbb{T}$ *is an exp-log difference operator.*

Proof. Let $\mathfrak{G}$ be a grid-based subset of $\mathbb{T}$ with $\mathfrak{G}\subseteq\{\mathfrak{m}_1,\ldots,\mathfrak{m}_n\}^*\ \mathfrak{n}$, for certain monomials $\mathfrak{m}_1,\ldots,\mathfrak{m}_n\prec 1$ and $\mathfrak{n}$ in $\mathfrak{T}$. Then the family $\mathcal{F}^*$ with $\mathcal{F}=(\delta(\mathfrak{m}_i))_{1\leqslant i\leqslant n}$ is grid-based, by proposition 2.14(*c*). It follows that $\delta\colon\mathfrak{T}\rightarrow\hat{\mathbb{T}}$ is grid-based, since $(\delta(\mathfrak{v}))_{\mathfrak{v}\in\mathfrak{G}}\subsetneq\mathcal{F}^*\ \delta(\mathfrak{n})$. By proposition 2.16, the extension of δ to $\mathbb{T}$ is a strong difference operator. If $f\in\mathbb{T}^{\prec}$, then $\delta(\mathfrak{m})\prec 1$ for all $\mathfrak{m}\in\operatorname{supp} f$, whence $\delta(f)=\sum f_{\mathfrak{m}}\,\delta(\mathfrak{m})\prec 1$. This proves that δ is asymptotic and, given $f\in\mathbb{T}^{\neq}$, it also follows that $\delta(f)\sim\delta(\tau_f)=c_f\,\delta(\mathfrak{d}_f)$. In particular, if $f>0$, then $\delta(f)>0$. This completes the proof of (*a*).

Now assume that $\delta(\log\mathfrak{m})=\log\delta(\mathfrak{m})$ for all $\mathfrak{m}\in\mathfrak{T}$. In order to prove (*b*), it obviously suffices to show that $\delta(\log f)=\log\delta(f)$ for all $f\in\mathbb{T}^{>}$. Now each $f\in\mathbb{T}^{>}$ may be decomposed as $f=c\,\mathfrak{m}\,(1+\varepsilon)$, with $c\in C^{>}$, $\mathfrak{m}\in\mathfrak{T}$ and $\varepsilon\prec 1$. For each $k\in\mathbb{N}^{>}$, we have $\delta(\frac{(-1)^{k-1}}{k}\,\varepsilon^k)=\frac{(-1)^{k-1}}{k}\,\delta(\varepsilon)^k$. Hence, $\delta(\log(1+\varepsilon))=\log(1+\delta(\varepsilon))$, by strong linearity. We conclude that

$$\begin{aligned}
\delta(\log f) &= \delta(\log c)+\delta(\log\mathfrak{d}_f)+\delta(\log(1+\varepsilon))\\
&= \log c+\log\delta(\mathfrak{d}_f)+\log(1+\delta(\varepsilon))\\
&= \log(c\,\delta(\mathfrak{d}_f)\,(1+\delta(\varepsilon)))\\
&= \log\delta(c\,\mathfrak{d}_f\,(1+\varepsilon))\\
&= \log\delta(f).
\end{aligned}$$

□

Proposition 5.10. *Let* $g\in\mathbb{T}^{>,\succ}$. *Then there exists a unique strong exp-log difference operator* $\circ_g$ *on* $\mathbb{T}$ *with* $\circ_g(x)=g$. *This difference operator is asymptotic and increasing.*

Proof. We will show by induction over $p\in\mathbb{N}$ that there exists a unique strong exp-log difference operator $\circ_g$ from $C_p[[[x]]]=C[[\mathfrak{T}_p]]$ into $\mathbb{T}$ with $\circ_g x=g$, and we will show that this difference operator is asymptotic and increasing.

For $p=0$, the axioms **Δ3** and **Δ4** imply that

$$\circ_g(x^{\alpha_0}\cdots\log_q^{\alpha_q}x)=g^{\alpha_0}\cdots\log_q^{\alpha_q}g$$

for all monomials $x^{\alpha_0}\cdots\log_q^{\alpha_q}x\in\mathfrak{T}_0$. If $x^{\alpha_0}\cdots\log_q^{\alpha_q}x\succ 1$, i.e. $\alpha_0=\cdots=\alpha_{i-1}=0$ and $\alpha_i>0$ for some i, we also get

$$\circ_g(x^{\alpha_0}\cdots\log_q^{\alpha_q}x)\in\mathbb{T}^{>,\succ},$$

since

$$\log_{i+1}^{\alpha_{i+1}}g\cdots\log_q^{\alpha_q}g\prec\!\!\prec\log_i^{\alpha_i}g\in\mathbb{T}^{>,\succ}.$$

This completes the proof in the case when $p=0$, by lemma 5.9.

If $p>0$, then the induction hypothesis states that there exists a unique strong exp-log difference operator $\circ_g: C_{p-1}[[x]] \to \mathbb{T}$ with $\circ_g(x)=g$, and $\circ_g$ is asymptotic and increasing. In view of **Δ4**, any extension of $\circ_g$ to $C_p[[x]]$ should therefore satisfy $\circ_g(\mathrm{e}^f)=\mathrm{e}^{\circ_g(f)}$ for all $\mathrm{e}^f \in \mathfrak{T}_p = \exp C_{p-1}[[x]]_{\succ}$. On the other hand, when defining $\circ_g$ in this way on $\mathfrak{T}_p$, we have

$$\circ_g(\mathrm{e}^{f_1}\mathrm{e}^{f_2}) = \mathrm{e}^{\circ_g(f_1+f_2)} = \mathrm{e}^{\circ_g(f_1)}\mathrm{e}^{\circ_g(f_2)} = \circ_g(\mathrm{e}^f)\circ_g(\mathrm{e}^g)$$

for all $\mathrm{e}^{f_1}, \mathrm{e}^{f_2} \in \mathfrak{T}_p$. Similarly,

$$\begin{aligned}(f=f_{\succ} \wedge \mathrm{e}^f \succ 1) &\Rightarrow (f>0 \wedge f \succ 1)\\ &\Rightarrow \circ_g(f) \in \mathbb{T}^{>,\succ}\\ &\Rightarrow \circ_g(\mathrm{e}^f) = \mathrm{e}^{\circ_g(\mathrm{e}^f)} \in \mathbb{T}^{>,\succ}\end{aligned}$$

for all $\mathrm{e}^f \in \mathfrak{T}_p$. This completes the proof in the general case, by lemma 5.9. □

Proposition 5.11.

a) $f\circ(g\circ h) = (f\circ g)\circ h$, *for all* $f\in\mathbb{T}$ *and* $g,h\in\mathbb{T}^{>,\succ}$.
b) $(f\circ g)' = g'(f'\circ g)$, *for all* $f\in\mathbb{T}$ *and* $g\in\mathbb{T}^{>,\succ}$.
c) Let $f,\delta\in\mathbb{T}$ *be such that* $\delta\prec x$ *and* $\mathfrak{m}^\dagger\delta\prec 1$ *for all* $\mathfrak{m}\in\operatorname{supp} f$. *Then*

$$f\circ(x+\delta) = f + f'\delta + \tfrac{1}{2}f''\delta^2 + \cdots \tag{5.4}$$

Proof. Property (a) follows from proposition 5.10 and the fact that $(\circ_h)\circ(\circ_g)$ and $\circ_{g\circ h}$ are both strong exponential difference operators which map x to $g\circ h$.

Let Φ be the set of $f\in\mathbb{T}$, for which $(f\circ g)' = g'(f'\circ g)$. We have $x\in\Phi$ and Φ is stable under grid-based summation, since the mappings $f\mapsto(f\circ g)'$ and $g'(f'\circ g)$ are both strongly linear. Φ is also stable under exponentiation and logarithm: if $f\in\Phi$, then

$$\begin{aligned}(\mathrm{e}^f\circ g)' &= (\mathrm{e}^{f\circ g})'\\ &= (f\circ g)'\mathrm{e}^{f\circ g}\\ &= g'(f'\circ g)\,\mathrm{e}^{f\circ g}\\ &= g'((f'\mathrm{e}^f)\circ g)\\ &= g'((\mathrm{e}^f)'\circ g)\end{aligned}$$

and $f>0$ implies

$$\begin{aligned}((\log f)\circ g)' &= (\log(f\circ g))'\\ &= g'(f'\circ g)/f\circ g\\ &= g'((\log f)'\circ g).\end{aligned}$$

This proves (b), since the smallest subset Φ of $\mathbb{T}$ which satisfies the above properties is $\mathbb{T}$ itself.

As to (c), we first have to prove that the right hand side of (5.4) is well-defined. Let $\delta \prec x$ be a transseries in $\mathbb{T}$ and denote by $\mathbb{T}^\flat$ the set of transseries f, such that $\mathfrak{m}^\dagger\ \delta \prec 1$ for all $\mathfrak{m} \in \operatorname{supp} f$. Given a transmonomial $\mathfrak{m}$, we have

$$\mathfrak{m}^\dagger \delta \prec 1 \Leftrightarrow (\log \mathfrak{m})' \prec 1/\delta \Leftrightarrow \log \mathfrak{m} \prec \int 1/\delta \Leftrightarrow \mathfrak{m} \prec\!\!\prec \mathrm{e}^{\int 1/\delta},$$

since $1/\delta \succ 1/x$. We infer that

$$\mathbb{T}^\flat = \{f \in \mathbb{T}: \forall \mathfrak{m} \in \operatorname{supp} f, \mathfrak{m} \prec\!\!\prec \mathrm{e}^{\int 1/\delta}\}.$$

Let us show that $\mathbb{T}^\flat$ is stable under differentiation. By the strong linearity of the differentiation, it suffices to prove that $\mathfrak{m}' \in \mathbb{T}^\flat$, for all transmonomials $\mathfrak{m}$ with $\mathfrak{m} \prec\!\!\prec \mathrm{e}^{\int 1/\delta}$. If $\mathfrak{m} \preccurlyeq\!\!\preccurlyeq x$, then $\mathfrak{n} \preccurlyeq\!\!\preccurlyeq x \prec\!\!\prec \mathrm{e}^{\int 1/\delta}$, for all $\mathfrak{n} \in \operatorname{supp} \mathfrak{m}'$. If $\mathfrak{m} \succ\!\!\succ x$, then $\mathfrak{n}/\mathfrak{m} \prec\!\!\prec \mathfrak{m}$ for all $\mathfrak{n} \in \operatorname{supp} \mathfrak{m}'$, whence $\mathfrak{n} \asymp\!\!\asymp \mathfrak{m} \prec\!\!\prec \mathrm{e}^{\int 1/\delta}$.

Now consider a transbasis $\mathfrak{B} = (\mathfrak{b}_1, ..., \mathfrak{b}_n)$, such that $\mathfrak{b}_1 = \log_p x, ..., x \in \mathfrak{B}$ and $\mathfrak{b}_1, ..., \mathfrak{b}_n \in \mathbb{T}^\flat$. By theorem 4.15($b$), any $f \in \mathbb{T}^\flat$ can be expanded with respect to such a transbasis. Let

$$\mathfrak{D} = \operatorname{supp} \mathfrak{b}_1^\dagger \cup \cdots \cup \operatorname{supp} \mathfrak{b}_n^\dagger \prec \tfrac{1}{\delta},$$

so that $\operatorname{supp} f' \subseteq (\operatorname{supp} f)\, \mathfrak{D} \subseteq \mathfrak{B}^C$, for all $f \in C[\![\mathfrak{B}^C]\!]$. Now let $f \in C[\![\mathfrak{B}^C]\!]$, $l \in \mathbb{N}$, and consider the family $\mathcal{T}_l$ of all terms

$$\tau_{\mathfrak{v},(\mathfrak{m}_1,\mathfrak{n}_1)\cdots(\mathfrak{m}_l,\mathfrak{n}_l)} = \tfrac{1}{l!}\,(f_{\mathfrak{v}}\, \mathfrak{v})\, (\mathfrak{v}_{\mathfrak{m}_1}^\dagger\, \mathfrak{m}_1)\, (\delta_{\mathfrak{n}_1}\, \mathfrak{n}_1) \cdots ((\mathfrak{v}\, \mathfrak{m}_1 \cdots \mathfrak{m}_{l-1})_{\mathfrak{m}_l}^\dagger\, \mathfrak{m}_l)\, (\delta_{\mathfrak{n}_l}\, \mathfrak{n}_l).$$

Then

$$\tfrac{1}{l!}\, f^{(l)}\, \delta^l = \sum \mathcal{T}_l.$$

Moreover, setting $\mathcal{T} = \coprod_{l \in \mathbb{N}} \mathcal{T}_l$, we have

$$\operatorname{mon} \mathcal{T} \subsetsim \operatorname{mon}(f)\, (\operatorname{mon}(\mathfrak{D})\, \operatorname{mon}(\delta))^*,$$

so $\mathcal{T}$ is grid-based, by proposition 2.14(c). Since $\mathcal{T}$ refines the family $(\frac{1}{l!}\, f^{(l)}\, \delta^l)_{l \in \mathbb{N}}$, it follows that the Taylor series in (5.4) is well-defined. For a similar reason, the mapping $\mathfrak{B}^C \to \mathbb{T};\ \mathfrak{v} \mapsto \sum_{l \geqslant 0} \frac{1}{l!}\, \mathfrak{v}^{(l)}\, \delta^l$ is grid-based, so the mapping $C[\![\mathfrak{B}^C]\!] \to \mathbb{T};\ f \mapsto \sum_{l \geqslant 0} \frac{1}{l!}\, f^{(l)}\, \delta^l$ is actually strongly linear.

Now let Φ be the subset of $\mathbb{T}^\flat$ of all f, such that (5.4) holds. Clearly, $x \in \Phi$ and Φ is stable under strongly linear combinations. We claim that Φ is also stable under exponentiation and logarithm. Indeed, assume that $f \in \Phi$ and $\mathrm{e}^f \in \mathbb{T}^\flat$. Then $1/f' \asymp \mathfrak{d}_{\mathrm{e}^f}/\mathfrak{d}_{\mathrm{e}^f}' \succ \delta$, $f'/f'' \succ \delta$, $f''/f''' \succ \delta$, ..., since $\mathfrak{d}_{\mathrm{e}^f}, f', f'', \ldots \in \mathbb{T}^\flat$. Hence $f^{(n)}\, \delta^n \prec 1$ for all $n \geqslant 1$, which allows us to expand

$$\begin{aligned} A &= (\mathrm{e}^f) \circ (x + \delta) = \mathrm{e}^{f + f'\delta + \frac{1}{2} f'' \delta^2 + \cdots} \\ &= \mathrm{e}^f\, (1 + \delta + \frac{1}{2} f''\, \delta^2 + \cdots + (\delta + \frac{1}{2} f''\, \delta^2 + \cdots)^2 + \ldots). \end{aligned}$$

We have to show that A coincides with

$$\begin{aligned} B &= \mathrm{e}^f + (\mathrm{e}^f)'\,\delta + \tfrac{1}{2}\,(\mathrm{e}^f)''\,\delta^2 \\ &= \mathrm{e}^f\,(1 + f'\,\delta + \tfrac{1}{2}\,((f')^2 + f'')\,\delta^2 + \cdots). \end{aligned}$$

But this follows from the fact that we may see $A = B$ as a formal identity in the ring $C[[\mathrm{e}^f, \delta, f', f'', \ldots]]$. Indeed, A and B satisfy the same differential equation

$$\begin{aligned} \frac{\partial A}{\partial \delta} &= \left(f' + f''\,\delta + \frac{1}{2}\,f'''\,\delta^2 + \cdots\right) A \\ &= f'\,A + \left(\frac{\partial A}{\partial f'}\,f'' + \frac{\partial A}{\partial f''}\,f''' + \cdots\right)\delta; \\ \frac{\partial B}{\partial \delta} &= f'\,B + \left(\frac{\partial B}{\partial f'}\,f'' + \frac{\partial B}{\partial f''}\,f''' + \cdots\right)\delta, \end{aligned}$$

and $[\delta^0]A = [\delta^0]B = \mathrm{e}^f$. Similarly, one may show that Φ is stable under logarithm. This proves (c), since the smallest subset of $\mathbb{T}^\flat$, which contains x and which is stable under strongly linear combinations, exponentiation and logarithm, is $\mathbb{T}^\flat$ itself. □

Exercise 5.13. Let $f \in \mathbb{T}$ and $g \in \mathbb{T}^{>,\succ}$.

a) Prove that the exponentiality of $f \circ g$ equals the sum of the exponentialities of f and g.
b) Prove that the exponential height resp. logarithmic depth of $f \circ g$ is bounded by the sum of the exponential heights resp. logarithmic depths of f and g.
c) Improve the bound in (b) by taking into account the exponentialities of f and g.

Exercise 5.14. Let $f, h \in \mathbb{T}$ and $g \in \mathbb{T}^{>,\succ}$ be such that $h \prec g$. Under which condition do we have

$$f \circ (g + h) = f \circ g + (f' \circ g)\,h + \frac{1}{2}\,(f'' \circ g)\,h^2 + \cdots ?$$

Exercise 5.15. Let $f \in \mathbb{T}$ and let $\mathcal{D}$ a grid-based family of transseries, such that $\mathfrak{m}^\dagger\,\delta \prec 1$, for all $\mathfrak{m} \in \operatorname{supp} f$ and $\delta \in \mathcal{D}$. prove that

$$f \circ \left(x + \sum \mathcal{D}\right) = \sum_{\delta_1 \cdots \delta_l \in \mathcal{D}^*} \frac{1}{l!}\,f^{(l)}\,\delta_1 \cdots \delta_l.$$

Exercise 5.16. Let $\mathfrak{m}$ be a transmonomial in $\mathbb{T}$ and $g \in \mathbb{T}^{>,\succ}$ a transseries, such that $\mathfrak{m} \circ g \succ\!\!\succ x$ and $\mathfrak{n} \prec\!\!\prec \log \mathfrak{m} \circ g$ for all $\mathfrak{n} \in \operatorname{supp} g$. Prove that $\mathfrak{m} \circ g$ is a transmonomial.

Exercise 5.17. Show that $\mathbb{R}\{\{x\}\}$ is stable under composition.

Exercise 5.18. Let $\mathfrak{A} = (\mathfrak{a}_1, \ldots, \mathfrak{a}_m)$ and $\mathfrak{B} = (\mathfrak{b}_1, \ldots, \mathfrak{b}_n)$ be two transbases and consider two series $f \in C[[\mathfrak{a}_1; \ldots, \mathfrak{a}_m]]$ and $g \in C[[\mathfrak{b}_1; \ldots; \mathfrak{b}_n]]^{>,\succ}$. Construct a transbasis for $f \circ g$ of size $\leqslant m + n$.

5.4 Functional inversion

5.4.1 Existence of functional inverses

Theorem 5.12. *Any* $g \in \mathbb{T}^{>,\succ}$ *admits a functional inverse* $g^{\mathrm{inv}} \in \mathbb{T}^{>,\succ}$ *with*

$$g^{\mathrm{inv}} \circ g = g \circ g^{\mathrm{inv}} = x.$$

Proof. Without loss of generality, one may assume that $g = x + \varepsilon$, where $\varepsilon \prec 1$ is exponential. Indeed, it suffices to replace g by $\log_{l-p} \circ g \circ \exp_l$ for sufficiently large l, where p is the exponentiality of g. Let $\mathfrak{B} = (\mathfrak{b}_1 = \mathrm{e}^x, \ldots, \mathfrak{b}_n)$ be a plane transbasis for ε. We will prove that g admits a functional inverse of the form $f = x + \delta$, where $\delta \prec 1$ can be expanded with respect to a plane transbasis $(\mathfrak{a}_1, \ldots, \mathfrak{a}_n)$ which satisfies

$$\begin{aligned} \mathfrak{a}_n &= \mathfrak{b}_n \circ (x + \delta_0) \\ \mathfrak{a}_{n-1} &= \mathfrak{b}_{n-1} \circ (x + \delta_{0,0}) \\ &\vdots \end{aligned}$$

Let us first assume that the constant coefficient ε_0 of ε in $\mathfrak{b}_n$ vanishes. Then proposition 5.11(c) implies that

$$K f := f \circ (x + \varepsilon) - f = f' \varepsilon + \frac{1}{2} f'' \varepsilon^2 + \cdots \tag{5.5}$$

for any $f \in C[[x; \mathfrak{b}_1; \ldots; \mathfrak{b}_n]]$. In particular, for every $\mathfrak{m} \in x^C \mathfrak{b}_1^C \cdots \mathfrak{b}_n^C$, we have

$$\operatorname{supp} \frac{K \mathfrak{m}}{\mathfrak{m}} \subseteq \mathfrak{K} := (\{x^{-1}, \mathfrak{b}_1^{\dagger}, \ldots, \mathfrak{b}_n^{\dagger}\} \operatorname{supp} \varepsilon)^*.$$

Now the functional inverse of g is given by

$$\begin{aligned} g^{\mathrm{inv}} &= x - K x + K^2 x - K^3 x + \cdots. \\ &= \sum_{(\mathfrak{k}_1, \ldots, \mathfrak{k}_l) \in \mathfrak{K}^*} (-1)^l (K x)_{\mathfrak{k}_1} (K \mathfrak{k}_1)_{\mathfrak{k}_2} \cdots (K \mathfrak{k}_{l-1})_{\mathfrak{k}_l} (x \mathfrak{k}_1 \cdots \mathfrak{k}_l) \end{aligned}$$

Since $K x = \varepsilon \in C[[\mathfrak{b}_1; \ldots; \mathfrak{b}_n]]$ and K maps $C[[\mathfrak{b}_1; \ldots; \mathfrak{b}_n]]$ into itself, we conclude that $g^{\mathrm{inv}} = x + \delta$, with $\delta \in C[[\mathfrak{b}_1; \ldots; \mathfrak{b}_n]]_{\prec}$.

The general case is proved by induction over n. If $n = 1$, then we must have $\varepsilon_0 = 0$, so we are done. So assume that $n > 1$. By the induction hypothesis, there exists a functional inverse $\tilde{f} = x + \tilde{\delta}$ for $\tilde{g} = x + \tilde{\varepsilon} = x + \varepsilon_0$, such that $\tilde{\delta} \in C[[\mathfrak{a}_1; \ldots; \mathfrak{a}_{n-1}]]_{\prec}$, where

$$\begin{aligned} \mathfrak{a}_{n-1} &= \mathfrak{b}_{n-1} \circ (x + \tilde{\delta}_0) \\ \mathfrak{a}_{n-2} &= \mathfrak{b}_{n-2} \circ (x + \tilde{\delta}_{0,0}) \\ &\vdots \end{aligned}$$

Now

$$g \circ \tilde{f} = x + (g - \tilde{g}) \circ \tilde{f} \in C[[\mathfrak{a}_1; \ldots; \mathfrak{a}_n]],$$

where $\mathfrak{a}_n = \mathfrak{b}_n \circ \tilde{f}$, and $((g-\tilde{g})\circ\tilde{f})_0 = 0$. It follows that $g\circ\tilde{f}$ has a functional inverse of the form $(g\circ\tilde{f})^{\mathrm{inv}} = x+\eta$ with $\eta\in C[\![\mathfrak{a}_1;\ldots;\mathfrak{a}_n]\!]$ and $\eta_0 = 0$. We conclude that $g^{\mathrm{inv}} = \tilde{f}\circ(g\circ\tilde{f})^{\mathrm{inv}}$ is a functional inverse of g and we have

$$g^{\mathrm{inv}} = \tilde{f}\circ(x+\eta) = \tilde{f} + \tilde{f}'\,\eta + \frac{1}{2}\,\tilde{f}''\,\eta^2 + \cdots \in C[\![\mathfrak{a}_1;\ldots;\mathfrak{a}_n]\!]. \qquad \square$$

5.4.2 The Translagrange theorem

We define a scalar product on $\mathbb{T}$ by

$$\langle f, g\rangle = (f g)_{\asymp}.$$

Given transseries $M, N\in\mathbb{T}$ and $f\in\mathbb{T}^{>,\succ}$, let us denote

$$f_{[M,N]} = \langle M\circ f, N\rangle.$$

When taking transmonomials for M and N, then the coefficients $f_{[M,N]}$ describe the post-composition operator with f. More precisely, for all $\mathfrak{m}$, $\mathfrak{n}\in\mathfrak{T}$ we have

$$(\mathfrak{m}\circ f)_{\mathfrak{n}} = f_{[\mathfrak{m},\mathfrak{n}^{-1}]}.$$

Theorem 5.13. *Let $M, N, \varepsilon \prec 1$ be exponential transseries and $f = x+\varepsilon$. Then $g = f^{\mathrm{inv}}$ satisfies*

$$g_{[M,N']} = -f_{[N,M']}.$$

Proof. Since $h_{\asymp} = (\int h)_x$ for all $h\in\mathbb{T}$, we have

$$\begin{aligned} g_{[M,N']} = \langle M\circ g, N'\rangle &= [\textstyle\int (M\circ g)\, N']_x;\\ f_{[N,M']} = \langle N\circ f, M'\rangle &= [\textstyle\int (N\circ f)\, M']_x. \end{aligned}$$

Since $[\int (N\circ f)\,M'] - [\int (N\circ f)\,M']_x\, x$ and $g - x$ are exponential, we have

$$[\textstyle\int (N\circ f)\, M']_x = [(\int (N\circ f)\, M')\circ g]_x.$$

Using the rule $(\int h)\circ g = \int (h\circ g)\, g'$, it follows that

$$[\textstyle\int (N\circ f)\, M']_x = [\int N\,(M'\circ g)\, g']_x = [\int N\,(M\circ g)']_x.$$

Now integration by parts yields

$$g_{[M,N]} + f_{[N,M']} = [\textstyle\int (M\circ g)\, N']_x + [\int N\,(M\circ g)']_x = [N\,(M\circ g)]_x$$

But $[N\,(M\circ g)]_x = 0$, since $N\,(M\circ g)$ is exponential. $\square$

The theorem generalizes to the case when M, N and ε are no longer exponential, by applying the following rule a finite number of times:

$$f_{[M,N]} = (\log\circ f\circ\exp)_{[M\circ\exp, N\circ\exp]}.$$

Corollary 5.14. *Let $M, N, \varepsilon \prec 1$ be transseries of depths $\leqslant l$ and $f = x + \varepsilon$. Then $g = f^{\mathrm{inv}}$ satisfies*

$$g_{[M, N'/\log_l']} = -f_{[N, M'/\log_l']}.$$

Exercise 5.19. Let $g = x + \varepsilon$ where ε is exponential and let K be as in (5.5).

a) Show that we do not always have $g^{\mathrm{inv}} = x - Kx + K^2 x + \cdots$.
b) Give a necessary and sufficient condition for which

$$g^{\mathrm{inv}} = x - Kx + K^2 x + \cdots.$$

Exercise 5.20.

a) A classical theorem of Liouville [Lio37, Lio38] states that $(x \log x)^{\mathrm{inv}}$ is not an exp-log function. Show that there exists no exp-log function f with $f \asymp (\log x \log\log x)^{\mathrm{inv}}$ (see [Har11] for a variant of this problem).
b) Show that there exists no exp-log function f with $f \asymp \mathrm{e}^{\int \mathrm{e}^{x^2}}$. Hint: use exercise 5.11.
c) Assume that $g \in \mathbb{T}^{>,\succ}$ is not an exp-log function. Show that there exists an $n \in \mathbb{N}$, such that there exists no exp-log function f with $f \asymp \exp_n g$.

Exercise 5.21. Show that $\mathbb{R}\{\{x\}\}$ is stable under functional inversion.

Exercise 5.22. Classify the convex subgroups of $(\mathbb{T}^{>,\succ}, \circ)$. Hint: G is a convex subgroup of $\mathbb{T}^{>,\succ}$ if and only if its contraction $\operatorname{con} G$ is a convex subgroup.

Exercise 5.23. Show that Lagrange's inversion formula is a special case of theorem 5.13.

Exercise 5.24. Show that theorem 5.13 still holds when $M = x$ and N is exponential.

Exercise 5.25. Let M, N be transseries and let $f \in \mathbb{T}^{>,\succ}$ be a transseries of level 0. Show that for all sufficiently large l, the inverse $g = f^{\mathrm{inv}}$ satisfies

$$g_{[M,N]} = -f_{[\int (N \log_l'), M'/\log_l']}.$$

If one allows $l = \omega$, then show that the formula holds for transseries of arbitrary levels.

6

Grid-based operators

Besides multiplication and strong summation, we have introduced other interesting operations on the field of transseries in the previous chapter, like differentiation, integration, composition and functional inversion. In this chapter we will perform a theoretical study of an even larger class of operations on transseries, which contains the above elementary operations, but also many natural combinations of them.

This theoretical study is carried out best in the context of "grid-based modules". Let C be a ring. In chapter 2, we defined a grid-based algebra to be a strong C-algebra of the form $C[[\mathfrak{M}]]$, where $\mathfrak{M}$ is a monomial monoid. An arbitrary subset $\mathfrak{S}$ of $\mathfrak{M}$ is called a *monomial set* and the set $C[[\mathfrak{S}]]$ of strong linear combinations of elements in $\mathfrak{S}$ a *grid-based module*.

In section 6.1, we start by generalizing the notion of strongly linear mappings from chapter 2 to the multilinear case. Most natural elementary operations like multiplication, differentiation, right composition, etc. can then be seen as either linear or bilinear "grid-based operators". In section 6.3, we next introduce the general concept of a grid-based operator. Roughly speaking, such an operator is a mapping $\Phi: C[[\mathfrak{M}]] \to C[[\mathfrak{N}]]$ which admits a "generalized Taylor series expansion"

$$\Phi = \Phi_0 + \Phi_1 + \Phi_2 + \cdots,$$

such that there exists a d-linear grid-based operator

$$\check{\Phi}_d: C[[\mathfrak{M}]]^d \to C[[\mathfrak{N}]]$$

with

$$\Phi_d(f) = \check{\Phi}_d(f, \ldots, f)$$

for each d. If $C \supseteq \mathbb{Q}$, then such Taylor series expansions are unique and we will show that the $\check{\Phi}_d$ may be chosen to be symmetric.

Multilinear grid-based operators may both be reinterpreted as general grid-based operators and linear grid-based operators using the "syntactic sugar isomorphisms"

$$\begin{aligned} C[[\mathfrak{M}_1 \amalg \cdots \amalg \mathfrak{M}_m]] &\cong C[[\mathfrak{M}_1]] \times \cdots \times C[[\mathfrak{M}_m]] \\ C[[\mathfrak{M}_1 \times \cdots \times \mathfrak{M}_m]] &\cong C[[\mathfrak{M}_1]] \otimes \cdots \otimes C[[\mathfrak{M}_m]] \end{aligned}$$

The first isomorphism also provides a notion of grid-based operators in several variables.

As promised, many operations can be carried with grid-based operators: they can be composed and one may define a natural strong summation on the space of grid-based operators $\Phi: C[[\mathfrak{M}]] \to C[[\mathfrak{N}]]$. An explicit strong basis of "symmetric atomic operators" for this space will be established in section 6.4.2. Last but not least, we will prove several implicit function theorems for grid-based operators in section 6.5. These theorems will be a key ingredient for the resolution of differential (and more general functional equations) in the next chapters.

6.1 Multilinear grid-based operators

6.1.1 Multilinear grid-based operators

Let $M_1, \ldots, M_m$ and N be strong modules over a ring C. A mapping

$$\Phi: M_1 \times \cdots \times M_m \to N$$

is said to be *strongly multilinear*, if for all $\mathcal{F}_1 \in \mathscr{S}(M_1), \ldots, \mathcal{F}_m \in \mathscr{S}(M_m)$, we have $M(\mathcal{F}_1, \ldots, \mathcal{F}_m) \in \mathscr{S}(N)$ and

$$\Phi\Big(\sum \mathcal{F}_1, \ldots, \sum \mathcal{F}_m\Big) = \sum \Phi(\mathcal{F}_1, \ldots, \mathcal{F}_m).$$

If M_1, ..., M_m and N are grid-based modules, then we also say that Φ is a *multilinear grid-based operator*.

Example 6.1. Given monomial monoids $\mathfrak{M}$ and $\mathfrak{N}$, all strongly linear mappings $L: C[[\mathfrak{M}]] \to C[[\mathfrak{N}]]$ are multilinear grid-based operators. Denoting $\mathbb{S} = C[[\mathfrak{M}]]$, we have in particular the following important types of linear grid-based operators:

1. *Left multiplication operators* $\times_f: \mathbb{S} \to \mathbb{S}, g \mapsto f g$, with $f \in \mathbb{S}$.
2. *Strong derivations* $d: \mathbb{S} \to \mathbb{S}$. If $\mathbb{S}$ admits R-powers, then such derivations should also satisfy $d f^{\lambda} = \lambda (d f) f^{\lambda - 1}$, whenever f^{λ} is well-defined for $f \in \mathbb{S}$ and $\lambda \in R$.
3. *Strong integrations*; these are partial, strongly linear right inverses $I: \mathbb{S} \to \mathbb{S}$ of strong derivations $d: \mathbb{S} \to \mathbb{S}$, i.e. $d I = \mathrm{Id}$.

4. *Strong difference operators* $\delta \colon \mathbb{S} \to \mathbb{S}$. If $\mathbb{S}$ admits R-powers, then such difference operators should also satisfy $\delta f^\lambda = (\delta f)^\lambda$, whenever f^λ is well-defined for $f \in \mathbb{S}$ and $\lambda \in R$).
5. *Strong summation operators*; these are partial, strongly linear right inverses $\Sigma \colon \mathbb{S} \to \mathbb{S}$ of finite difference operators, i.e. $(\delta - \mathrm{Id})\,\Sigma = \mathrm{Id}$, for some strong difference operator $\delta \colon \mathbb{S} \to \mathbb{S}$.

Example 6.2. Given a monomial monoid $\mathfrak{M}$, the multiplication $\cdot \colon C[[\mathfrak{M}]]^2 \to C[[\mathfrak{M}]]$ and the scalar product $C[[\mathfrak{M}]]^2 \to C;\ (f, g) \mapsto \langle f, g \rangle = (f\,g)_{\asymp}$ are strongly bilinear mappings.

Example 6.3. Compositions

$$\begin{aligned} \Psi \circ (\Phi_1, \ldots, \Phi_n) \colon \prod_{i=1}^{n} \prod_{j=1}^{m_i} M_{i,j} &\longrightarrow V; \\ ((f_{i,j})_{1 \leqslant j \leqslant m_n})_{1 \leqslant i \leqslant n} &\longmapsto \Psi(\Phi_1(f_{1,1}, \ldots, f_{1,m_1}), \\ &\qquad \ldots, \\ &\qquad \Phi_n(f_{n,1}, \ldots, f_{n,m_n})) \end{aligned}$$

of multilinear grid-based operators

$$\begin{aligned} \Psi \colon N_1 \times \cdots \times N_n &\longrightarrow V \\ \Phi_i \colon M_{i,1} \times \cdots \times M_{i,m_i} &\longrightarrow N_i \qquad (i = 1, \ldots, n) \end{aligned}$$

are again multilinear grid-based operators.

Example 6.4. The m-linear grid-based operators of the form $\Phi \colon C[[\mathfrak{M}_1]] \times \cdots \times C[[\mathfrak{M}_m]] \to C[[\mathfrak{N}]]$ form a C-module. For instance, if $d \colon \mathbb{S} \to \mathbb{S}$ is a strong derivation, where $\mathbb{S} = C[[\mathfrak{M}]]$, then *strong differential operators* of the form

$$L = L_r\, d^r + \cdots + L_0$$

are linear grid-based operators. In section 6.4.1, we will see that we may actually define strong summations on spaces of grid-based operators.

6.1.2 Operator supports

Let $\Phi \colon C[[\mathfrak{M}_1]] \times \cdots \times C[[\mathfrak{M}_m]] \to C[[\mathfrak{N}]]$ be an m-linear grid-based operator, such that $\mathfrak{M}_1, \ldots, \mathfrak{M}_m$ and $\mathfrak{N}$ are all subsets of a common monomial group $\mathfrak{G}$. Then the *operator support* of L is defined by

$$\operatorname{supp} \Phi = \bigcup_{(\mathfrak{m}_1, \ldots, \mathfrak{m}_m) \in \mathfrak{M}_1 \times \cdots \times \mathfrak{M}_m} \operatorname{supp} \frac{\Phi(\mathfrak{m}_1, \ldots, \mathfrak{m}_m)}{\mathfrak{m}_1 \cdots \mathfrak{m}_m}.$$

The operator support is the smallest subset of $\mathfrak{G}$, such that

$$\operatorname{supp} \Phi(f_1, \ldots, f_m) \subseteq (\operatorname{supp} \Phi)\,(\operatorname{supp} f_1) \cdots (\operatorname{supp} f_m), \tag{6.1}$$

for all $(f_1, ..., f_m) \in C[\![\mathfrak{M}_1]\!] \times \cdots \times C[\![\mathfrak{M}_m]\!]$. Given $\mathfrak{S}_1 \subseteq \mathfrak{M}_1, ..., \mathfrak{S}_m \subseteq \mathfrak{M}_m$, we also denote

$$\operatorname{supp}_{\mathfrak{S}_1 \times \cdots \times \mathfrak{S}_m} \Phi = \operatorname{supp} \Phi_{|C[\![\mathfrak{S}_1]\!] \times \cdots \times C[\![\mathfrak{S}_m]\!]}.$$

Example 6.5. We have

$$\begin{aligned} \operatorname{supp} \cdot &= \{1\}; \\ \operatorname{supp} \Psi \circ \prod_{i=1}^{n} \Phi_i &\subseteq (\operatorname{supp} \Psi)(\operatorname{supp} \Phi_1) \cdots (\operatorname{supp} \Phi_n), \end{aligned}$$

for multilinear operators $\Phi_k: C[\![\mathfrak{M}_1]\!] \times \cdots \times C[\![\mathfrak{M}_m]\!] \to C[\![\mathfrak{N}_k]\!]$ $(k = 1, ..., n)$ and $\Psi: C[\![\mathfrak{N}_1]\!] \times \cdots \times C[\![\mathfrak{N}_n]\!] \to C[\![\mathfrak{V}]\!]$.

Exercise 6.1. Let $L_1, ..., L_k: C[\![\mathfrak{M}]\!] \to C[\![\mathfrak{M}]\!]$ be infinitesimal linear grid-based operators (i.e. $\operatorname{supp} L_i \prec 1$ for $i = 1, ..., k$).

a) Show that $f(L_1, ..., L_k)$ is well-defined for non-commutative series $f \in C\langle\langle z_1, ..., z_n\rangle\rangle$.
b) Determine the largest subspace of $\mathbb{T} = C[\![\![x]\!]\!]$ on which e^{∂^2} is a well-defined bijection.

Exercise 6.2.

a) Is a multilinear grid-based operator necessarily a multilinear well-based operator?
b) Show that $C[[\mathfrak{M}^{\preccurlyeq}]]^* \cong C[[\mathfrak{M}^{\succcurlyeq}]]$ for well-based series, if $\mathfrak{M}$ is totally ordered. Here $C[[\mathfrak{M}^{\preccurlyeq}]]^*$ denotes the strong dual of $C[[\mathfrak{M}^{\preccurlyeq}]]$.
c) Show that (*b*) does not hold for grid-based series. How to characterize $C[\![\mathfrak{M}]\!]^*$?

Exercise 6.3.

a) Let $\mathbb{T}^\flat = C[\![\mathfrak{T}^\flat]\!] = \mathbb{T}_{\prec\!\!\prec e^x}$ be the set of transseries $f \in \mathbb{T}$ with $\mathfrak{m} \prec\!\!\prec e^x$ for all $\mathfrak{m} \in \operatorname{supp} f$ and consider the space $\mathscr{D}_{\mathbb{T}^\flat}$ of operators

$$L = \sum_{n \in \mathbb{N}} L_n \partial^n \in \mathbb{T}^\flat[[\partial]], \tag{6.2}$$

such that $\bigcup_{n \in \mathbb{N}} \operatorname{supp} L_n$ is a grid-based. Show that $\mathscr{D}_{\mathbb{T}^\flat}$ operates on $\mathbb{T}^\flat$ and that $\mathscr{D}_{\mathbb{T}^\flat}$ is stable under composition.
b) Let $\mathbb{T}^\flat = C[\![\mathfrak{T}^\flat]\!] = \mathbb{T}_{\preccurlyeq\!\!\prec e^x}$ and consider the space $\mathscr{D}_{\mathbb{T}^\flat}$ of operators (6.2), such that $(L_n)_{n \in \mathbb{N}}$ is a grid-based family. Show that $\mathscr{D}_{\mathbb{T}^\flat}$ operates on $\mathbb{T}^\flat$ and that $\mathscr{D}_{\mathbb{T}^\flat}$ is stable under composition.

6.2 Strong tensor products

It is often useful to consider multilinear mappings

$$M_1 \times \cdots \times M_m \to N$$

as linear mappings

$$M_1 \otimes \cdots \otimes M_m \to N.$$

A similar thing can be done in the strongly linear setting. We will restrict ourselves to the case when $M_1, \ldots, M_m$ are grid-based modules, in which case the tensor product has a particularly nice form:

Proposition 6.6. *Let $\mathfrak{M}_1, \ldots, \mathfrak{M}_m$ be monomial sets and denote*

$$\mathfrak{M} = \mathfrak{M}_1 \times \cdots \times \mathfrak{M}_m.$$

Consider the mapping

$$\begin{aligned} \mu: C[[\mathfrak{M}_1]] \times \cdots \times C[[\mathfrak{M}_m]] &\longrightarrow C[[\mathfrak{M}]] \\ (f_1, \ldots, f_m) &\longmapsto \sum_{\mathfrak{m} \in \mathfrak{M}} f_{1,\mathfrak{m}_1} \cdots f_{m,\mathfrak{m}_m} (\mathfrak{m}_1, \ldots, \mathfrak{m}_m) \end{aligned}$$

This mapping is well-defined and strongly multilinear. Moreover, for every strongly multilinear mapping

$$\Phi: C[[\mathfrak{M}_1]] \times \cdots \times C[[\mathfrak{M}_m]] \to N$$

into an arbitrary strong C-module, there exists a unique strongly linear mapping

$$L: C[[\mathfrak{M}]] \to N,$$

such that $\Phi = L \circ \mu$.

Lemma 6.7. *Let $\mathcal{F}$ be a grid-based family of monomials in $\mathfrak{M}$. Then there exist grid-based families $\mathcal{G}_1 \in \mathscr{F}(\mathfrak{M}_1), \ldots, \mathcal{G}_m \in \mathscr{F}(\mathfrak{M}_m)$ with $\mathcal{F} \precsim \mathcal{G}_1 \times \cdots \times \mathcal{G}_m$.*

Proof. Let $\mathfrak{S}_k$ be the projection of $\mathfrak{S} = \bigcup_{f \in \mathcal{F}} \operatorname{supp} f$ on $\mathfrak{M}_k$, for $k = 1, \ldots, m$. We have $\mathfrak{S}_k \subseteq \mathfrak{e}_{k,1}^{\mathbb{N}} \cdots \mathfrak{e}_{k,p_k}^{\mathbb{N}} \{\mathfrak{f}_{k,1}, \ldots, \mathfrak{f}_{k,q_k}\}$ for certain $\mathfrak{e}_{k,l} \prec 1$ and $\mathfrak{f}_{k,l}$. Given $\mathfrak{m} \in \mathfrak{S}_k$, we will denote

$$\deg \mathfrak{m} = \min \{i_1 + \cdots + i_{p_k} : \mathfrak{m} = \mathfrak{e}_{k,1}^{i_1} \cdots \mathfrak{e}_{k,p_k}^{i_{p_k}} \mathfrak{f}_{k,j}\}.$$

Given $\mathfrak{m} \in \mathfrak{M}$, we define its multiplicity by

$$\mu(\mathfrak{m}) = \operatorname{card} (f \in \mathcal{F}: f_{\mathfrak{m}} \neq 0).$$

Given $\mathfrak{m}_k \in \mathfrak{S}_k$, let

$$\begin{aligned} \mu_k(\mathfrak{m}_k) = \max \{ & \mu(\mathfrak{m}_1, \ldots, \mathfrak{m}_m): \\ & \forall i \in \{1, \ldots, m\}, \mathfrak{m}_i \in \mathfrak{S}_i \wedge \deg \mathfrak{m}_i \leqslant \deg \mathfrak{m}_k\}. \end{aligned}$$

Then for all $(\mathfrak{m}_1, \ldots, \mathfrak{m}_m) \in \mathfrak{S}$, we have

$$\begin{aligned} \mu(\mathfrak{m}_1, \ldots, \mathfrak{m}_m) &\leqslant \max \{\mu_1(\mathfrak{m}_1), \ldots, \mu_m(\mathfrak{m}_m)\} \\ &\leqslant \mu_1(\mathfrak{m}_1) \cdots \mu_m(\mathfrak{m}_m). \end{aligned}$$

Hence

$$\mathcal{F} \precsim \mathcal{G}_1 \times \cdots \times \mathcal{G}_m$$

for $\mathcal{G}_k = (\mathfrak{m}_k)_{\mathfrak{m}_k \in \mathfrak{S}_k, i \in \{1, \ldots, \mu_k(\mathfrak{m}_k)\}}$ $(k = 1, \ldots, m)$. □

Proof of proposition 6.6. Given grid-based subsets $\mathfrak{G}_k \subseteq \mathfrak{M}_k$ with $k = 1, \ldots, m$, the set $\mathfrak{G}_1 \times \cdots \times \mathfrak{G}_m$ is clearly a grid-based subset of $\mathfrak{M}$. This implies that μ is well-defined. More generally, given grid-based families of terms $\mathcal{T}_k \in \mathscr{F}(C\,\mathfrak{M}_k)$ $(k = 1, \ldots, m)$, the family $\mu(\mathcal{T}_1, \ldots, \mathcal{T}_m) \in \mathscr{F}(C\,\mathfrak{M})$ is again grid-based. Now consider arbitrary grid-based families $\mathcal{F}_k \in \mathscr{S}(C[[\mathfrak{M}_k]])$ and let $\mathcal{T}_k = \operatorname{term} \mathcal{F}_k$, for $k = 1, \ldots, m$. Then

$$\begin{aligned}\mu\Big(\sum \mathcal{F}_1, \ldots, \sum \mathcal{F}_m\Big) &= \mu\Big(\sum \mathcal{T}_1, \ldots, \sum \mathcal{T}_m\Big)\\ &= \sum \mu(\mathcal{T}_1, \ldots, \mathcal{T}_m)\\ &= \sum \mu(\mathcal{F}_1, \ldots, \mathcal{F}_m).\end{aligned}$$

This shows that μ is multilinear.

Inversely, if $\mathfrak{G}$ is a grid-based subset of $\mathfrak{M}$, then its projections $\pi_k(\mathfrak{G})$ on $\mathfrak{M}_k$ for $j = 1, \ldots, m$ are again grid-based, and we have

$$\mathfrak{G} \subseteq \pi_1(\mathfrak{G}) \times \cdots \times \pi_m(\mathfrak{G}).$$

Consequently, given a strongly multilinear mapping

$$\Phi: C[[\mathfrak{M}_1]] \times \cdots \times C[[\mathfrak{M}_m]] \to N,$$

the mapping

$$\begin{aligned} L: C[[\mathfrak{M}]] &\longrightarrow N\\ \sum_{\mathfrak{m} \in \mathfrak{M}} f_{\mathfrak{m}}\, \mathfrak{m} &\longmapsto \sum_{\mathfrak{m} \in \mathfrak{M}} f_{\mathfrak{m}}\, \Phi(\mathfrak{m})\end{aligned}$$

is well-defined. Moreover, if $\mathcal{F} \in \mathscr{S}(C[[\mathfrak{M}]])$, then the above lemma implies that there exist $\mathcal{G}_k \in \mathscr{F}(\mathfrak{M}_k)$ $(k = 1, \ldots, m)$ with $\operatorname{mon} \mathcal{F} \precsim \mathcal{G}_1 \times \cdots \times \mathcal{G}_m$, whence

$$L(\operatorname{mon} \mathcal{F}) \precsim \Phi(\mathcal{G}_1, \ldots, \mathcal{G}_m).$$

It follows that $L(\operatorname{mon} \mathcal{F})$, $L(\operatorname{term} \mathcal{F})$ and $L(\mathcal{F})$ are summable families in N. Finally, using strong associativity, we have

$$\begin{aligned} L\Big(\sum \operatorname{term} \mathcal{F}\Big) &= L\Bigg(\sum_{\mathfrak{m} \in \mathfrak{M}} \Bigg(\sum_{c\mathfrak{m} \in \operatorname{term} \mathcal{F}} c\Bigg) \mathfrak{m}\Bigg)\\ &= \sum_{\mathfrak{m} \in \mathfrak{M}} \Bigg(\sum_{c\mathfrak{m} \in \operatorname{term} \mathcal{F}} c\Bigg) \Phi(\mathfrak{m})\\ &= \sum L(\operatorname{term} \mathcal{F}).\end{aligned}$$

We conclude that $L(\sum \mathcal{F}) = \sum L(\mathcal{F})$. □

We call $C[[\mathfrak{M}_1]] \otimes \cdots \otimes C[[\mathfrak{M}_m]] = C[[\mathfrak{M}_1 \times \cdots \times \mathfrak{M}_m]]$ (together with the mapping μ) the strong tensor product of $C[[\mathfrak{M}_1]]$, ..., $C[[\mathfrak{M}_m]]$. An immediate consequence of proposition 6.6 is the principle of extension by strong multilinearity:

Corollary 6.8. *Let $\mathfrak{M}_1$, ..., $\mathfrak{M}_m$ and $\mathfrak{N}$ be monomial monoids and assume that φ is a mapping, such that*

$$(\varphi(\mathfrak{m}_1, \ldots, \mathfrak{m}_m))_{(\mathfrak{m}_1, \ldots, \mathfrak{m}_m) \in \mathfrak{G}_1 \times \cdots \times \mathfrak{G}_m}$$

is a grid-based family for any grid-based subsets $\mathfrak{G}_1 \subseteq \mathfrak{M}_1, \ldots, \mathfrak{G}_m \subseteq \mathfrak{M}_m$. Then there exists a unique strongly multilinear mapping

$$\Phi: C[\![\mathfrak{M}_1]\!] \times \cdots \times C[\![\mathfrak{M}_m]\!] \to C[\![\mathfrak{N}]\!]$$

with $\Phi_{|\mathfrak{M}_1 \times \cdots \times \mathfrak{M}_m} = \varphi$.

Proof. Using extension by strong linearity, there exists a unique strongly linear mapping $L: C[\![\mathfrak{M}_1 \times \cdots \times \mathfrak{M}_m]\!] \to C[\![\mathfrak{N}]\!]$, with $L_{|\mathfrak{M}_1 \times \cdots \times \mathfrak{M}_m} = \varphi$. Then $\Phi = L \circ \mu$ is the unique mapping we are looking for. □

Exercise 6.4. When do we have $\mathscr{L}(C[\![\mathfrak{M}]\!], C[\![\mathfrak{N}]\!]) \cong C[\![\mathfrak{M}]\!]^* \otimes C[\![\mathfrak{N}]\!]$, where $\mathscr{L}(C[\![\mathfrak{M}]\!], C[\![\mathfrak{N}]\!])$ denotes the space of strongly linear mappings from $C[\![\mathfrak{M}]\!]$ into $C[\![\mathfrak{N}]\!]$?

Exercise 6.5.

a) Generalize proposition 6.6 to the case of well-based series.

b) Show that a well-based family $(f_i)_{i \in I} \in C[[\mathfrak{M}]]^I$ corresponds to an element of $C[[I \times \mathfrak{M}]]$.

c) Define a family $\mathcal{F} \in \mathscr{F}(C[\![\mathfrak{M}]\!])$ to be super-grid-based $\mathcal{F} \approx (f_\mathfrak{i})_{\mathfrak{i} \in \mathfrak{I}}$ with $\mathfrak{I} \subseteq z^{\mathbb{N}^n}$ and $f = \sum_{(\mathfrak{i},\mathfrak{m})} f_{\mathfrak{i},\mathfrak{m}} (\mathfrak{i}, \mathfrak{m}) \in C[\![\mathfrak{I} \times \mathfrak{M}]\!]$. Show that $C[\![\mathfrak{M}]\!]$ is a strong C-algebra for super-grid-based summation.

d) Give an example of a grid-based family which is not super-grid-based.

Exercise 6.6. Show that tensor products exist in the general strongly linear setting (see also exercise 2.20). Hint:

a) Let M_1, ..., M_m be strong modules. Consider the set F of all mappings f: $M_1 \times \cdots \times M_m \to C$, whose support is contained in a set $S_1 \times \cdots \times S_m$ such that each S_i is a summable subset of M_i. Construct a natural embedding $\nu: M_1 \times \cdots \times M_m \to F$ and give F the structure of a strong C-module.

b) Let Z be the strong submodule of F, which is generated by all elements of the form

$$\Big(\sum_{i_1 \in I_1} \lambda_{i_1} x_{i_1}, \ldots, \sum_{i_m \in I_m} \lambda_{i_m} x_{i_m}\Big) - \sum_{\substack{i_1 \in I_1 \\ \vdots \\ i_m \in I_m}} \lambda_{i_1} \cdots \lambda_{i_m} (x_{i_1}, \ldots, x_{i_m}),$$

where the I_k are mutually disjoint. Then the strong quotient

$$M_1 \otimes \cdots \otimes M_m = F/Z$$

with $\mu = \pi_{F/Z} \circ \nu$ satisfies the universal property of the strong tensor product.

6.3 Grid-based operators

6.3.1 Definition and characterization

Let $\mathfrak{M}$ and $\mathfrak{N}$ be monomial sets. A mapping $\Phi: C[[\mathfrak{M}]] \to C[[\mathfrak{N}]]$ is said to be a *grid-based operator* if there exists a family $(\check{\Phi}_i)_{i\in\mathbb{N}}$ of multilinear grid-based operators $\check{\Phi}_i: C[[\mathfrak{M}]]^i \to C[[\mathfrak{N}]]$, such that for all $\mathcal{F}\in\mathscr{S}(C[[\mathfrak{M}]])$, the family $(\check{\Phi}_i(f_1,\ldots,f_i))_{i\in\mathbb{N},f_1,\ldots,f_i\in\mathcal{F}}$ is grid-based, and

$$\Phi\left(\sum \mathcal{F}\right) = \sum_{\substack{i\in\mathbb{N}\\ f_1,\ldots,f_i\in\mathcal{F}}} \check{\Phi}_i(f_1,\ldots,f_i). \tag{6.3}$$

We call $(\check{\Phi}_i)_{i\in\mathbb{N}}$ a *multilinear family* for Φ. Considering the family of a single element $f\in C[[\mathfrak{M}]]$, the formula (6.3) reduces to

$$\begin{aligned} \Phi(f) &= \sum_{i\in\mathbb{N}} \Phi_i(f), \qquad \text{with} \\ \Phi_i(f) &\hat{=} \check{\Phi}_i(f) = \check{\Phi}_i(f,\ldots,f). \end{aligned} \tag{6.4}$$

Assuming that $C\supseteq\mathbb{Q}$, each Φ_i is uniquely determined and we call it the *homogeneous part* of degree i of Φ:

Proposition 6.9. *Let $\Phi: C[[\mathfrak{M}]]\to C[[\mathfrak{N}]]$ be a grid-based operator and let $\check{\Phi}_i: C[[\mathfrak{M}]]^i\to C[[\mathfrak{N}]]$ be multilinear grid-based operators, such that* (6.4) *holds for all $f\in C[[\mathfrak{M}]]$. If $C\supseteq\mathbb{Q}$ and $\Phi=0$, then $\Phi_i=0$ for each $i\in\mathbb{N}$.*

Proof. We observe that it suffices to prove that $\Phi_i=0$ for each $i\in\mathbb{N}$, since the $\check{\Phi}_i$ are symmetric and $C\supseteq\mathbb{Q}$ is torsion-free. Assume the contrary and let $f\in C[[\mathfrak{M}]]$ be such that $\Phi_i(f)\neq 0$ for some i. Choose

$$\mathfrak{m}\in\mathfrak{S}=\bigcup_{i\in\mathbb{N}} \operatorname{supp}\Phi_i(f)\neq\varnothing.$$

Since $(\Phi_i(f))_{i\in\mathbb{N}}$ is a grid-based family, there exist only a finite number of indices i, such that $\mathfrak{m}\in\operatorname{supp}\Phi_i(f)$. Let $i_1<\cdots<i_n$ be those indices.

Let $c_k=\Phi_{i_k}(f)_{\mathfrak{m}}$ for all $k\in\{1,\ldots,n\}$. For any $l\in\{1,\ldots,n\}$, we have $\Phi_{i_k}(l\,f)_{\mathfrak{m}}=l^{i_k}c_k$, by multilinearity. On the other hand,

$$\Phi(l\,f)_{\mathfrak{m}}=\Phi_{i_1}(l\,f)_{\mathfrak{m}}+\cdots+\Phi_{i_n}(l\,f)_{\mathfrak{m}}=0$$

for each l, so that

$$\begin{pmatrix} 1 & \cdots & 1\\ \vdots & & \vdots\\ n^{i_1} & \cdots & n^{i_n}\end{pmatrix}\begin{pmatrix} c_1\\ \vdots\\ c_n\end{pmatrix}=0.$$

The matrix on the left hand side admits an inverse with rational coefficients (indeed, by the sign rule of Descartes, a real polynomial $\alpha_1 x^{i_1}+\cdots+\alpha_n x^{i_n}$ cannot have n distinct positive zeros unless $\alpha_1=\cdots=\alpha_n=0$). Since $C\supseteq\mathbb{Q}$, it follows that $c_1=\cdots=c_n=0$. This contradiction completes the proof. □

Proposition 6.10. *Let* $\Phi: C[[\mathfrak{M}]] \to C[[\mathfrak{N}]]$ *be a grid-based operator and assume that* $C \supseteq \mathbb{Q}$. *Then there exist a unique multilinear family* $(\check{\Phi}_i)_{i\in\mathbb{N}}$ *for* Φ, *such that each* $\check{\Phi}_i$ *is symmetric.*

Proof. Let $(\tilde{\Phi}_i)_{i\in\mathbb{N}}$ be an arbitrary multilinear family for Φ. Then the $\check{\Phi}_i$ defined by

$$\check{\Phi}_i(f_1, \ldots, f_i) = \frac{1}{i!} \sum_{\sigma \in \mathfrak{S}_i} \tilde{\Phi}_i(f_{\sigma(1)}, \ldots, f_{\sigma(i)}).$$

form a multilinear family of symmetric operators for Φ. Moreover, each $\check{\Phi}_i$ is determined uniquely in terms of Φ_i by

$$\check{\Phi}_i(f_1, \ldots, f_i) = \frac{1}{i!} \sum_{J \subseteq \{1,\ldots,i\}} (-1)^{i-|J|}\, \Phi_i\!\left(\sum_{j\in J} f_j\right).$$

We conclude by proposition 6.9. □

Assume that $\mathfrak{M}$ and $\mathfrak{N}$ are subsets of a common monomial group $\mathfrak{G}$. If we have $C \supseteq \mathbb{Q}$ and Φ and $(\check{\Phi}_i)_{i\in\mathbb{N}}$ are as in proposition 6.10, then we call

$$\operatorname{supp} \Phi = \operatorname{supp} \check{\Phi}_0 \cup \operatorname{supp} \check{\Phi}_1 \cup \operatorname{supp} \check{\Phi}_2 \cup \cdots$$

the *operator support* of Φ. For all $f \in C[[\mathfrak{M}]]$, we have

$$\operatorname{supp} \Phi(f) \subseteq (\operatorname{supp} \Phi)\,(\operatorname{supp} f)^*.$$

Notice also that $\operatorname{supp} \Phi_i = \operatorname{supp} \check{\Phi}_i$ for all i.

6.3.2 Multivariate grid-based operators and compositions

In a similar way that we have the natural isomorphism

$$C[[\mathfrak{M}_1 \times \cdots \times \mathfrak{M}_m]] \;\cong\; C[[\mathfrak{M}_1]] \otimes \cdots \otimes C[[\mathfrak{M}_m]],$$

for tensor products, we also have a natural isomorphism

$$\begin{aligned} C[[\mathfrak{M}_1 \amalg \cdots \amalg \mathfrak{M}_m]] &\longrightarrow C[[\mathfrak{M}_1]] \times \cdots \times C[[\mathfrak{M}_m]], \\ f &\longmapsto \left(\sum_{\mathfrak{m}\in\mathfrak{M}_1} f_{\mathfrak{m}}\, \mathfrak{m}, \ldots, \sum_{\mathfrak{m}\in\mathfrak{M}_m} f_{\mathfrak{m}}\, \mathfrak{m}\right) \end{aligned}$$

for Cartesian products. This allows us to reinterpret mappings "in several series" $C[[\mathfrak{M}_1]] \times \cdots \times C[[\mathfrak{M}_m]] \to N$ as mappings "in one series" $C[[\mathfrak{M}_1 \amalg \cdots \amalg \mathfrak{M}_m]] \to N$. In particular, any multilinear grid-based operator $\Phi: C[[\mathfrak{M}_1]] \times \cdots \times C[[\mathfrak{M}_m]] \to C[[\mathfrak{N}]]$ can be seen as a grid-based operator in from $C[[\mathfrak{M}_1 \amalg \cdots \amalg \mathfrak{M}_m]]$ into $C[[\mathfrak{N}]]$. More generally, the natural isomorphism may be used in order to extend the notion of grid-based operators to mappings $C[[\mathfrak{M}_1]] \times \cdots \times C[[\mathfrak{M}_m]] \to C[[\mathfrak{N}]]$.

Let $\Phi: C[[\mathfrak{M}]] \to C[[\mathfrak{N}]]$ and $\Psi: C[[\mathfrak{N}]] \to C[[\mathfrak{V}]]$ be two grid-based operators. Then $\Psi \circ \Phi$ is again a grid-based operator. Indeed, let $(\check{\Phi}_i)_{i\in\mathbb{N}}$ and $(\check{\Psi}_j)_{j\in\mathbb{N}}$ be multilinear families for Φ and Ψ. Then for all $\mathcal{F} \in \mathscr{S}(C[[\mathfrak{M}]])$, we have

$$\Psi \circ \Phi\left(\sum \mathcal{F}\right) = \Psi\left(\sum_{\substack{i\in\mathbb{N}\\ f_1,\ldots,f_i\in K}} \check{\Phi}_i(f_1,\ldots,f_i)\right)$$

$$= \sum_{\substack{j\in\mathbb{N}\\ i_1,\ldots,i_j\in\mathbb{N}\\ f_{1,1},\ldots,f_{1,i_1}\in\mathcal{F}\\ \vdots\\ f_{j,1},\ldots,f_{j,i_j}\in\mathcal{F}}} \begin{array}{l}\check{\Psi}_j(\check{\Phi}_{i_1}(f_{1,1},\ldots,f_{1,i_1}),\\ \ldots,\\ \check{\Phi}_{i_j}(f_{j,1},\ldots,f_{j,i_j}))\end{array}$$

so that the $(\widecheck{\Psi\circ\Phi})_l$ defined by

$$(\widecheck{\Psi\circ\Phi})_l = \sum_{\substack{j\in\mathbb{N}\\ i_1+\cdots+i_j=l}} \check{\Psi}_j \circ (\check{\Phi}_{i_1},\ldots,\check{\Phi}_{i_j})$$

form a multilinear family for $\Psi \circ \Phi$.

Exercise 6.7. Assume that $C \supseteq \mathbb{Q}$ and let $\Phi: C[[\mathfrak{M}]] \to C[[\mathfrak{N}]]$ be a grid-based operator. Is it true that for any $\mathfrak{S} \subsetneq \operatorname{supp} \Phi$ there exists an $f \in C[[\mathfrak{M}]]$ with $\operatorname{supp} \Phi(f) \nsubseteq \mathfrak{S}\,(\operatorname{supp} f)^*$?

Exercise 6.8. Define the "derivative" of a grid-based operator $\Phi: C[[\mathfrak{M}]] \to C[[\mathfrak{N}]]$.

Exercise 6.9.

a) Characterize the intervals $\mathfrak{I}$ of the set of infinitesimal transmonomials $\mathfrak{T}_{\prec}$ (i.e. for all $\mathfrak{m}, \mathfrak{n} \in \mathfrak{I}$ and $\mathfrak{v} \in \mathfrak{T}$, we have $\mathfrak{m} \preccurlyeq \mathfrak{v} \preccurlyeq \mathfrak{n} \Rightarrow \mathfrak{v} \in \mathfrak{I}$), such that for all $g \in x + C[[\mathfrak{I}]]$, the operator $\circ_g$ is a grid-based operator on $C[[\mathfrak{I}]]$.
b) With $\mathfrak{I}$ as in (a), show that the operators $C[[\mathfrak{I}]]^2 \to C[[\mathfrak{I}]]$; $(\varepsilon, \delta) \mapsto (x+\varepsilon)\circ(x-\delta) - x$ and $C[[\mathfrak{I}]] \to C[[\mathfrak{I}]]; \varepsilon \mapsto (x+\varepsilon)^{\mathrm{inv}} - x$ are grid-based.

6.4 Atomic decompositions

6.4.1 The space of grid-based operators

Let $\mathscr{L}(M_1, \ldots, M_m, N)$ be the space of strongly multilinear operators $\Phi: M_1 \times \cdots \times M_m \to N$. Then $\mathscr{L}(M_1, \ldots, M_m, N)$ is clearly a C-module. More generally, a family $(\Phi_i)_{i\in I}$ of elements in $\mathscr{L}(M_1, \ldots, M_m, N)$ is said to be *summable*, if for all $\mathcal{F}_1 \in \mathscr{S}(M_1), \ldots, \mathcal{F}_m \in \mathscr{S}(M_m)$, we have

$$\coprod_{i\in I} \Phi_i(\mathcal{F}_1,\ldots,\mathcal{F}_m) \in \mathscr{S}(N).$$

In that case, we define the sum $\sum_{i\in I} \Phi_i \in \mathscr{L}(M_1, ..., M_m, N)$ by

$$\sum_{i\in I} \Phi_i: (f_1, ..., f_m) \longmapsto \sum_{i\in I} \Phi_i(f_1, ..., f_m).$$

This gives $\mathscr{L}(M_1, ..., M_m, N)$ the structure of a strong C-module.

Similarly, let $\mathscr{G}(C[[\mathfrak{M}]], C[[\mathfrak{N}]])$ denote the space of grid-based operators $\Phi: C[[\mathfrak{M}]] \to C[[\mathfrak{N}]]$. This space is clearly a C-module. A family $(\Phi_j)_{j\in J} \in \mathscr{G}(C[[\mathfrak{M}]], C[[\mathfrak{N}]])^J$ is said to be *summable*, if for all $\mathcal{F} \in \mathscr{S}(C[[\mathfrak{M}]])$, the family

$$(\check{\Phi}_{j,i}(f_1, ..., f_i))_{j\in J, i\in \mathbb{N}, (f_1, ..., f_i)\in \mathcal{F}^i}$$

is a grid-based family. In that case, the sum

$$\sum_{j\in J} \Phi_j: f \mapsto \sum_{j\in J} \Phi_j(f)$$

is a grid-based operator and $\mathscr{G}(C[[\mathfrak{M}]], C[[\mathfrak{N}]])$ is a strong C-module for this summation. In particular, we have

$$\Phi = \Phi_0 + \Phi_1 + \Phi_2 + \cdots \tag{6.5}$$

for all $\Phi \in \mathscr{G}(C[[\mathfrak{M}]], C[[\mathfrak{N}]])$. We call (6.5) the *decomposition* of Φ *into homogeneous parts*.

6.4.2 Atomic decompositions

Let $\mathfrak{M}_1, ..., \mathfrak{M}_m$ and $\mathfrak{N}$ be monomials sets. Given $\mathfrak{m}_1 \in \mathfrak{M}_1, ..., \mathfrak{m}_m \in \mathfrak{M}_m$ and $\mathfrak{n} \in \mathfrak{N}$, the operator

$$\Omega_{\mathfrak{m}_1, ..., \mathfrak{m}_m, \mathfrak{n}}: C[[\mathfrak{M}_1]] \times \cdots \times C[[\mathfrak{M}_m]] \longrightarrow C[[\mathfrak{N}]]$$

with

$$\Omega_{\mathfrak{m}_1, ..., \mathfrak{m}_m, \mathfrak{n}}(f_1, ..., f_m) = c\, f_{1,\mathfrak{m}_1} \cdots f_{m,\mathfrak{m}_m}\, \mathfrak{n}$$

is an m-linear grid-based operator. Operators of this form, which are said to be *atomic*, form a strong basis of $\mathscr{L}(C[[\mathfrak{M}_1]], ..., C[[\mathfrak{M}_m]], C[[\mathfrak{N}]])$, since any operator $\Phi \in \mathscr{L}(C[[\mathfrak{M}_1]], ..., C[[\mathfrak{M}_m]], C[[\mathfrak{N}]])$ may be uniquely decomposed as

$$\Phi = \sum_{\substack{\mathfrak{m}_1 \in \mathfrak{M}_1, ..., \mathfrak{m}_m \in \mathfrak{M}_m \\ \mathfrak{n} \in \mathfrak{N}}} \Phi(\mathfrak{m}_1, ..., \mathfrak{m}_m)_{\mathfrak{n}}\, \Omega_{\mathfrak{m}_1, ..., \mathfrak{m}_m, \mathfrak{n}}. \tag{6.6}$$

We call (6.6) the *atomic decomposition* of Φ. More generally, an *atomic family* is a summable family $\mathcal{A} = (c_\alpha\, \Omega_\alpha)_{\alpha\in\mathcal{A}}$, with $c_\alpha \in C$ and $\Omega_\alpha = \Omega_{\mathfrak{i}_{\alpha,1}, ..., \mathfrak{i}_{\alpha,m}, \mathfrak{o}_\alpha}$, where $\mathfrak{i}_{\alpha,1}, ..., \mathfrak{i}_{\alpha,m} \in \mathfrak{M}$ and $\mathfrak{o}_\alpha \in \mathfrak{N}$.

Assume now that $C \supseteq \mathbb{Q}$. Given a grid-based operator $\Phi: C[[\mathfrak{M}]] \to C[[\mathfrak{N}]]$, let the $\check{\Phi}_i$ be as in proposition 6.10. Then we have

$$\Phi = \sum_{\mathfrak{m}_1 \cdots \mathfrak{m}_i \in \mathfrak{M}^*, \mathfrak{n} \in \mathfrak{N}} \check{\Phi}_i(\mathfrak{m}_1, ..., \mathfrak{m}_i)_{\mathfrak{n}}\, \widehat{\Omega_{\mathfrak{m}_1, ..., \mathfrak{m}_i, \mathfrak{n}}} \tag{6.7}$$

and we call this formula the *atomic decomposition* of Φ. More generally, a family $\mathcal{A} = (c_\alpha\, \Omega_\alpha)_{\alpha \in \mathcal{A}}$, where $c_\alpha \in C$ and $\Omega_\alpha = \Omega_{\mathfrak{i}_{\alpha,1},\ldots,\mathfrak{i}_{\alpha,|\alpha|},\mathfrak{o}_\alpha}$, is called an *atomic family*, if the family $\hat{\mathcal{A}} = (c_\alpha\, \hat{\Omega}_\alpha)_{\alpha \in \mathcal{A}}$ is summable in $\mathscr{G}(C[[\mathfrak{M}]], C[[\mathfrak{N}]])$.

Since the $\check{\Phi}_i$ in (6.7) are symmetric, the atomic decomposition is slightly redundant. Let $\sim$ be the equivalence relation on $\mathfrak{M}^*$, such that $\mathfrak{m}_1 \cdots \mathfrak{m}_i \sim \mathfrak{n}_1 \cdots \mathfrak{n}_j$ if and only if $j = i$ and there exists a permutation of indices α, such that $\mathfrak{n}_i = \mathfrak{m}_{\alpha(i)}$ for all i. Given $\hat{\mathfrak{m}} \in \mathfrak{M}^*/\sim$, $\mathfrak{m}_1 \cdots \mathfrak{m}_m \in \hat{\mathfrak{m}}$ and $\mathfrak{n} \in \mathfrak{N}$, we define

$$\Omega_{\hat{\mathfrak{m}},\mathfrak{n}} = \widehat{\Omega_{\mathfrak{m}_1,\ldots,\mathfrak{m}_i,\mathfrak{n}}}.$$

Clearly, $\Omega_{\hat{\mathfrak{m}},\mathfrak{n}}$ does not depend on the choice of $\mathfrak{m}_1 \cdots \mathfrak{m}_m \in \hat{\mathfrak{m}}$ and operators of the form $\Omega_{\hat{\mathfrak{m}},\mathfrak{n}}$ will be called *symmetric atomic operators*. Setting

$$\check{\Phi}(\hat{\mathfrak{m}}) = \sum_{\mathfrak{m}_1 \cdots \mathfrak{m}_i \in \hat{\mathfrak{m}}} \check{\Phi}_i(\mathfrak{m}_1, \ldots, \mathfrak{m}_i),$$

for all $\hat{\mathfrak{m}} \in \mathfrak{M}^*/\sim$, the decomposition

$$\Phi = \sum_{\hat{\mathfrak{m}} \in \mathfrak{M}^*/\sim,\, \mathfrak{n} \in \mathfrak{N}} \check{\Phi}(\hat{\mathfrak{m}})_\mathfrak{n}\, \Omega_{\hat{\mathfrak{m}},\mathfrak{n}}$$

is unique. We call it the *symmetric atomic decomposition* of Φ.

6.4.3 Combinatorial interpretation of atomic families

Consider an atomic family $\mathcal{A}$ with $\Omega_\alpha: C[[\mathfrak{M}]]^{|\alpha|} \to C[[\mathfrak{N}]]$ for each $\alpha \in \mathcal{A}$. We may interpret the Ω_α as combinatorial boxes with *inputs* $\mathfrak{i}_{\alpha,1}, \ldots, \mathfrak{i}_{\alpha,|\alpha|} \in \mathfrak{M}$ and *output* $\mathfrak{o}_\alpha \in \mathfrak{N}$. We define a partial ordering on $\mathcal{A}$ by $\alpha \prec \alpha' \Leftrightarrow \mathfrak{o}_\alpha \prec \mathfrak{o}_{\alpha'}$. Given a subset $\mathfrak{S}$ of $\mathfrak{M}$, we denote by $\mathcal{A}_\mathfrak{S}$ the atomic family of all $\alpha \in \mathcal{A}$ with $\{\mathfrak{i}_{\alpha,1}, \ldots, \mathfrak{i}_{\alpha,|\alpha|}\} \subseteq \mathfrak{S}$. Finally, given a monomial set $\mathfrak{M}$, we denote by $\mathcal{D}_\mathfrak{M}$ the atomic family $(\Omega_{\mathfrak{m},\mathfrak{m}})_{\mathfrak{m} \in \mathfrak{M}}$, so that $\sum \mathcal{D}_\mathfrak{M}$ is the identity operator on $C[[\mathfrak{M}]]$.

Remark 6.11. A convenient way to check whether a family $\mathcal{A} = (c_\alpha\, \Omega_\alpha)_{\alpha \in \mathcal{A}}$ is atomic is to prove that for each grid-based subset $\mathfrak{S} \subseteq \mathfrak{M}$ we have

1. The set $\mathfrak{o}_{\mathcal{A}_\mathfrak{S}}$ is grid-based.
2. For each $\mathfrak{n} \in \mathfrak{N}$, there exist only a finite number of $\alpha \in \mathcal{A}_\mathfrak{S}$ with $\mathfrak{o}_\alpha = \mathfrak{n}$.

Consider two atomic families $\mathcal{A}$ and $\mathcal{B}$, where $\Omega_\alpha: C[[\mathfrak{N}]]^{|\alpha|} \to C[[\mathfrak{V}]]$ and $\Omega_\beta: C[[\mathfrak{M}]]^{|\beta|} \to C[[\mathfrak{N}]]$ for all $\alpha \in \mathcal{A}$ and $\beta \in \mathcal{B}$. We define their composition to be the family $(c_\varsigma\, \Omega_\varsigma)_{\varsigma \in \mathcal{A} \circ \mathcal{B}}$ with formal index set

$$\begin{aligned}\mathcal{A} \circ \mathcal{B} = \{&\alpha \circ (\beta_1, \ldots, \beta_{|\alpha|}):\\ &\alpha \in \mathcal{A} \wedge \beta_1, \ldots, \beta_{|\alpha|} \in \mathcal{B} \wedge \mathfrak{o}_{\beta_1} = \mathfrak{i}_{\alpha,1} \wedge \cdots \wedge \mathfrak{o}_{\beta_{|\alpha|}} = \mathfrak{i}_{\alpha,|\alpha|}\}\end{aligned}$$

and

$$\begin{aligned} c_{\alpha \circ (\beta_1, \ldots, \beta_{|\alpha|})} &= c_\alpha\, c_{\beta_1} \cdots c_{\beta_{|\alpha|}};\\ \Omega_{\alpha \circ (\beta_1, \ldots, \beta_{|\alpha|})} &= \Omega_{\mathfrak{i}_{\beta_1,1}, \ldots, \mathfrak{i}_{\beta_1,|\beta_1|}, \ldots, \mathfrak{i}_{\beta_{|\alpha|},1}, \ldots, \mathfrak{i}_{\beta_{|\alpha|},|\beta_{|\alpha|}|}, \mathfrak{o}_\alpha}.\end{aligned}$$

We may see the $\alpha \circ (\beta_1, ..., \beta_{|\alpha|})$ as combinatorial structures, such that the outputs $\mathfrak{o}_{\beta_k}$ of the β_k coincide with the inputs $\mathfrak{i}_{\alpha,k}$ of α (see figure 6.1). A similar computation as at the end of section 6.3.2 yields:

Proposition 6.12. *Let $\mathcal{A}$ and $\mathcal{B}$ be two atomic families as above. Then $\mathcal{A} \circ \mathcal{B}$ is again an atomic family and*

$$\sum \widehat{\mathcal{A} \circ \mathcal{B}} = \left(\sum \hat{\mathcal{A}}\right) \circ \left(\sum \hat{\mathcal{B}}\right). \qquad \square$$

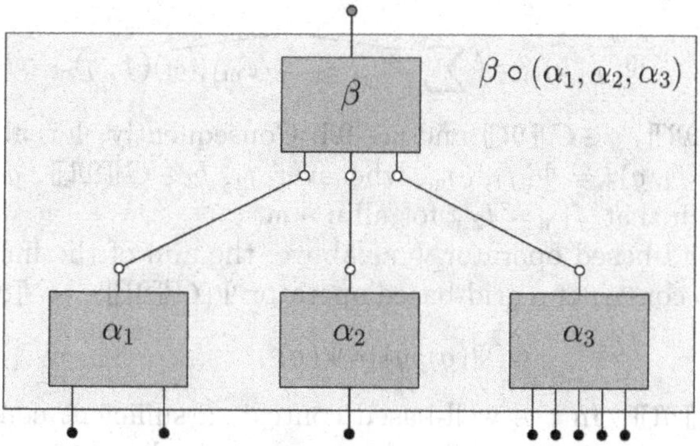

Fig. 6.1. Combinatorial interpretation of the composition of atomic operators.

Exercise 6.10. Show that the mapping

$$\begin{aligned} \circ_{L_1,...,L_k}: C\langle\langle z_1, ..., z_k\rangle\rangle &\to \mathscr{L}(C[[\mathfrak{M}]], C[[\mathfrak{M}]]) \\ f &\mapsto f(L_1, ..., L_k) \end{aligned}$$

from exercise 6.1 is a strong C-algebra morphism.

Exercise 6.11. Show that $\mathscr{L}(M_1, ..., M_m, N)$ and $\mathscr{L}(M_1 \otimes \cdots \otimes M_m, N)$ are naturally isomorphic as sets. Show that this natural isomorphism also preserves the strong C-module structure.

Exercise 6.12. Show that an atomic family $\mathcal{A}$ is summable, if and only if $\mathcal{A}_{\mathfrak{S}}$ is grid-based for every grid-based subset $\mathfrak{S} \subseteq \mathfrak{M}$.

Exercise 6.13. Generalize the theory from sections 6.3 and 6.4 to the well-based setting.

6.5 Implicit function theorems

Let $\mathfrak{M}$ and $\mathfrak{N}$ be monomial sets which are contained in a common monomial monoid. Consider a grid-based operator

$$\begin{aligned} \Phi: C[[\mathfrak{M}]] \times C[[\mathfrak{N}]] &\longrightarrow C[[\mathfrak{M}]] \\ (f, g) &\longmapsto \Phi(f, g) \end{aligned}$$

together with its atomic decomposition $\Phi = \sum \mathcal{A}$. We say that

- Φ is *strictly extensive* in f if $\mathfrak{o}_\alpha \prec \mathfrak{i}_{\alpha,k}$ whenever $\mathfrak{i}_{\alpha,k} \in \mathfrak{M}$.
- Φ is *extensive* in f *with multipliers* in a set $\mathfrak{E}$, if $\mathfrak{o}_\alpha \in \mathfrak{i}_{\alpha,k}\ \mathfrak{E}$ whenever $\mathfrak{i}_{\alpha,k} \in \mathfrak{M}$.
- Φ is *contracting* in f if $\Phi(f_2, g) - \Phi(f_1, g) \preccurlyeq f_2 - f_1$ for all $f_1, f_2 \in C[[\mathfrak{M}]]$ and $g \in C[[\mathfrak{N}]]$. Here we write $f \preccurlyeq g$ if for all $\mathfrak{m} \in \operatorname{supp} f$, there exists an $\mathfrak{n} \in \operatorname{supp} g$ with $\mathfrak{m} \prec \mathfrak{n}$.

If Φ is strictly extensive in f, then we have in particular

$$\Phi(f, g)_\mathfrak{m} = (\sum \widehat{\mathcal{A}_{\{\mathfrak{m}' \in \mathfrak{m} : \mathfrak{m}' \succ \mathfrak{m}\} \amalg \mathfrak{N}}})(f, g)_\mathfrak{m}$$

for all $f \in C[[\mathfrak{M}]]$, $g \in C[[\mathfrak{N}]]$ and $\mathfrak{m} \in \mathfrak{M}$. Consequently, Φ is also contracting in f, since $\Phi(f_2, g)_\mathfrak{m} = \Phi(f_1, g)_\mathfrak{m}$, whenever $f_1, f_2 \in C[[\mathfrak{M}]]$, $g \in C[[\mathfrak{N}]]$ and $\mathfrak{m} \in \mathfrak{M}$ are such that $f_{1,\mathfrak{n}} = f_{2,\mathfrak{n}}$ for all $\mathfrak{n} \succ \mathfrak{m}$.

Given a grid-based operator Φ as above, the aim of the implicit function theorems is to construct a grid-based operator $\Psi: C[[\mathfrak{N}]] \to C[[\mathfrak{M}]]$, such that

$$\Phi(\Psi(g), g) = \Psi(g) \tag{6.8}$$

for all $g \in C[[\mathfrak{N}]]$. In the well-based context, a sufficient condition for the existence (and uniqueness) of such an operator is the strict extensiveness of Φ in f. In the grid-based context we need additional conditions in order to preserve the grid-based property. In this section, we present three possible choices for these extra conditions, which lead each to a grid-based implicit function theorem.

6.5.1 The first implicit function theorem

Theorem 6.13. *Consider a grid-based operator*

$$\begin{aligned} \Phi: C[[\mathfrak{M}]] \times C[[\mathfrak{N}]] &\to C[[\mathfrak{M}]] \\ (f, g) &\mapsto \Phi(f, g), \end{aligned}$$

which is extensive in f with multipliers in a grid-based set $\mathfrak{E} \prec 1$. Then for each $g \in C[[\mathfrak{N}]]$, there exists a unique $\Psi(g)$ which satisfies (6.8) *and the operator $\Psi: C[[\mathfrak{N}]] \to C[[\mathfrak{M}]]$ is grid-based. Furthermore, for all $g \in C[[\mathfrak{N}]]$, we have*

$$\operatorname{supp} \Psi(g) \subseteq (\operatorname{supp} \Phi(0, g))\, \mathfrak{E}^*.$$

If $C \supseteq \mathbb{Q}$, then we also have

$$\operatorname{supp} \Psi \subseteq (\operatorname{supp} \Phi)^+.$$

Proof. Let $\Phi = \sum \mathcal{A}$ be the atomic decomposition of Φ. Consider the family $\mathcal{B} = \amalg_{d \in \mathbb{N}} \mathcal{B}_d$, where the $\mathcal{B}_d$ are recursively defined by

$$\begin{aligned} \mathcal{B}_0 &= \mathcal{A}_\mathfrak{N} \\ \mathcal{B}_{d+1} &= (\mathcal{A} \setminus \mathcal{A}_\mathfrak{N}) \circ (\mathcal{B}_d \amalg \mathcal{D}_\mathfrak{N}) \end{aligned}$$

See figure 6.2 for the illustration of a member of $\mathcal{B}$. We claim that $\mathcal{B}$ is an atomic family. Indeed, let $\mathfrak{S} \subseteq \mathfrak{N}$ be a grid-based set. Let us prove by induction over d that

$$\operatorname{supp} \mathfrak{o}_\varsigma \subseteq \mathfrak{S}\, \mathfrak{E}^d \tag{6.9}$$

for all $\varsigma \in \mathcal{B}_{d,\mathfrak{S}}$. This is clear if $d = 0$. If $d \geqslant 1$, then we may write $\varsigma = \alpha \circ (\beta_1, \ldots, \beta_d)$, where $\mathfrak{i}_{\alpha,k} = \mathfrak{o}_{\beta_k} \in \mathfrak{M}$ for at least one k. By the induction hypothesis, we have $\operatorname{supp} \mathfrak{o}_{\beta_k} \subseteq \mathfrak{S}\, \mathfrak{E}^{d-1}$, so that $\mathfrak{o}_\varsigma \in \mathfrak{o}_{\beta_k} \mathfrak{E} \subseteq \mathfrak{S}\, \mathfrak{E}^d$. This shows that $\bigcup_{\varsigma \in \mathcal{B}_\mathfrak{S}} \mathfrak{o}_\varsigma \subseteq \mathfrak{S}\, \mathfrak{E}^*$. Moreover, given $\mathfrak{m} \in \mathfrak{S}\, \mathfrak{E}^*$, there are only a finite number of d with $\mathfrak{m} \in \mathfrak{S}\, \mathfrak{E}^d$. It follows that $\mathcal{B}$ is an atomic family, by remark 6.11 and the fact that each $\mathcal{B}_d$ is atomic.

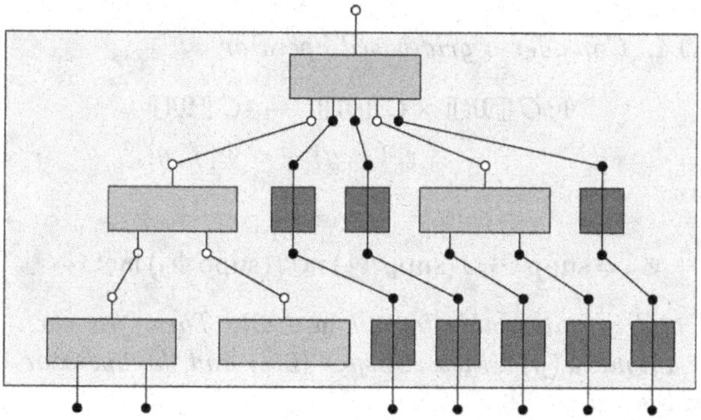

Fig. 6.2. Illustration of a member of $\mathcal{B}_3$. The white dots correspond to elements of $\mathfrak{M}$ and the black dots to elements of $\mathfrak{N}$. The light boxes belong to $\mathcal{A}$ and the dark ones to $\mathcal{D}_\mathfrak{N}$.

Now consider the grid-based operator

$$\Psi = \sum \mathcal{B} \colon C[[\mathfrak{M}]] \times C[[\mathfrak{N}]] \to C[[\mathfrak{M}]].$$

Identifying $C[[\mathfrak{M}]] \times C[[\mathfrak{N}]]$ and $C[[\mathfrak{M} \amalg \mathfrak{N}]]$ via the natural isomorphism, we have

$$(\Psi(g), g) = \Psi(g) + g = (\sum \mathcal{B} \amalg \mathcal{D}_\mathfrak{N})(g),$$

for all $g \in C[[\mathfrak{N}]]$. Similarly, for all $(f, g) \in C[[\mathfrak{M}]] \times C[[\mathfrak{N}]]$, we have

$$\Phi_{\text{rest}}(f, g) = \Phi(f, g) - \Phi(0, g) = (\sum \mathcal{A} \setminus \mathcal{A}_\mathfrak{N})(f, g).$$

Applying proposition 6.12, we conclude that

$$\begin{aligned}
\Psi(g) &= (\sum \mathcal{B}_0)(g) + (\sum \mathcal{B} \setminus \mathcal{B}_0)(g) \\
&= (\sum \mathcal{A}_\mathfrak{N})(g) + (\sum (\mathcal{A} \setminus \mathcal{A}_\mathfrak{N}) \circ (\mathcal{B} \amalg \mathcal{D}_\mathfrak{N}))(g) \\
&= \Phi(0, g) + \Phi_{\text{rest}}(\Psi(g), g) \\
&= \Phi(\Psi(g), g),
\end{aligned}$$

for all $g \in C[[\mathfrak{N}]]$. As to the uniqueness of $\Psi(g)$, assume that $f_1, f_2 \in C[[\mathfrak{N}]]$ are such that $\Phi(f_1, g) = f_1$ and $\Phi(f_2, g) = f_2$. Then we have

$$\Phi(f_2, g) - \Phi(f_1, g) = f_2 - f_1 \prec\!\!\prec f_2 - f_1,$$

which is only possible if $f_2 = f_1$.

Let us finally prove the bounds on the supports. The first one follows directly from (6.9). The second one follows from the fact that the operator support of an element in $\mathcal{B}$ is the product of the operator supports of all combinatorial boxes on the nodes of the corresponding tree. □

6.5.2 The second implicit function theorem

Theorem 6.14. *Consider a grid-based operator*

$$\begin{aligned} \Phi: C[[\mathfrak{M}]] \times C[[\mathfrak{N}]] &\rightarrow C[[\mathfrak{M}]] \\ (f, g) &\mapsto \Phi(f, g), \end{aligned}$$

such that

$$\mathfrak{E}_{\mathfrak{m}} = \operatorname{supp} \Phi_1 \cup (\operatorname{supp} \Phi_2)\, \mathfrak{m} \cup (\operatorname{supp} \Phi_3)\, \mathfrak{m}^2 \cup \cdots$$

is grid-based and infinitesimal for all $\mathfrak{m} \in \mathfrak{M}$. Then, for each $g \in C[[\mathfrak{N}]]$, there exists a unique $\Psi(g)$ which satisfies (6.8) *and the operator $\Psi: C[[\mathfrak{N}]] \rightarrow C[[\mathfrak{M}]]$ is grid-based.*

Proof. Let $g \in C[[\mathfrak{N}]]$, with support $\mathfrak{S} = \operatorname{supp} g$. There exist finite sets $\mathfrak{F}$ and $\mathfrak{D} \prec 1$, such that $\mathfrak{S} \subseteq \mathfrak{F}\mathfrak{D}^*$. Let

$$\mathfrak{E} = \left(\bigcup_{\mathfrak{m} \in \mathfrak{F}} \mathfrak{E}_{\mathfrak{m}} \right)^{+} \mathfrak{D}^*$$

Then we have $\mathfrak{E} \prec 1$ and

$$\mathfrak{E} \supseteq \bigcup_{\mathfrak{m} \in \mathfrak{F}(\mathfrak{D} \cup \mathfrak{E})^*} \mathfrak{E}_{\mathfrak{m}}.$$

We now observe that $\Phi(\cdot, g)$ maps $C[[\mathfrak{F}(\mathfrak{D} \cup \mathfrak{E})^*]]$ into itself, so we may apply theorem 6.13 to this mapping with the same $\mathfrak{E}$. This proves the existence and uniqueness of $\Psi(g)$. With similar notations as in theorem 6.13, it also follows that $\mathcal{B}$ is again a grid-based atomic family, so that $\Psi = \sum \hat{\mathcal{B}}$ is a grid-based operator. □

6.5.3 The third implicit function theorem

Theorem 6.15. *Consider a grid-based operator*

$$\begin{aligned} \Phi: C[[\mathfrak{M}]] \times C[[\mathfrak{N}]] &\rightarrow C[[\mathfrak{M}]] \\ (f, g) &\mapsto \Phi(f, g), \end{aligned}$$

which is strictly extensive in f. Assume that

$$\mathfrak{G} = \operatorname{supp} \Phi_0 \cup \operatorname{supp} \Phi_1 \cup \cdots$$

is grid-based and $\mathfrak{G} \prec 1$. Then for each $g \in C[[\mathfrak{N}]]$, there exists a unique $\Psi(g)$ which satisfies (6.8) *and the operator $\Psi: C[[\mathfrak{N}]] \to C[[\mathfrak{M}]]$ is grid-based.*

Proof. With the notations of the proof of theorem 6.13, let us first show that $\mathcal{B}_{\mathfrak{S}}$ is a well-based family for every grid-based set $\mathfrak{S} \subseteq \mathfrak{N}$. For each $\alpha \in \mathcal{A}$, let $\bar{\alpha} = \mathfrak{o}_\alpha / (\mathfrak{i}_{\alpha,1} \cdots \mathfrak{i}_{\alpha,|\alpha|}) \in \mathfrak{G}$. To each $\beta \in \mathcal{B}_{\mathfrak{S}}$, we associate a tree $\bar{\beta} \in (\mathfrak{S} \amalg \mathfrak{G})^\top$, by setting $\bar{\beta} = \mathfrak{o}_\beta$ if $\beta \in \mathcal{D}_{\mathfrak{N}} \amalg \mathcal{B}_0$, and

$$\overline{\alpha \circ (\beta_1, \ldots, \beta_{|\alpha|})} =$$

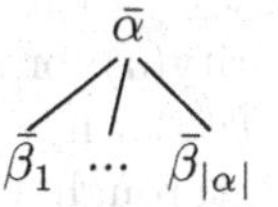

for $\alpha \circ (\beta_1, \ldots, \beta_{|\alpha|}) \in \mathcal{B} \setminus \mathcal{B}_0$. Since Φ is strictly extensive in f, this mapping is strictly increasing. Furthermore, the inverse image of each tree in $(\mathfrak{S} \amalg \mathfrak{G})^\top$ is finite and $(\mathfrak{S} \amalg \mathfrak{G})^\top$ is well-based by Higman's theorem. This together implies that $\mathcal{B}_{\mathfrak{S}}$ is well-based.

Let us show that $\mathcal{B}_{\mathfrak{S}}$ is actually a grid-based. For each tree $\bar{\beta} \in (\mathfrak{S} \amalg \mathfrak{G})^\top$, let $\mathfrak{o}_{\bar{\beta}} = \prod_{a \in \bar{\beta}} l(a)$, so that $\mathfrak{o}_{\bar{\beta}} = \mathfrak{o}_\beta$ for all $\beta \in \mathcal{B}$. Now consider

$$\mathfrak{T} = \{(\bar{\alpha}, (\bar{\beta}_1, \ldots, \bar{\beta}_l)) \in \mathfrak{G} \times (\mathfrak{S} \amalg \mathfrak{G})^{\top *} : \mathfrak{o}_{\bar{\alpha}} \mathfrak{o}_{\bar{\beta}_1} \cdots \mathfrak{o}_{\bar{\beta}_l} \prec 1\}.$$

Let $\mathfrak{F}$ be the finite subset of $\preccurlyeq$-maximal elements of $\mathfrak{T}$. Notice that we may naturally interpret elements

$$(\bar{\alpha}, (\bar{\beta}_1, \ldots, \bar{\beta}_l)) \in \mathfrak{G} \times (\mathfrak{S} \amalg \mathfrak{G})^{\top *}$$

as trees

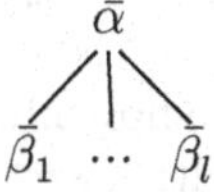

$$\in (\mathfrak{S} \amalg \mathfrak{G})^\top.$$

Given a grid-based set $\mathfrak{A}$ and $\mathfrak{m} \in \mathfrak{A}$, let us denote

$$\operatorname{res}_{\mathfrak{A}} \mathfrak{m} = \left\{\frac{\mathfrak{n}}{\mathfrak{m}} : \mathfrak{n} \in \mathfrak{A}, \mathfrak{n} \prec \mathfrak{m}\right\}.$$

Consider

$$\mathfrak{E} = \left(\mathfrak{G} \cup \{\mathfrak{o}_{\bar{\varsigma}} : \bar{\varsigma} \in \mathfrak{F}\} \cup \bigcup_{\substack{\bar{\varsigma} \in \mathfrak{F} \\ \mathfrak{l} \in l(\bar{\varsigma})}} \operatorname{res}_{\mathfrak{S} \amalg \mathfrak{G}} \mathfrak{l}\right)^* \setminus \{1\}.$$

We claim that $\mathfrak{E}$ satisfies the hypothesis of theorem 6.13.

Indeed, consider $\varsigma = \alpha \circ (\beta_1, \ldots, \beta_{|\alpha|}) \in \mathcal{B}_{\mathfrak{S}} \cap \mathcal{B}_d$ and let us show by induction over d that $\mathfrak{o}_\varsigma \in \mathfrak{i}_{\varsigma,k} \mathfrak{E}$ for every k with $\mathfrak{o}_{\beta_k} \in \mathfrak{M}$. Now

$$\bar{\varsigma} = (\bar{\alpha}, (\bar{\beta}_1, \ldots, \bar{\beta}_{k-1}, \bar{\beta}_{k+1}, \ldots, \bar{\beta}_{|\alpha|})) \preccurlyeq \bar{\varsigma}'$$

for some $\bar{\varsigma}' \in \mathfrak{F}$. In other words, there exists an embedding $\varphi: \bar{\varsigma}' \to \bar{\varsigma}$ which fixes the root. Consider a factorization $\varphi = \psi \circ \varphi'$ of this embedding through a tree $\bar{\varpi}$ with $\mathfrak{o}_{\bar{\varpi}} \in \mathfrak{i}_{\varsigma,k}\, \mathfrak{E}$, such that $a \in \operatorname{im} \varphi'$ for all $a \in \bar{\varpi}$ with $l(\psi(a)) \neq l(a)$, and such that

$$\delta_\psi = \operatorname{card}\{b \in \bar{\varsigma} : \forall a \in \bar{\varpi}, b = \psi(a) \Rightarrow l(b) \neq l(a)\}$$

is minimal. Assume for contradiction that $\delta_\psi \neq 0$. We distinguish three cases:

Case 1. $l(\psi(a)) \neq l(a)$ for some $a \in \bar{\varpi}$.
Consider the tree $\bar{\varpi}'$ with the same nodes as $\bar{\varpi}$ and $l_{\bar{\varpi}'}(b) = l_{\bar{\varpi}}(b)$ if $b \neq a$ and $l_{\bar{\varpi}'}(a) = l_{\bar{\varsigma}}(\psi(a))$. Then we may factor $\psi = \xi \circ \psi'$ through $\bar{\varpi}'$ with $\delta_\xi = \delta_\psi - 1$ and $\mathfrak{o}_{\bar{\varpi}'} \in \mathfrak{o}_{\bar{\varpi}}\, \mathfrak{E} \subseteq \mathfrak{i}_{\varsigma,k}\, \mathfrak{E}$.

Case 2. $\operatorname{arity}(\psi(a)) > \operatorname{arity}(a)$ for some $a \in \bar{\varpi}$.
Let $\bar{\kappa}$ be a child of $\psi(a)$ whose root is not in the image of ψ. Then we may factor $\psi = \xi \circ \psi'$ through a tree $\bar{\varpi}'$ which is obtained by adding $\bar{\kappa}$ as a child to a at the appropriate place, in such a way that $\delta_\xi = \delta_\psi - \operatorname{card} \bar{\kappa}$. Moreover, since $\kappa \in \mathcal{B}_0 \cup \cdots \cup \mathcal{B}_{d-1}$, the induction hypothesis implies that $\mathfrak{o}_{\bar{\kappa}} \in \mathfrak{E}$, so that $\mathfrak{o}_{\bar{\varpi}'} = \mathfrak{o}_{\bar{\varpi}}\, \mathfrak{o}_{\bar{\kappa}} \in \mathfrak{i}_{\varsigma,k}\, \mathfrak{E}$.

Case 3. we are not in cases 1 and 2.
Since $\delta_\psi \neq 0$, there exists a $b \in \bar{\varsigma} \setminus \operatorname{im} \psi$ with a successor $c = \psi(a)$. Let $\bar{\kappa}_1, \ldots, \bar{\kappa}_p$ be the children of b, so that c is the root of $\bar{\kappa}_i$ for some i. Consider the tree $\bar{\varpi}'$ which is obtained by substituting the subtree $\bar{\lambda}$ of $\bar{\varpi}$ with root a by

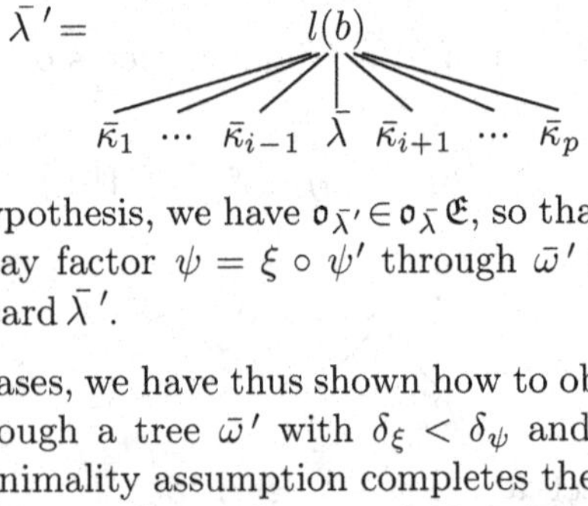

By the induction hypothesis, we have $\mathfrak{o}_{\bar{\lambda}'} \in \mathfrak{o}_{\bar{\lambda}}\, \mathfrak{E}$, so that $\mathfrak{o}_{\bar{\varpi}'} \in \mathfrak{o}_{\bar{\varpi}}\, \mathfrak{E} \subseteq \mathfrak{i}_{\varsigma,k}\, \mathfrak{E}$. Furthermore, we may factor $\psi = \xi \circ \psi'$ through $\bar{\varpi}'$ in such a way that $\delta_\xi = \delta_\psi + \operatorname{card} \bar{\lambda} - \operatorname{card} \bar{\lambda}'$.

In each of these three cases, we have thus shown how to obtain a factorization $\varphi = \xi \circ (\psi' \circ \varphi')$ through a tree $\bar{\varpi}'$ with $\delta_\xi < \delta_\psi$ and $\mathfrak{o}_{\bar{\varpi}'} \in \mathfrak{i}_{\varsigma,k}\, \mathfrak{E}$. This contradiction of the minimality assumption completes the proof of our claim. We conclude the proof by applying theorem 6.13 and by noticing that $\mathcal{B}$ is grid-based, so that $\Psi = \sum \hat{\mathcal{B}}$ is a grid-based operator. □

Exercise 6.14. Give an example of a contracting mapping which is not strictly extensive.

Exercise 6.15. In the first implicit function theorem, show that the condition that f has multipliers in a grid-based set $\mathfrak{E} \prec 1$ cannot be omitted. Hint: consider the equation $f(x) = x + f(\sqrt{x})$.

Exercise 6.16. Give an example where the second implicit function theorem may be applied, but not the first. Also give an example where the third theorem may be applied, but not the second.

Exercise 6.17. Prove the following implicit function theorem for well-based series:

> Let $\Phi: C[[\mathfrak{M}]] \times C[[\mathfrak{N}]] \to C[[\mathfrak{M}]]$; $(f, g) \mapsto F(f, g)$ be a well-based operator which is extensive in f. Then for each $g \in C[[\mathfrak{N}]]$, there exists a unique $\Psi(g)$ which satisfies (6.8) and the operator $\Phi: C[[\mathfrak{N}]] \to C[[\mathfrak{M}]]$ is well-based.

6.6 Multilinear types

One obtains interesting subclasses of grid-based operators by restricting the homogeneous parts to be of a certain type. More precisely, let $\mathfrak{M}$ be a monomial monoid and let $\mathscr{T}$ be a set of strongly multilinear mappings $\Phi: C[\![\mathfrak{M}]\!]^{|\Phi|} \to C[\![\mathfrak{M}]\!]$. We say that $\mathscr{T}$ is a *multilinear type* if

MT1. The constant mapping $\{0\} \mapsto f$ is in $\mathscr{T}$, for each $f \in C[\![\mathfrak{M}]\!]$.

MT2. The projection mapping π_i: $C[\![\mathfrak{M}]\!]^k \to C[\![\mathfrak{M}]\!]$ is in $\mathscr{T}$, for each $i \in \{1, ..., k\}$.

MT3. The multiplication mapping $\cdot: C[\![\mathfrak{M}]\!]^2 \to C[\![\mathfrak{M}]\!]$ is in $\mathscr{T}$.

MT4. If $\Psi, \Phi_1, ..., \Phi_{|\Psi|} \in \mathscr{T}$, then $\Psi \circ (\Phi_1, ..., \Phi_{|\Psi|}) \in \mathscr{T}$.

Given subsets $\mathfrak{V}_1, ..., \mathfrak{V}_v, \mathfrak{W}_1, ..., \mathfrak{W}_w$ of $\mathfrak{M}$, we say that a strongly multilinear mapping

$$\Phi: C[\![\mathfrak{V}_1]\!] \times \cdots \times C[\![\mathfrak{V}_v]\!] \to C[\![\mathfrak{W}_1]\!] \times \cdots \times C[\![\mathfrak{W}_w]\!]$$

is an *atom of type* $\mathscr{T}$, if for $i = 1, ..., w$, there exists a mapping $\Phi_i: C[\![\mathfrak{M}]\!]^v \to C[\![\mathfrak{M}]\!]$ in $\mathscr{T}$, such that $\pi_i \circ \Phi$ coincides with the restriction of the domain and image of Φ_i to $C[\![\mathfrak{V}_1]\!] \times \cdots \times C[\![\mathfrak{V}_v]\!]$ resp. $C[\![\mathfrak{W}_i]\!]$. We say that Φ is *of type* $\mathscr{T}$, if Φ is the sum of a grid-based family of atoms of type $\mathscr{T}$. A grid-based operator

$$\Phi: C[\![\mathfrak{V}_1]\!] \times \cdots \times C[\![\mathfrak{V}_v]\!] \to C[\![\mathfrak{W}_1]\!] \times \cdots \times C[\![\mathfrak{W}_w]\!]$$

is said to be of type $\mathscr{T}$, if $\check{\Phi}_i$ is of type $\mathscr{T}$ for all i.

Example 6.16. For any set $\mathscr{S}$ of grid-based operators $C[\![\mathfrak{M}]\!] \to C[\![\mathfrak{M}]\!]$, there exists a smallest multilinear type $\mathscr{T} = \langle \mathscr{S} \rangle$ which contains $\mathscr{S}$. Taking $\mathbb{T} = C[\![\mathfrak{M}]\!]$ to be the field of grid-based transseries, interesting special cases are obtained when taking $\mathscr{S} = \{\partial\}$ or $\mathscr{S} = \{\int\}$. Grid-based operators of type $\langle\{\partial\}\rangle$ resp. $\langle\{\int\}\rangle$ are called *differential* resp. *integral grid-based operators.*

Exercise 6.18. Show that compositions of grid-based operators of type $\mathscr{T}$ are again of type $\mathscr{T}$.

Exercise 6.19. State and prove the implicit function theorems from the previous section for grid-based operators of a given type $\mathscr{T}$.

Exercise 6.20. For which subfields of $\mathbb{T}$ and $g \in \mathbb{T}^{>,\succ}$ do the grid-based operators of types $\langle\{\circ_g\}\rangle$ and $\langle\{\partial\}\rangle$ coincide?

7

Linear differential equations

Let $L = L_r \partial_r + \cdots + L_0 \in \mathbb{T}[\partial]$ be a linear differential operator with transseries coefficients and $g \in \mathbb{T}$. In this chapter, we study the linear differential equation

$$L f = g. \tag{7.1}$$

In our grid-based context, it is convenient to study the equation (7.1) in the particular case when L_0, ..., L_r and g can be expanded w.r.t. a plane transbasis $\mathfrak{B}$. In order to solve the equation $f^{(r)} = 1$, we necessarily need to consider solutions in $C[x]$. Therefore, we will regard L as an operator on $C[\![x^{\mathbb{N}} \mathfrak{B}^C]\!] = C[x][\![\mathfrak{B}^C]\!]$. Assuming that we understand how to solve (7.1) for $L \in C[\![\mathfrak{B}^C]\!][\partial]$ and $f, g \in C[\![x^{\mathbb{N}} \mathfrak{B}^C]\!]$ and assuming that we understand how this resolution depends on $\mathfrak{B}$ and upward shiftings, the incomplete transbasis theorem will enable us to solve (7.1) in the general case.

A first step towards the resolution of (7.1) is to find candidates for dominant terms of solutions f. It turns out that the dominant monomial of $L f$ only depends on the dominant term of f, except if $\tau_f \in C^{\neq} \mathfrak{H}_L$, where $\mathfrak{H}_L$ is a finite set of "irregular" monomials. The corresponding mapping $T_L: \tau_f \mapsto \tau_{L f}$ is called the trace of L, and its properties will be studied in section 7.3. In particular, we will show that T_L is invertible.

In section 7.4 we will show that the invertibility of the trace implies the existence of a strong right inverse L^{-1} of L. Moreover, the constructed right inverse is uniquely determined by the fact that $(L^{-1} g)_{\mathfrak{h}} = 0$ for all $\mathfrak{h} \in \mathfrak{H}_L$ (for which we call it "distinguished"). Furthermore, we may associate to each $\mathfrak{h} \in \mathfrak{H}_L$ a solution $h^{\mathfrak{h}} = \mathfrak{h} - L^{-1} L \mathfrak{h} \sim \mathfrak{h}$ to the homogeneous equation $L h = 0$ and these solutions form a "distinguished basis" of the space H_L of all solutions.

Now finding all solutions to (7.1) it equivalent to finding one particular solution $f = L^{-1} g$ and the space H_L of solutions to the homogeneous equation. Solving the homogeneous equation $L h = 0$ is equivalent to solving the Riccati equation

$$R_L(f) = 0, \tag{7.2}$$

which is an algebraic differential equation in $f = h^\dagger$ (see section 7.2). In section 7.5, we will show that (7.2) is really a "deformation" of the algebraic equation $L_r f^r + \cdots + L_0 = 0$, so we apply a deformation of the Newton polygon method from chapter 3 to solve it. In fact, we will rather solve the equation "modulo $o(1)$", which corresponds to finding the dominant monomials in $\mathfrak{H}_L$ of solutions to the homogeneous equation (see section 7.6).

Of course, an equation like $f'' + f = 0$ does not admit any non-trivial solutions in the transseries. In order to guarantee that the solution space H_L of the homogeneous equation has dimension r, we need to consider transseries solutions with complex coefficients and oscillating monomials. In section 7.7 we will briefly consider the resolution of (7.1) in this more general context. In section 7.8 we will also show that, as a consequence of the fact that $\dim H_L = r$, we may factor L as a product of linear operators.

7.1 Linear differential operators

7.1.1 Linear differential operators as series

Let $\mathbb{T} = C[[\mathfrak{T}]] = C[[[x]]]$ be the field of grid-based transseries in x over a real-closed exp-log field of constants C. In what follows, it will often be convenient to regard linear differential operators $L = L_r \partial^r + \cdots + L_0 \in \mathbb{T}[\partial]$ as elements of $C[\partial][[\mathfrak{T}]]$. In particular, each non-zero operator L admits a dominant monomial

$$\mathfrak{d}_L = \max_{\preccurlyeq} \{\mathfrak{d}_{L_0}, \ldots, \mathfrak{d}_{L_r}\}$$

and a dominant coefficient

$$c_L = L_{\mathfrak{d}_L} = L_{r,\mathfrak{d}_L} \partial^r + \cdots + L_{0,\mathfrak{d}_L} \in C[\partial],$$

for which we will also use the alternative notation

$$L_* = c_L.$$

Similarly, the asymptotic relations $\preccurlyeq$, $\prec$, $\preccurlyeq^\sharp$, $\prec^\sharp$, etc. extend to $\mathbb{T}[\partial]$. In order to avoid confusion with the support of L as an operator, the support of L as a series will be denoted by $\operatorname{supp}_{\mathrm{ser}} L$.

Proposition 7.1. *Given $K, L \in \mathbb{T}[\partial]^{\neq}$ with $L \asymp 1$, we have*

$$c_{KL} = c_K c_L.$$

Proof. Without loss of generality, one may assume that $K \asymp 1$, modulo division of K by $\mathfrak{d}_K$. Then

$$KL = c_K c_L + \sum_{0 \leqslant i,j} \sum_{0 \leqslant k \leqslant i} \sum_{\substack{\mathfrak{m} \preccurlyeq 1, \mathfrak{n} \preccurlyeq 1 \\ \mathfrak{m} \prec 1 \vee \mathfrak{n} \prec 1}} \binom{k}{i} K_{i,\mathfrak{m}} L_{j,\mathfrak{n}} \mathfrak{m} \mathfrak{n}^{(i-k)} \partial^{k+j}.$$

Now each term in the big sum at the right hand side is infinitesimal. □

7.1.2 Multiplicative conjugation

Given a linear differential operator $L \in \mathbb{T}[\partial]$ and a non-zero transseries h, there exists a unique linear differential operator $L_{\times h}$, such that

$$L_{\times h}(f) = L(h f)$$

for all f. We call $L_{\times h}$ a *multiplicative conjugate* of L. Its coefficients are given by

$$L_{\times h,i} = \sum_{j \geqslant i} \binom{j}{i} L_j h^{(j-i)}. \tag{7.3}$$

Notice that $L_{\times h_1 h_2} = L_{\times h_1, \times h_2}$ for all $h_1, h_2 \in \mathbb{T}^{\neq}$.

Proposition 7.2. *If $h \succ\!\!\succ x$, then*

$$L_{\times h} \asymp_h h L.$$

Proof. From $h \succ\!\!\succ x$ it follows that $h^{(i)} \preccurlyeq_h h$ for all i. Then (7.3) implies $L_{\times h} \preccurlyeq_h h L$. Conversely, we have

$$L = L_{\times h, /h} \preccurlyeq_h h^{-1} L_{\times h}. \qquad \square$$

7.1.3 Upward shifting

In order to reduce the study of a general linear differential equation $L f = g$ over the transseries to the case when the coefficients are exponential, we define the *upward shifting* $L\uparrow$ and *downward shifting* $L\downarrow$ of L to be the unique operators with

$$\begin{aligned} (L\uparrow)(f\uparrow) &= (L f)\uparrow \\ (L\downarrow)(f\downarrow) &= (L f)\downarrow \end{aligned}$$

for all f. In other words, the resolution of $L f = g$ is equivalent to the resolution of $(L\uparrow)(f\uparrow) = g\uparrow$. The coefficients of $L\uparrow$ and $L\downarrow$ are explicitly given by

$$(L\uparrow)_i = \sum_{j \geqslant i} s_{j,i} \mathrm{e}^{-jx} (L_j\uparrow), \tag{7.4}$$

$$(L\downarrow)_i = \sum_{j \geqslant i} S_{j,i} x^i (L_j\downarrow), \tag{7.5}$$

where the $s_{j,i}, S_{j,i} \in \mathbb{Z}$ are Stirling numbers of the first resp. kind, which are determined by

$$\begin{aligned} f(\log x)^{(j)} &= \sum_{i=0}^{j} s_{j,i} x^{-j} f^{(i)}(\log x). \\ (f(\mathrm{e}^x))^{(j)} &= \sum_{i=0}^{j} S_{j,i} \mathrm{e}^{ix} f^{(i)}(\mathrm{e}^x). \end{aligned}$$

Upward and downward shifting are compatible with multiplicative conjugation in the sense that

$$\begin{aligned} L_{\times h}\uparrow &= (L\uparrow)_{\times h\uparrow} \\ L_{\times h}\downarrow &= (L\downarrow)_{\times h\downarrow} \end{aligned}$$

for all $h \in \mathbb{T}^{\neq}$. We will denote by $\uparrow_l$ resp. $\downarrow_l$ the l-th iterates of $\uparrow$ and $\downarrow$.

Exercise 7.1. Let $g \in \mathbb{T}^{>,\succ}$ and $L \in \mathbb{T}[\partial]$.

a) Show that there exists a unique $L_{\circ g} \in \mathbb{T}[\partial]$ with

$$L_{\circ g}(f \circ g) = L(f) \circ g$$

for all $f \in \mathbb{T}$.

b) Give an explicit formula for $L_{\circ g, i}$ for each $i \in \mathbb{N}$.

c) Show that $L \mapsto L_{\circ g}$ is a ring homomorphism.

Exercise 7.2. Let $\varphi \in \mathbb{T}^{\neq}$ and $\tilde{\partial} = \varphi\, \partial$. Denote by ${}^{\varphi}\mathbb{T}$ the field $\mathbb{T}$ with differentiation $\tilde{\partial}$.

a) Show that each $L \in ({}^{\varphi}\mathbb{T})[\tilde{\partial}]$ can be reinterpreted as an operator $L^{\varphi} \in \mathbb{T}[\partial]$.

b) Given $L \in \mathbb{T}[\partial]$, let $\sigma_{\varphi}(L) \in ({}^{\varphi}\mathbb{T})[\tilde{\partial}]$ be the result of the substitution of $\tilde{\partial}$ for ∂ in L. If $\int \varphi^{-1} \in \mathbb{T}^{>,\succ}$, then show that $\sigma_{\varphi}(L)^{\varphi} = L_{\circ \int \varphi^{-1}}$.

Exercise 7.3. Let $g \in \mathbb{T}^{>,\succ}$ and $\varphi = 1/g'$, so that $(\mathbb{T}, \partial) \cong (\mathbb{T} \circ g, \varphi\partial); f \mapsto f \circ g$.

a) Given $L \in \mathscr{D}_{\mathbb{T}_{\prec\!\!\prec e^x}}$ (see exercise 6.3), let $\sigma_{\varphi}(L) = \sum_{n \in \mathbb{N}} L_n (\varphi\, \partial)^n$. Show that $\sigma_{\varphi}(L)$ naturally operates on $\mathbb{T}_{\prec\!\!\prec e^g}$. Also show that the space $\mathscr{D}_{\mathbb{T}_{\prec\!\!\prec e^g}}$ of all such operators only depends on $\mathfrak{d}_g$.

b) Same question, but for $L \in \mathscr{D}_{\mathbb{T}_{\preceq\!\!\prec e^x}}$.

c) Under which condition on g can the operator $\tilde{L} = \sigma_{\varphi}(L)$ in either of the above questions be rewritten as an operator of the form $\sum_{n \in \mathbb{N}} \tilde{L}_n\, \partial^n$?

Exercise 7.4. Let $\mathbb{T}^{\flat} = C[[\mathfrak{T}^{\flat}]] \notin \{C, \mathbb{T}\}$ be a flat subspace of $\mathbb{T}$.

a) Extend the definition of $\mathscr{D}_{\mathbb{T}^{\flat}}$ in exercises 6.3 and 7.3 to the present case.

b) Let $\mathbb{T}^{\flat_1} \subseteq \mathbb{T}^{\flat_2}$ be two flat subspaces of $\mathbb{T}$ of the above type. Characterize $\mathscr{D}_{\mathbb{T}^{\flat_1}} \cap \mathscr{D}_{\mathbb{T}^{\flat_2}}$.

Exercise 7.5. Let $g \in \mathbb{T}^{>,\succ}$.

a) Determine $\varphi \in \mathbb{T}$ so that $\circ_g = e^{\varphi\partial}$.

b) Given $\lambda \in C$, construct the λ-th iterate $g^{\circ\lambda}$ of g.

c) Determine the maximal flat subspace $\mathbb{T}^{\flat} = C[[\mathfrak{T}^{\flat}]]$ of $\mathbb{T}$ such that $\circ_g \in \mathscr{D}_{\mathbb{T}^{\flat}}$.

Exercise 7.6. Let $g_1, \ldots, g_k \in x + \mathbb{T}_{\prec\!\!\prec e^x, \preccurlyeq}$. Consider an operator

$$L = \sum_{i=1}^{k} \sum_{j=0}^{r_i - 1} A_{i,j} \circ_{g_i} \partial^j,$$

where $A_{i,j} \in \mathbb{T}_{\prec\!\!\prec e^x}$.

a) Show that $L \in \mathscr{D}_{\mathbb{T}_{\prec\!\!\prec e^x}}$ and let $L_0, L_1, \ldots$ be such that $L = \sum_{n \in \mathbb{N}} L_n\, \partial^n$.

b) Assuming that $L \neq 0$, show that there exists a $\nu < r_1 + \cdots + r_k$ with $\mathfrak{d}_{L_\nu} = \max_{n \in \mathbb{N}} \mathfrak{d}_{L_n}$.

Exercise 7.7. Let $\mathbb{T}^\flat$ be as in exercise 7.3(*a*) or (*b*), and $\lambda \in C$.

a) Given $E = \sum_{n \in \mathbb{N}} E_n (\varphi \partial)^n \in \mathscr{D}_{\mathbb{T}^\flat}$ with $\mathfrak{d}_E = \max_{\preccurlyeq} \mathfrak{d}_{E_n} \preccurlyeq 1$ and $E_0 \prec 1$. Show that

$$\begin{aligned}
\log(1+E) &= \sum_{n \geqslant 1} \frac{(-1)^{n+1}}{n} E^n \in \mathscr{D}_{\mathbb{T}^\flat} \\
\exp E &= \sum_{n \geqslant 0} \frac{1}{n!} E^n \in \mathscr{D}_{\mathbb{T}^\flat} \\
(1+E)^\lambda &= \sum_{n \in \mathbb{N}} \binom{\lambda}{n} E^n = \exp(\lambda \log(1+E)) \in \mathscr{D}_{\mathbb{T}^\flat}
\end{aligned}$$

are well-defined.

b) Let $\varphi \in \mathbb{T}^{\flat, \succcurlyeq, >}$, $\ell = \log \varphi$, $K = \varphi + E$ and $L = \log(1 + \varphi^{-1} E)$. Show that

$$\begin{aligned}
\log(K) &= \ell + L + \frac{1}{2}[\ell, L] + \frac{1}{12}[\ell, [\ell, L]] + \frac{1}{12}[L, [L, \ell]] + \cdots \\
K^\lambda &= \exp(\lambda \log(K))
\end{aligned}$$

are well-defined.

c) Given a transmonomial $\mathfrak{m} \in \mathfrak{T}$ with $\mathfrak{m} \succ 1$ and $\mathfrak{m} \succ\!\!\succ x$, show that

$$\partial^\lambda(\mathfrak{m}) = \mathfrak{m}\,(\mathfrak{m}^{-1} \partial\, \mathfrak{m})^\lambda(1)$$

is well-defined. Extend the definition of ∂^λ to $\mathbb{T}_{\succ, \succ\!\!\succ x}$ and show that ∂^{-1} corresponds to the distinguished integration.

7.2 Differential Riccati polynomials

7.2.1 The differential Riccati polynomial

Given a transseries $f \in \mathbb{T}$, we may rewrite the successive derivatives of $f^{(i)}$ as

$$f^{(i)} = U_i(f^\dagger)\, f, \tag{7.6}$$

where the $U_i \in \mathbb{Z}\{F\}$ are universal differential polynomials given by

$$\begin{aligned}
U_0 &= 1 \\
U_{i+1} &= F U_i + U_i'.
\end{aligned}$$

For instance:

$$\begin{aligned}
U_0 &= 1 \\
U_1 &= F \\
U_2 &= F^2 + F' \\
U_3 &= F^3 + 3 F F' + F'' \\
U_4 &= F^4 + 6 F^2 F' + 4 F F'' + 3 (F')^2 + F''' \\
&\ \vdots
\end{aligned}$$

In particular, for each linear differential operator $L = L_r \partial^r + \cdots + L_0 \in \mathbb{T}[\partial]$, there exists a unique differential polynomial $R_L = L_r U_r + \cdots + L_0 U_0 \in \mathbb{T}\{F\}$ such that

$$L(f) = R_L(f^\dagger)\, f \tag{7.7}$$

for all $f \in \mathbb{T}$. We call R_L the *differential Riccati polynomial* associated to L. Notice that R_L is uniquely determined by the polynomial

$$R_{L,\mathrm{alg}} = L_r F^r + \cdots + L_0 \in \mathbb{T}[F],$$

which is called the *algebraic part* of R_L.

7.2.2 Properties of differential Riccati polynomials

Let $P \in \mathbb{T}\{F\}$ be a differential polynomial with transseries coefficients. Like in the case of differential operators, we may consider P as a series in $C\{F\}[[\mathfrak{T}]]$, where $\mathfrak{T}$ denotes the set of transmonomials. Given $\varphi \in \mathbb{T}$ we also define $P_{+\varphi}$ to be the unique differential polynomial in $\mathbb{T}\{F\}$, such that

$$P_{+\varphi}(f) = P(\varphi + f)$$

for all $f \in \mathbb{T}$. We call $P_{+\varphi}$ an *additive conjugate* of P. Additive conjugates of the differential Riccati polynomials correspond to multiplicative conjugates of the corresponding linear differential operators:

Proposition 7.3. *For all L and $\varphi \in \mathbb{T}^{\neq}$, we have*

$$R_{L,+\varphi^\dagger} = R_{\varphi^{-1} L_{\times\varphi}}. \tag{7.8}$$

Proof. For all $f \in \mathbb{T}$, we have

$$(\varphi^{-1} L_{\times\varphi})(f) = \varphi^{-1} L(\varphi f) = \varphi^{-1} R_L(f^\dagger + \varphi^\dagger)\, \varphi f = R_{L,+\varphi^\dagger}(f),$$

so (7.8) follows from the uniqueness property of differential Riccati polynomials. □

Given a linear differential operator $L = L_r \partial^r + \cdots + L_0 \in \mathbb{T}[\partial]$, we call

$$L' = r\, L_r \partial^{r-1} + \cdots + L_1 \in \mathbb{T}[\partial]$$

the *derivative* of L.

Proposition 7.4. *For all $L \in \mathbb{T}[\partial]$, we have*

$$R_{L'} = \frac{\partial R_L}{\partial F}; \tag{7.9}$$

$$R_{L',\mathrm{alg}} = R'_{L,\mathrm{alg}}. \tag{7.10}$$

Proof. We claim that $\frac{\partial U_i}{\partial F} = i\, U_{i-1}$ for all $i \geqslant 1$. Indeed, $\frac{\partial U_1}{\partial F} = 1$ and, using induction,

$$\begin{aligned}
\frac{\partial U_{i+1}}{\partial F} &= U_i + F \frac{\partial U_i}{\partial F} + \frac{\partial^2 U_i}{\partial F \partial F} F' + \cdots + \frac{\partial^2 U_i}{\partial F^{(i-1)} \partial F} F^{(i)} \\
&= U_i + i\, F U_{i-1} + i \frac{\partial U_{i-1}}{\partial F} F' + \cdots + i \frac{\partial U_{i-1}}{\partial F^{(i-1)}} F^{(i)} \\
&= U_i + i\, F U_{i-1} + i\, U'_{i-1} \\
&= (i+1)\, U_i
\end{aligned}$$

for all $i \geqslant 2$. Our claim immediately implies (7.9) and (7.10). □

Corollary 7.5. *For all* $L = L_r \partial^r + \cdots + L_0 \in \mathbb{T}[\partial]$ *and* $\varphi \in \mathbb{T}$, *we have*

$$R_{L,+\varphi} = \frac{1}{r!} R_{L^{(r)}}(\varphi)\, U_r + \cdots + R_L(\varphi)\, U_0; \qquad (7.11)$$

$$R_{L,+\varphi,\mathrm{alg}} = \frac{1}{r!} R_{L^{(r)}}(\varphi)\, F^r + \cdots + R_L(\varphi). \qquad (7.12)$$

Exercise 7.8. Prove that

$$U_n(F+G) = \sum_{i=0}^{n} \binom{n}{i} U_i(F)\, U_{n-i}(G).$$

Exercise 7.9. Show that

$$R_{L\uparrow} = R_L \uparrow_{\times \mathrm{e}^{-x}};$$

$$R_{L\downarrow} = R_L \downarrow_{\times x},$$

where $R_L \uparrow$ and $R_L \downarrow$ are defined in section 8.2.3.

7.3 The trace of a linear differential operator

Let $L: C[[\mathfrak{M}]] \to C[[\mathfrak{N}]]$ be a linear grid-based operator. A term $\tau = c\,\mathfrak{m} \in C^{\neq}\,\mathfrak{M}$ is said to be *regular for* L, if Lf is regular for all $f \in C[[\mathfrak{M}]]$ with $\tau(f) = \tau$ and if $\tau(L\,f)$ does not depend on the choice of such an f. In particular, a monomial in $\mathfrak{M}$ is said to be regular for L if it is regular as a term. We will denote by $\mathfrak{R}_L \subseteq \mathfrak{M}$ the set of all regular monomials for L and by $\mathfrak{H}_L \subseteq \mathfrak{M}$ the set of *irregular* monomials. The mapping

$$T_L: C^{\neq}\,\mathfrak{R}_L \longrightarrow C^{\neq}\,\mathfrak{N}$$
$$\tau \longmapsto \tau(L\tau)$$

is called the *trace* of L. For all $\tau_1, \tau_2 \in C^{\neq}\,\mathfrak{R}_L$, we have

$$\tau_1 \preccurlyeq \tau_2 \Longrightarrow T_L(\tau_1) \preccurlyeq T_L(\tau_2). \qquad (7.13)$$

Given a linear differential equation $Lf = g$ over the transseries $\mathbb{T}$ with $g \neq 0$, finding a term τ with $T_L(\tau) = \tau_g$ corresponds to finding a good candidate for the first term of a solution. In the next section we will show that this first term may indeed be completed into a full solution.

7.3.1 The trace relative to plane transbases

Let $L \in C[[\mathfrak{B}^C]][\partial]$ be a linear differential operator, where $\mathfrak{B} = (\mathfrak{b}_1, \ldots, \mathfrak{b}_n)$ is a plane transbasis. We will consider L as a grid-based operator on $C[[x^{\mathbb{N}}\,\mathfrak{B}^C]]$, so that its trace $T_L = T_{L;\mathfrak{B}}$ is a mapping from $x^{\mathbb{N}}\,\mathfrak{B}^C \setminus \mathfrak{H}_L$ into $x^{\mathbb{N}}\,\mathfrak{B}^C$.

Proposition 7.6. *Given $x^i\mathfrak{m}\in x^{\mathrm{N}}\mathfrak{B}^C$, we have*

$$x^i\mathfrak{m}\in\mathfrak{H}_L \Longleftrightarrow L_{\times\mathfrak{m},*}(x^i)=0.$$

Proof. Modulo replacing L by $\mathfrak{d}(L_{\times\mathfrak{m}})^{-1}L_{\times\mathfrak{m}}$, we may assume without loss of generality that $\mathfrak{m}=1$ and $L\asymp 1$. Let j be minimal with $L_{*,j}\neq 0$, so that $L_*(x^i)=0$ if and only if $i<j$.

Now $i<j$ implies $L(x^i)=(L-L_*)(x^i)\prec_{\mathrm{e}^x}1$. Furthermore, $L_{\times\mathrm{e}^{-\alpha x}}(1)\asymp\mathrm{e}^{-\alpha x}$ for all but the finite number of α such that $L_*(\mathrm{e}^{-\alpha x})=0$. It follows that $L(x^i)\prec L(\mathrm{e}^{-\alpha x})$ for a sufficiently small $\alpha>0$, whence $x^i\in\mathfrak{H}_L$.

If $i\geqslant j$, then $L_*(x^i)\asymp x^{i-j}$. Given $\mathfrak{n}\in x^{\mathrm{N}}\mathfrak{B}^C$ with $\mathfrak{n}\prec x^i$, we have either $\mathfrak{n}\prec_{\mathrm{e}^x}1$ or $\mathfrak{n}=x^k$ with $k<i$. In the first case, $L(\mathfrak{n})=L_{\times\mathfrak{n}}(1)\preccurlyeq L_{\times\mathfrak{n}}\prec_{\mathrm{e}^x}1$. In the second case, we have either $k<j$ and $L(\mathfrak{n})\prec_{\mathrm{e}^x}1$ or $k\geqslant j$ and $L(\mathfrak{n})\asymp x^{k-j}\prec x^{i-j}$. So we always have $L(\mathfrak{n})\prec x^{i-j}$. Hence $x^i\in\mathfrak{H}_L$, by strong linearity. □

Proposition 7.7. *For every $\mathfrak{m}\in\mathfrak{B}^C$ there exists a unique $\mathfrak{n}\in\mathfrak{B}^C$ with $L_{\times\mathfrak{n}}\asymp\mathfrak{m}$.*

Proof. Let $\mathfrak{m}\in\mathfrak{B}^C$ and consider $\mathfrak{v}=\mathfrak{m}/\mathfrak{d}_L=\mathfrak{b}_1^{\alpha_1}\cdots\mathfrak{b}_n^{\alpha_n}$. We will prove the proposition by induction over the maximal i such that $\alpha_i\neq 0$. If such an i does not exist, then we have nothing to prove. Otherwise, proposition 7.2 implies

$$\tilde{\mathfrak{v}}:=\frac{\mathfrak{m}}{\mathfrak{d}(L_{\times\mathfrak{b}_i^{\alpha_i}})}\asymp_{\mathfrak{b}_i}\frac{\mathfrak{m}}{\mathfrak{b}_i^{\alpha_i}\mathfrak{d}(L)}=\mathfrak{b}_1^{\alpha_1}\cdots\mathfrak{b}_{i-1}^{\alpha_{i-1}}.$$

It follows that $\tilde{\mathfrak{v}}=\mathfrak{b}_1^{\tilde{\alpha}_1}\cdots\mathfrak{b}_{i-1}^{\tilde{\alpha}_{i-1}}$ for certain $\tilde{\alpha}_1,\ldots,\tilde{\alpha}_{i-1}$. By the induction hypothesis, there exists an $\tilde{\mathfrak{n}}$ with $L_{\times\mathfrak{b}_i^{\alpha_i},\times\tilde{\mathfrak{n}}}\asymp\mathfrak{m}$. Hence $L_{\times\mathfrak{n}}\asymp\mathfrak{m}$ for $\mathfrak{n}=\tilde{\mathfrak{n}}\,\mathfrak{b}_i^{\alpha_i}$. Furthermore, given $\mathfrak{e}\in\mathfrak{B}^C\setminus\{1\}$, we have $L_{\times\mathfrak{n}\mathfrak{e}}\asymp_{\mathfrak{e}}\mathfrak{m}\,\mathfrak{e}\not\asymp_{\mathfrak{e}}\mathfrak{m}$. This proves the uniqueness of $\mathfrak{n}$. □

Proposition 7.8. *The trace T_L of L is invertible.*

Proof. Let $\tau=c\,x^i\,\tilde{\mathfrak{m}}\in x^{\mathrm{N}}\mathfrak{B}^C$. By the previous proposition, there exists a unique $\tilde{\mathfrak{n}}$ with $L_{\times\tilde{\mathfrak{n}}}\asymp\tilde{\mathfrak{m}}$. Modulo the replacement of L by $\tilde{\mathfrak{m}}^{-1}L_{\times\tilde{\mathfrak{n}}}$ we may assume without loss of generality that $\tilde{\mathfrak{m}}=\tilde{\mathfrak{n}}=1$. Let j be minimal with $L_{*,j}\neq 0$. Then

$$L(x^{i+j})=\sum_{k\geqslant j}L_{*,k}\frac{\partial^k x^{i+j}}{\partial x^k}+o_{\mathrm{e}^x}(1)=\frac{(i+j)!}{i!}L_{*,j}x^i+o(x^i).$$

Setting

$$\upsilon=\frac{c\,i!}{(i+j)!\,L_{*,j}}x^{i+j},$$

we thus have $T_L(\upsilon)=\tau$. Notice that proposition 7.6 implies $x^{i+j}\notin\mathfrak{H}_L$. □

Example 7.9. Let $\mathfrak{B}=(\mathrm{e}^x,\mathrm{e}^{\mathrm{e}^x})$ and consider the operator

$$L=\mathrm{e}^{-2x}\partial^3-2\,\mathrm{e}^{-x}\partial^2+\partial+1.$$

Given $\mathfrak{m} = \mathrm{e}^{a\mathrm{e}^x + bx}$ and $K = \mathfrak{m}^{-1} L_{\times \mathfrak{m}}$, we have

$$\begin{aligned} K \;=\; & \mathrm{e}^{-2x}\,\partial^3 + \\ & \left((3a-2)\,\mathrm{e}^{-x} + 3b\,\mathrm{e}^{-2x}\right)\partial^2 + \\ & \left(1 - 4a + 3a^2 + (6ab - 4b + 3a)\,\mathrm{e}^{-x} + 3b^2\,\mathrm{e}^{-2x}\right)\partial + \\ & (a^3 - 2a^2 + a)\,\mathrm{e}^x + 3a^2 b - 4ab + b + 3a^2 - 2a + 1 + \\ & (3ab^2 - 2b^2 + 3ab + a)\,\mathrm{e}^{-x} + b^3\,\mathrm{e}^{-2x}. \end{aligned}$$

Now the following cases may occur:

Case	$\mathfrak{d}_K$	c_K	$x^i\,\mathfrak{m} \in \mathfrak{H}_L$	$T_L(x^i\,\mathfrak{m})$
$\alpha \notin \{0,1\}$	e^x	$a\,(a-1)^2$	no	$a\,(a-1)^2\,\mathrm{e}^x\,x^i\,\mathfrak{m}$
$\alpha = 0, \beta \neq -1$	1	$b+1+\partial$	no	$(b+1)\,x^i\,\mathfrak{m}$
$\alpha = 0, \beta = -1$	1	∂	iff $i=0$	$i\,x^{i-1}\,\mathfrak{m}$ (if $i \neq 0$)
$\alpha = 1$	1	2	no	$2\,x^i\,\mathfrak{m}$

7.3.2 Dependence of the trace on the transbasis

Let $L \in C[[\mathfrak{B}^C]][\partial]$, where $\mathfrak{B}$ is a plane transbasis and let us study the dependence of the trace $T_{L;\mathfrak{B}} = T_L$ of L on $\mathfrak{B}$. Given a plane supertransbasis $\hat{\mathfrak{B}}$ of $\mathfrak{B}$, proposition 7.6 implies that $\mathfrak{H}_{L;\hat{\mathfrak{B}}} \cap x^{\mathbb{N}}\,\mathfrak{B}^C = \mathfrak{H}_{L;\mathfrak{B}}$ and $T_{L;\hat{\mathfrak{B}}}$ clearly coincides with $T_{L;\mathfrak{B}}$ on $C^{\neq}\,x^{\mathbb{N}}\,\mathfrak{B}^C \setminus \mathfrak{H}_{L;\mathfrak{B}}$. Similarly, if $\hat{\mathfrak{B}}$ is a second transbasis such that $C[[x^{\mathbb{N}}\,\hat{\mathfrak{B}}^C]]$ and $C[[x^{\mathbb{N}}\,\mathfrak{B}^C]]$ coincide as subsets of $\mathbb{T}$, then $\mathfrak{H}_{L;\hat{\mathfrak{B}}} = (\mathfrak{d} \circ T_I)(\mathfrak{H}_{L;\mathfrak{B}})$ and $T_{L;\hat{\mathfrak{B}}} \circ T_I = T_I \circ T_{L;\mathfrak{B}}$, where $I\colon C[[x^{\mathbb{N}}\,\hat{\mathfrak{B}}^C]] \to C[[x^{\mathbb{N}}\,\mathfrak{B}^C]]$ denotes the "identification mapping".

Proposition 7.10. *Let* $\hat{\mathfrak{B}} = (\mathrm{e}^x, \mathfrak{b}_1\uparrow, \ldots, \mathfrak{b}_n\uparrow)$. *Then* $\mathfrak{H}_{L\uparrow;\hat{\mathfrak{B}}} = \mathfrak{H}_{L,\mathfrak{B}}\uparrow$ *and* $T_{L\uparrow;\hat{\mathfrak{B}}}(\mathfrak{v}\uparrow) = T_{L;\mathfrak{B}}(\mathfrak{v})\uparrow$ *for all* $\mathfrak{v} \in C^{\neq}\,(x^{\mathbb{N}}\,\mathfrak{B}^C \setminus \mathfrak{H}_{L;\mathfrak{B}})$.

Proof. We clearly have

$$T_{L\uparrow;\hat{\mathfrak{B}}}(\mathfrak{v}\uparrow) = \tau(L\uparrow(\mathfrak{v}\uparrow)) = \tau(L(\mathfrak{v})\uparrow) = \tau(L(\mathfrak{v}))\uparrow = T_{L;\mathfrak{B}}(\mathfrak{v})\uparrow$$

for all $\mathfrak{v} \in C^{\neq}\,(x^{\mathbb{N}}\,\mathfrak{B}^C \setminus \mathfrak{H}_{L;\mathfrak{B}})$. Given

$$\mathfrak{n} = (\log x)^j\,x^i\,\mathfrak{m} \in (\log x)^{\mathbb{N}}\,x^C\,\mathfrak{B}^C,$$

let us show that $\mathfrak{n} \in \mathfrak{H}_{L;\mathfrak{B}} \Leftrightarrow \mathfrak{n}\uparrow \in \mathfrak{H}_{L\uparrow;\hat{\mathfrak{B}}}$. Modulo replacing L by $\mathfrak{d}(L_{\times\mathfrak{m}})^{-1}L_{\times\mathfrak{m}}$, we may assume without loss of generality that $\mathfrak{m} = 1$ and $L \asymp 1$.

Assume that $\mathfrak{n} \in \mathfrak{H}_{L;\mathfrak{B}}$, so that $j = 0$, $i \in \mathbb{N}$ and $L_*(x^i) = 0$. Then $L = L_* + o_{\mathrm{e}^x}(1)$ implies $L\uparrow = L_*\uparrow + o_{\mathrm{e}^{\mathrm{e}^x}}(1)$ and $L\uparrow_{\times \mathrm{e}^{ix}} = L_*\uparrow_{\times \mathrm{e}^{ix}} + o_{\mathrm{e}^{\mathrm{e}^x}}(1)$. Hence $L\uparrow_{\times \mathrm{e}^{ix},*} = L_*\uparrow_{\times \mathrm{e}^{ix},*}$. Since $L_*(x^i) = 0$, we also observe that $L_{\times x^i,0,*} = 0$, whence $L\uparrow_{\times \mathrm{e}^{ix},0,*} = 0$. But this means in particular that

$$L\uparrow_{\times \mathrm{e}^{ix},*} = L_*\uparrow_{\times \mathrm{e}^{ix},*}(1) = L_*\uparrow_{\times \mathrm{e}^{ix},\mathfrak{d}(L_*\uparrow_{\times \mathrm{e}^{ix}}),0} = 0.$$

In other words, $\mathfrak{n}\uparrow \in \mathfrak{H}_{L\uparrow;\hat{\mathfrak{B}}}$.

Assume now that $\mathfrak{n} \notin \mathfrak{H}_{L;\mathfrak{B}}$ and let k be minimal with $L_{*,k} \neq 0$. Then $L_*(\mathfrak{n}) \asymp \mathfrak{n}^{(k)} \asymp_x x^{i-k} \neq 0$ so

$$L_* \uparrow_{\times e^{ix}}(x^j) \asymp \mathfrak{n}^{(k)} \uparrow \asymp_{e^x} e^{(i-k)x}.$$

On the other hand, $L_* \uparrow \asymp e^{-kx}$, whence $L_* \uparrow_{\times e^{ix}} \preccurlyeq e^{(i-k)x}$. This is only possible if $L_* \uparrow_{\times e^{ix}} \asymp e^{(i-k)x}$ and $L_* \uparrow_{\times e^{ix},*}(x^j) \neq 0$. In other words, $\mathfrak{n} \uparrow \notin \mathfrak{H}_{L\uparrow;\hat{\mathfrak{B}}}$. □

Proposition 7.11. *Let $\mathfrak{B}$ be a transbasis of level $1-l \leqslant 1$ containing $\log_{l-1} x$, ..., x and denote $\mathfrak{S}_{\mathfrak{B}} = (\log_l x)^{\mathbb{N}} \mathfrak{B}^C$. Let $L \in C[[\mathfrak{B}^C]]$ and let $\mathfrak{H}_{L;\mathfrak{B}}$ be the set of singular monomials of L as an operator on $C[[\mathfrak{S}_{\mathfrak{B}}]]$. Then*

$$\mathfrak{H}_{L;\mathfrak{B}} = \mathfrak{H}_L \cap \mathfrak{S}_{\mathfrak{B}}.$$

Proof. Clearly, $\mathfrak{H}_{L;\mathfrak{B}} \subseteq \mathfrak{H}_L \cap \mathfrak{S}_{\mathfrak{B}}$. Assume for contradiction that there exists an $\mathfrak{m} \in (\mathfrak{H}_L \cap \mathfrak{S}_{\mathfrak{B}}) \setminus \mathfrak{H}_{L;\mathfrak{B}}$. Then there exists an $\mathfrak{n} \in \mathfrak{T}$ with $\mathfrak{n} \prec \mathfrak{m}$ and $L\mathfrak{m} \preccurlyeq L\mathfrak{n}$. Let $\mathfrak{B}'$ be a super-transbasis of $\mathfrak{B}$ for $\mathfrak{n}$, of level $1 - l'$, and which contains $\log_{l-1} x, ..., x$. Setting $\hat{\mathfrak{B}} = \{\log_{l'-1} x, ..., \log_l x\}$, proposition 7.10 now implies

$$\mathfrak{H}_{L\uparrow_{l'};\mathfrak{B}'\uparrow_{l'}} \cap \mathfrak{S}_{\mathfrak{B}\uparrow_{l'}} = \mathfrak{H}_{L\uparrow_{l'};\hat{\mathfrak{B}}\uparrow_{l'}} \cap \mathfrak{S}_{\mathfrak{B}\uparrow_{l'}} = \mathfrak{H}_{L\uparrow_l;\mathfrak{B}\uparrow_l} \uparrow_{l'-l} = \mathfrak{H}_{L;\mathfrak{B}} \uparrow_{l'}.$$

Hence, $\mathfrak{m}\uparrow_{l'} \notin \mathfrak{H}_{L\uparrow_{l'};\mathfrak{B}'\uparrow_{l'}}$ so that $(L\mathfrak{m})\uparrow_{l'} = L\uparrow_{l'}(\mathfrak{m}\uparrow_{l'}) \succ L\uparrow_{l'}(\mathfrak{n}\uparrow_{l'}) = (L\mathfrak{n})\uparrow_{l'}$. This contradiction completes the proof. □

Proposition 7.12. *Let $L \in \mathbb{T}[\partial]^{\neq}$ be a linear differential operator on $\mathbb{T}$. Then the trace T_L of L is invertible.*

Proof. Given $\tau \in C^{\neq} \mathfrak{T}$, the incomplete transbasis theorem implies that there exists a transbasis $\mathfrak{B}$ for τ like in proposition 7.11. By proposition 7.8, there exists an $\upsilon\uparrow_l \in x^{\mathbb{N}} \mathfrak{B}\uparrow_l^C \setminus \mathfrak{H}_{L\uparrow_l;\mathfrak{B}\uparrow_l}$ with $T_{L\uparrow_l}(\upsilon\uparrow_l) = \tau\uparrow_l$. By proposition 7.11, we have $\upsilon \in C^{\neq} \mathfrak{R}_L$ and $T_L(\upsilon) = T_{L\uparrow_l}(\upsilon\uparrow_l)\downarrow_l = \tau$. □

7.3.3 Remarkable properties of the trace

Assume again that $L \in C[[\mathfrak{B}^C]][\partial]$, where $\mathfrak{B}$ is a plane transbasis.

Proposition 7.13. *The set*

$$\mathfrak{F} = \bigcup_{\mathfrak{m} \in \mathfrak{B}^C} \frac{\mathfrak{d}(L_{\times\mathfrak{m}})}{\mathfrak{m}}$$

is finite.

Proof. Considering $\lambda_1, ..., \lambda_n$ as indeterminates, the successive derivatives of $\mathfrak{m} = \mathfrak{b}_1^{\lambda_1} \cdots \mathfrak{b}_n^{\lambda_n}$ satisfy

$$\mathfrak{m}^{(i)} / \mathfrak{m} = U_i(\lambda_1 \mathfrak{b}_1^\dagger + \cdots + \lambda_n \mathfrak{b}_n^\dagger) \in C[\lambda_1, ..., \lambda_n][[\mathfrak{B}^C]],$$

where the U_i are as in (7.6). Consequently, we may see

$$\tilde{L} = \frac{L_{\times\mathfrak{m}}}{\mathfrak{m}} = \sum_{i=0}^{r} \sum_{j=i}^{r} \binom{j}{i} L_j U_{j-i}(\mathfrak{m}^\dagger) \partial^i$$

as an element of $C[\lambda_1, ..., \lambda_n][[\mathfrak{B}^C]][\partial]$ for each i.

Assume for contradiction that $\mathfrak{F}$ is infinite. Since $\mathfrak{F} \subseteq \operatorname{supp}_{\text{ser}} \tilde{L}$, there exists an infinite sequence $\mathfrak{v}_1 \succ \mathfrak{v}_2 \succ \cdots$ of elements in $\mathfrak{F}$. For each $\mathfrak{v}_i$, let $\mathfrak{n}_i = \mathfrak{b}_1^{\alpha_{i,1}} \cdots \mathfrak{b}_n^{\alpha_{n,i}}$ be such that $\mathfrak{v}_i = \mathfrak{d}(L_{\times \mathfrak{n}_i})/\mathfrak{n}_i$. Now each $\mathfrak{v}_i$ induces an ideal I_i of $C[\lambda_1, ..., \lambda_n]$, generated by all coefficients of $\tilde{L}_{\mathfrak{w}}$ with $\mathfrak{w} \succ \mathfrak{v}_i$. We have $Z_1 \subseteq Z_2 \subseteq \cdots$ and each $(\alpha_{i+1,1}, ..., \alpha_{i+1,n})$ is a zero of Z_i, but not of Z_{i+1}. It follows that $Z_1 \subsetneq Z_2 \subsetneq \cdots$, which contradicts the Noetherianity of $C[\lambda_1, ..., \lambda_n]$. □

Corollary 7.14. *There exist unique strongly linear mappings*

$$\Delta: C[[x^{\mathbb{N}} \mathfrak{B}^C \setminus \mathfrak{H}_L]] \longrightarrow C[[x^{\mathbb{N}} \mathfrak{B}^C]]$$
$$\Delta^{-1}: C[[x^{\mathbb{N}} \mathfrak{B}^C]] \longrightarrow C[[x^{\mathbb{N}} \mathfrak{B}^C \setminus \mathfrak{H}_L]]$$

which extend T_L and T_L^{-1}. Furthermore,

a) $\operatorname{supp} \Delta \subseteq \{1, ..., x^{-r}\} \mathfrak{F}$ *and* $\operatorname{supp} \Delta^{-1} \subseteq \{1, ..., x^r\} \mathfrak{F}^{-1}$.

b) $T_\Delta = T_L$ *and* $T_{\Delta^{-1}} = T_L^{-1}$. □

Proposition 7.15. *Given $K, L \in C[[\mathfrak{B}^C]][\partial]^{\neq}$, we have*

$$\mathfrak{H}_{KL} = \mathfrak{H}_L \amalg \mathfrak{d}(T_L^{-1}(\mathfrak{H}_K))$$

and

$$T_{KL} = T_K \circ T_L.$$

Proof. Let $\mathfrak{m} \in \mathfrak{R}_L \setminus \mathfrak{d}(T_L^{-1}(\mathfrak{H}_K))$. Then for all $\mathfrak{n} \prec \mathfrak{m}$, we have $L\mathfrak{n} \prec L\mathfrak{m}$ and $KL\mathfrak{n} \prec KL\mathfrak{m}$. By strong linearity, it follows that $KLf \prec KL\mathfrak{m}$ for all $f \in C[[x^{\mathbb{N}} \mathfrak{B}^C]]$ with $f \prec \mathfrak{m}$. This shows that $\mathfrak{m} \in \mathfrak{R}_{KL}$ and $\mathfrak{H}_{KL} \subseteq \mathfrak{H}_L \amalg \mathfrak{d}(T_L^{-1}(\mathfrak{H}_K))$.

Conversely, let $\mathfrak{m} \in \mathfrak{H}_L$ and assume that $L\mathfrak{m} \neq 0$. Then $L\mathfrak{n} \succ L\mathfrak{m}$ for all $\mathfrak{n} \prec \mathfrak{m}$ with $\mathfrak{n} \succ T_L^{-1}(\tau(L\mathfrak{m}))$. If $x^i \mathfrak{v} \in \mathfrak{H}_L$, then proposition 7.6 implies $i < r$ and $x^j \mathfrak{v} \in \mathfrak{H}_L$ for all $j < i$. Hence $T_L^{-1}(\tau(L\mathfrak{m})) \prec_{e^x} \mathfrak{m}$ and we may choose $\mathfrak{n}$ so that $\mathfrak{n} \notin T_L^{-1}(\mathfrak{H}_K)$. But then $KL\mathfrak{n} \succ KL\mathfrak{m}$ and $\mathfrak{m} \in \mathfrak{H}_{KL}$. If $\mathfrak{m} \in \mathfrak{H}_L$ satisfies $L\mathfrak{m} = 0$, then we clearly have $\mathfrak{m} \in \mathfrak{H}_{KL}$.

Similarly, let $\mathfrak{m} = x^i \mathfrak{v} \in \mathfrak{H}_K \cap \operatorname{im} T_L$ and denote $\tilde{\mathfrak{m}} = \mathfrak{d}(T_L^{-1}(\mathfrak{m}))$. Then $K\mathfrak{n} \succ K\mathfrak{m}$ for all $\mathfrak{n} \prec \mathfrak{m}$ with $K\mathfrak{m} \neq 0 \Rightarrow \mathfrak{n} \succ T_K^{-1}(\tau(K\mathfrak{m}))$. Moreover, we may choose $\mathfrak{n} \in \mathfrak{R}_K$ such that $\mathfrak{n} \succ (\operatorname{supp} L\, \tilde{\mathfrak{m}})_{\prec \mathfrak{v}}$ and $K(x^j\, \mathfrak{v}) \neq 0 \Rightarrow \mathfrak{n} \succ T_K^{-1}(\tau(K(x^j\, \mathfrak{v})))$ for all $j \leqslant i$. This ensures that $K\mathfrak{n} \succ KL\,\tilde{\mathfrak{m}}$. Denoting $\tilde{\mathfrak{n}} = \mathfrak{d}(T_L^{-1}(\mathfrak{n})) \prec \tilde{\mathfrak{m}}$, we conclude that $KL\,\tilde{\mathfrak{n}} \asymp K\mathfrak{n} \succ KL\,\tilde{\mathfrak{m}}$, whence $\tilde{\mathfrak{m}} \in \mathfrak{H}_{KL}$.

As to second identity, let $v \in C^{\neq} \mathfrak{R}_{KL}$. Then $Lv \sim T_L(v)$ and $T_L(v) \notin C^{\neq} \mathfrak{H}_{\mathfrak{K}}$ implies $KLv \sim K(T_L(v))$. Hence $T_{KL}(v) = \tau(KLv) = \tau(K(T_L(v)) = T_K(T_L(v))$. □

Exercise 7.10. Prove the propositions of section 7.3.3 for operators $L \in \mathbb{T}[\partial]$.

Exercise 7.11. Generalize the results from this section to the well-based setting.

Exercise 7.12. Let $L = 1 + \circ_{x+1} - 2\circ_{x+\pi} \in \mathscr{D}_{\mathbb{T}_{\prec\!\!\prec e^x}}$. Determine $\mathfrak{R}_L$.

7.4 Distinguished solutions

Let $\mathfrak{M}$ and $\mathfrak{N}$ be monomial sets, such that $\mathfrak{M}$ is totally ordered. Given a linear grid-based operator $L: C[[\mathfrak{M}]] \to C[[\mathfrak{N}]]$ and $g \in C[[\mathfrak{N}]]$, we say that $f \in C[[\mathfrak{M}]]$ is a *distinguished solution* to the equation

$$Lf = g, \tag{7.14}$$

if for any other solution $\hat{f} \in C[[\mathfrak{M}]]$, we have $f_{\mathfrak{d}(\hat{f}-f)} = 0$. Clearly, if a distinguished solution exists, then it is unique. A mapping $L^{-1}: C[[\mathfrak{N}]] \to C[[\mathfrak{M}]]$ is said to be a *distinguished right inverse* of L, if $L L^{-1} = \mathrm{Id}$ and $L^{-1} g$ is a distinguished solution solution to (7.14) for each $g \in C[[\mathfrak{N}]]$. A *distinguished solution* to the homogeneous equation

$$Lh = 0 \tag{7.15}$$

is a series $h \in C[[\mathfrak{M}]]$ with $c_h = 1$ and $h_{\mathfrak{d}(\hat{h})} = 0$ for all other solutions $\hat{h}$ with $\mathfrak{d}_{\hat{h}} \neq \mathfrak{d}_h$. A *distinguished basis* of the solution space H_L of (7.15) is a strong basis which consists exclusively of distinguished solutions. If it exists, then the distinguished basis is unique.

Remark 7.16. Distinguished solutions can sometimes be used for the renormalization of "divergent" solutions to differential equations; see [vdH01b] for details.

7.4.1 Existence of distinguished right inverses

Theorem 7.17. *Assume that the trace T_L is invertible and both T_L and T_L^{-1} extend to strongly linear mappings*

$$\begin{aligned} \Delta: C[[\mathfrak{R}_L]] &\to C[[\mathfrak{N}]] \\ \Delta^{-1}: C[[\mathfrak{N}]] &\to C[[\mathfrak{R}_L]]. \end{aligned}$$

Assume also that $\operatorname{supp} L$ *and* $\operatorname{supp} \Delta^{-1}$ *are grid-based. Then*

a) L admits a distinguished and grid-based right inverse

$$L^{-1}: C[[\mathfrak{N}]] \to C[[\mathfrak{R}_L]].$$

b) The elements $h^{\mathfrak{h}} = \mathfrak{h} - L^{-1} L\mathfrak{h}$ with $\mathfrak{h} \in \mathfrak{H}_L$ form a distinguished basis for H_L.

Proof. Let $R = L - \Delta$. Then the operator $R\Delta^{-1}$ is strictly extensive, and the operator $(\mathrm{Id} + R\Delta^{-1})\Delta$ coincides with L on $C[[\mathfrak{R}_L]]$. Now consider the functional

$$\Phi(f, g) = g - R\Delta^{-1} f.$$

By theorem 6.14, there exists a strongly linear operator

$$\Psi = (\mathrm{Id} + R\Delta^{-1})^{-1} = \mathrm{Id} - R\Delta^{-1} + (R\Delta^{-1})^2 + \cdots,$$

such that $\Phi(\Psi(g), g) = \Psi(g)$ for all $g \in C[[\mathfrak{N}]]$. Consequently,

$$L^{-1} = \Delta^{-1}(\mathrm{Id} + R\Delta^{-1})^{-1}: C[[\mathfrak{N}]] \to C[[\mathfrak{R}_L]]$$

is a strongly linear right inverse for L. Given $h \in H_L^{\neq}$, we also observe that $\mathfrak{d}_h \in \mathfrak{H}_L$; otherwise, $\tau_{Lh} = T_L(\tau_h) \neq 0$. Consequently, $f = L^{-1} g$ is the distinguished solution of (7.14) for all $g \in C[[\mathfrak{N}]]$. This proves (a).

As to (b), we first observe that

$$L h^{\mathfrak{h}} = L\mathfrak{h} - L L^{-1} L\mathfrak{h} = 0$$

for all $\mathfrak{h} \in \mathfrak{H}_L$. The solution $h^{\mathfrak{h}}$ is actually distinguished, since

$$\operatorname{supp} h^{\mathfrak{h}} \cap \mathfrak{H}_L \subseteq \{\mathfrak{h}\}$$

and $\mathfrak{d}_{\hat{h}} \in \mathfrak{H}_L$ for all $\hat{h} \in H_L$. In fact, we claim that

$$h^{\mathfrak{h}} \sim \mathfrak{h}. \tag{7.16}$$

Indeed, if $L^{-1} L\mathfrak{h} \succ \mathfrak{h}$, then we would have $\mathfrak{d}_{L^{-1}L\mathfrak{h}} \in \mathfrak{R}_L$, so

$$L\mathfrak{h} \prec L(L^{-1} L h) = L h,$$

which is impossible. Now let h be an arbitrary solution to (7.15) and consider

$$\begin{aligned} \tilde{h} &= \sum_{\mathfrak{h} \in \mathfrak{H}_L} h_{\mathfrak{h}}\, \mathfrak{h}. \\ \hat{h} &= \tilde{h} - L^{-1} L \tilde{h} = \sum_{\mathfrak{h} \in \mathfrak{H}_L} h_{\mathfrak{h}}\, h^{\mathfrak{h}}. \end{aligned}$$

Then we have $\hat{h}_{\mathfrak{h}} = h_{\mathfrak{h}}$ for all $\mathfrak{h} \in \mathfrak{H}_L$, by the distinguished property of the $h^{\mathfrak{h}}$ and (7.16). Consequently, $\hat{h} - h \in H_L \cap C[[\mathfrak{R}_L]] = \{0\}$. This proves (b). □

Corollary 7.18. *Let $\mathfrak{B} = (\mathfrak{b}_1, ..., \mathfrak{b}_n)$ be a plane transbasis and let $L \in C[[\mathfrak{B}^C]][\partial]$ be a linear differential operator on $C[[x^{\mathbb{N}} \mathfrak{B}^C]]$. Then L admits a distinguished right inverse L^{-1} and H_L admits a finite distinguished basis.*

Proof. In view of proposition 7.8 and corollary 7.14, we may apply theorem 7.17. By general differential algebra, we know that H_L is finite dimensional. □

Corollary 7.19. *Let $L \in \mathbb{T}[\partial]$ be a linear differential operator on $\mathbb{T}$. Then L admits a distinguished right inverse and H_L admits a finite distinguished basis.*

Proof. Given $g \in \mathbb{T}$, let us first prove that $Lf = g$ admits a distinguished solution. Let $\mathfrak{B}$ be a transbasis for g as in proposition 7.11 and consider $f = L\uparrow_l^{-1}(g\uparrow_l)\downarrow_l$. Then

$$Lf = L\uparrow_l(f\uparrow_l)\downarrow_l = L\uparrow_l(L\uparrow_l^{-1}(g\uparrow_l))\downarrow_l = g.$$

From proposition 7.11, it follows that

$$f_{\mathfrak{h}} = (L\uparrow_l^{-1}(g\uparrow_l))_{\mathfrak{h}\uparrow_l} = 0$$

for all

$$\mathfrak{h} \in \mathfrak{H}_L \cap \mathfrak{S}_{\mathfrak{B}} = (\mathfrak{H}_{L\uparrow_l} \cap \mathfrak{S}_{\mathfrak{B}\uparrow_l})\downarrow_l = \mathfrak{H}_{L\uparrow_l;\mathfrak{B}\uparrow_l}\downarrow_l.$$

Hence f is the distinguished solution to $Lf = g$. In particular, the construction of $L^{-1} g := f$ is independent of the choice of $\mathfrak{B}$. The operator L^{-1} is strongly linear, since each grid-based family in $\mathscr{F}(\mathbb{T})$ is also a family in $\mathscr{F}(C[[\mathfrak{S}_{\mathfrak{B}}]])$ for some $\mathfrak{B}$ as above, and L^{-1} is strongly linear on $C[[\mathfrak{S}_{\mathfrak{B}}]]$. □

Example 7.20. With L as in example 7.9, we have

$$L^{-1}\mathrm{e}^x = \tfrac{1}{2}\mathrm{e}^x + 1 - \tfrac{1}{2}x\,\mathrm{e}^{-x} + (x-1)\,\mathrm{e}^{-2x} + \left(-\tfrac{15}{4}x + \tfrac{43}{8}\right)\mathrm{e}^{-3x} + \cdots.$$

7.4.2 On the supports of distinguished solutions

Let $\mathfrak{B} = (\mathfrak{b}_1, \ldots, \mathfrak{b}_n)$ be a plane transbasis and let $L \in C[[\mathfrak{B}^C]][\partial]$ be a linear differential operator on $C[[x^{\mathbb{N}}\mathfrak{B}^C]]$ of order r.

Proposition 7.21. *The operator support of L^{-1} is bounded by*

$$\operatorname{supp} L^{-1} \subseteq \mathfrak{V}\,\mathfrak{W}^*,$$

where

$$\begin{aligned}
\mathfrak{V} &= \{1, \ldots, x^r\}\left\{\frac{\mathfrak{m}}{\mathfrak{d}(L_{\times\mathfrak{m}})}\,\middle|\,\mathfrak{m} \in \mathfrak{B}^C\right\};\\
\mathfrak{W} &= \{1, \ldots, x^r\}\left(\bigcup_{\mathfrak{m} \in \mathfrak{B}^C} \frac{\operatorname{supp_{ser}} L_{\times\mathfrak{m}}}{\mathfrak{d}(L_{\times\mathfrak{m}})} \setminus \{1\}\right) \cup \{x^{-1}, x^{-2}, \ldots\}
\end{aligned}$$

are grid-based sets and $\mathfrak{W} \prec 1$.

Proof. With the notations from the proof of theorem 7.17,

$$\begin{aligned}
\operatorname{supp} \Delta^{-1} &\subseteq \mathfrak{V};\\
\operatorname{supp}(R\,\Delta^{-1}) &\subseteq \mathfrak{W}.
\end{aligned}$$

It follows that

$$\begin{aligned}
\operatorname{supp} L^{-1} &= \operatorname{supp} \Delta^{-1}(\mathrm{Id} + R\,\Delta^{-1})^{-1}\\
&\subseteq (\operatorname{supp} \Delta^{-1})\,(\operatorname{supp}(R\,\Delta^{-1}))^*\\
&\subseteq \mathfrak{V}\,\mathfrak{W}^*.
\end{aligned}$$

Recall that $\mathfrak{V}$ is finite, by proposition 7.13. This also implies that $\mathfrak{W}$ is grid-based. □

Proposition 7.22. *Given $d \in \mathbb{N}$, let*

$$\begin{aligned}
C[[\mathfrak{B}^C]][x]_d &= \{f \in C[[\mathfrak{B}^C]][x] : \deg_x f \leqslant d\}\\
&\subseteq C[[x^{\mathbb{N}}\mathfrak{B}^C]] = C[x][[\mathfrak{B}^C]].
\end{aligned}$$

Setting $s = \operatorname{card} \mathfrak{H}_L \leqslant r$, *we have*

a) L *maps* $C[[\mathfrak{B}^C]][x]_d$ *into* $C[[\mathfrak{B}^C]][x]_d$.
b) L^{-1} *maps* $C[[\mathfrak{B}^C]][x]_d$ *into* $C[[\mathfrak{B}^C]][x]_{d+s}$.
c) $H_L \subseteq C[[\mathfrak{B}^C]][x]_s$.

Proof. For all $f = f_d x^d + \cdots + f_0 \in C[[\mathfrak{B}^C]][x]_d$, we have

$$\begin{aligned} L f &= L_{\times x^d} f_d + \cdots + L_{\times x} f_1 + L f_0 \\ &= ((L f_d) x^d + \cdots + L^{(d)} f_d) + \cdots + ((L f_1) x + L' f_1) + L f_0 \\ &= (L f_d) x^d + \cdots + (L^{(d-1)} f_d + \cdots + L f_1) x + (L^{(d)} f_d + \cdots + L f_0). \end{aligned}$$

This shows (a). As to (b), let $g \in C[[\mathfrak{B}^C]][x]_d$ and consider

$$\begin{aligned} \mathfrak{D} &= \{x^i \, \mathfrak{m} \in \mathfrak{R}_L : i \leqslant d + \operatorname{card} \{\mathfrak{h} \in \mathfrak{H}_L : \mathfrak{h} \succcurlyeq \mathfrak{m}\}\}; \\ \mathfrak{I} &= \{x^i \, \mathfrak{m} \in x^{\mathbb{N}} \mathfrak{B}^C : i \leqslant d + \operatorname{card} \{\mathfrak{h} \in \mathfrak{H}_L : L_{\times \mathfrak{h}} \succcurlyeq \mathfrak{m}\}\} \\ & \quad (\mathfrak{d} \circ L)(\mathfrak{D}). \end{aligned}$$

Then T_L is a bijection between $C^{\neq} \mathfrak{D}$ and $C^{\neq} \mathfrak{I}$ and L maps $C[[\mathfrak{D}]]$ into $C[[\mathfrak{I}]]$. By theorem 7.17, it follows that the restriction of L to $C[[\mathfrak{D}]]$ admits a distinguished right inverse, which necessarily coincides with the restriction of L^{-1} to $C[[\mathfrak{I}]]$. This proves (b), since $C[[\mathfrak{D}]] \subseteq C[[\mathfrak{B}^C]][x]_{d+s}$ and $C[[\mathfrak{I}]] \supseteq C[[\mathfrak{B}^C]][x]_d$. Moreover, for each element $h^{\mathfrak{h}}$ of the distinguished basis of H_L, we have $h^{\mathfrak{h}} = \mathfrak{h} + L^{-1} L h \in C[[\mathfrak{B}^C]][x]_s$. This proves ($c$). □

Exercise 7.13. Show that $T_{L^{-1}} = T_L^{-1}$.

Exercise 7.14. Show that we actually have $H_L \subseteq C[[\mathfrak{B}^C]][x]_{s-1}$ in proposition 7.22(c).

Exercise 7.15. Let $\mathfrak{B}$ and $\hat{\mathfrak{B}}$ be plane transbases in the extended sense of exercise 4.15. Given $L \in C[[\mathfrak{B}^C]][\partial]$, let $L_{;\mathfrak{B}}^{-1}$ denote the distinguished right inverse of L as an operator on $C[[x^{\mathbb{N}} \mathfrak{B}^C]]$.

a) Show that $L_{;\mathfrak{B}}^{-1}$ is the restriction of $L_{;\hat{\mathfrak{B}}}^{-1}$ to $C[[x^{\mathbb{N}} \mathfrak{B}^C]]$, if $\hat{\mathfrak{B}}$ is a supertransbasis of $\mathfrak{B}$.
b) If $C[[\mathfrak{B}^C]] = C[[\hat{\mathfrak{B}}^C]]$, then show that $L_{;\mathfrak{B}}^{-1} = L_{;\hat{\mathfrak{B}}}^{-1}$ if and only if $\hat{\mathfrak{B}}^C = \mathfrak{B}^C$.
c) If $\hat{\mathfrak{B}} = (e^x, \mathfrak{b}_1 \uparrow, \ldots, \mathfrak{b}_n \uparrow)$, then show that $L \uparrow_{;\hat{\mathfrak{B}}}^{-1} (g \uparrow) = L_{;\mathfrak{B}}^{-1} (g) \uparrow$ for all $g \in C[[x^{\mathbb{N}} \mathfrak{B}^C]]$.

Exercise 7.16. Let $\mathbb{T}^\flat = C[[\mathfrak{T}^\flat]] \ni x$ be a flat subspace of $\mathbb{T}$ and $\mathfrak{T}^\sharp$ the steep complement of $\mathfrak{T}^\flat$, so that $\mathbb{T} = \mathbb{T}^\flat[[\mathfrak{T}^\sharp]]$. Consider $L \in \mathbb{T}[\partial]$ as a strong operator on $\mathbb{T}^\flat[[\mathfrak{T}^\sharp]]$ (notice that L is not $\mathbb{T}^\flat$-linear). Let $\mathfrak{R}_L^\sharp$ be the set of monomials $\mathfrak{m}^\sharp \in \mathfrak{T}^\sharp$ such that $\mathfrak{d}^\sharp(L(\lambda^\sharp \mathfrak{m}^\sharp))$ does not depend on $\lambda^\sharp \in \mathbb{T}^{\flat,\neq}$ and such that the mapping $\lambda^\sharp \mapsto c^\sharp(L(\lambda^\sharp \mathfrak{m}^\sharp)); \mathbb{T}^{\flat,\neq} \to \mathbb{T}^{\flat,\neq}$ is invertible.

a) Exhibit an operator in $\mathbb{T}^\flat[\partial]$ which maps $\lambda^\sharp$ to $c^\sharp(L(\lambda^\sharp \mathfrak{m}^\sharp))$ and relate $\mathfrak{R}_L^\sharp$ and $\mathfrak{R}_L$.

b) Generalize theorem 7.17 to the setting of strongly additive operators and relate the distinguished right inverses of L as an operator on $\mathbb{T}$ and as an operator on $\mathbb{T}^\flat[[\mathfrak{T}^\sharp]]$.
c) Given a plane transbasis $\mathfrak{B}$, $L \in C[[\mathfrak{B}^C]][\partial]$ and $g \in C[[\mathfrak{B}^C]][x]$, give a concrete algorithm to compute the recursive expansion of $L_{;\mathfrak{B}}^{-1} g$.

Exercise 7.17. Let $L \in \mathbb{T}[\partial]^{\neq}$ and let $\mathfrak{m}$ be a transmonomial. Prove that

$$\begin{aligned}(L \times_{\mathfrak{m}})^{-1} &= \times_{\mathfrak{m}}^{-1} L^{-1}\\ (\times_{\mathfrak{m}} L)^{-1} &= L^{-1} \times_{\mathfrak{m}}^{-1}\end{aligned}$$

Exercise 7.18. Let $L \in \mathbb{T}[\partial]^{\neq}$ and $g \in \mathbb{T}^{>,\succ}$. When do we have

$$(L_{\circ g})^{-1} = (L^{-1})_{\circ g}?$$

Here $(L^{-1})_{\circ g}$ is the unique operator such that

$$(L^{-1})_{\circ g}(f \circ g) = (L^{-1} f) \circ g$$

for all f.

Exercise 7.19.

a) Show that $(K L)^{-1} = L^{-1} K^{-1}$ for $K = \partial^2 + e^{e^x}$ and $L = \partial^2 + 2\partial + e^x$.
b) Show that $(K L)^{-1} \neq L^{-1} K^{-1}$ for $K = \partial^2 - e^{e^x}$ and $L = \partial^2 + 2 e^x \partial + 1$.
c) Do we always have $(L L)^{-1} = L^{-1} L^{-1}$?

Exercise 7.20.

a) Prove that each non-zero $L \in \mathscr{D}_{\mathbb{T}_{\prec\!\!\prec e^x}}$ admits a distinguished right-inverse on $\mathbb{T}_{\prec\!\!\prec e^x}$.
b) Can $\mathfrak{H}_L$ be infinite?
c) Same questions for $L \in \mathscr{D}_{\mathbb{T}_{\underline{\prec\!\!\prec} e^x}}$.

Exercise 7.21. Consider an operator L as in exercise 7.6.

a) For any $g \in \mathbb{T}^{\neq}_{\prec\!\!\prec e^{x^2}}$, show that $g^{-1} L_{\times g}$ is an operator of the same kind.
b) Show that L admits a distinguished right-inverse on $\mathbb{T}_{\prec\!\!\prec e^{x^2}}$.
c) Assuming that $A_{i,j} \in \mathbb{T}_{\underline{\prec\!\!\prec} e^x}$, show that L admits a distinguished right-inverse on $\mathbb{T}_{\underline{\prec\!\!\prec} e^{x^2}}$.
d) Given $g \in \mathbb{T}_{\underline{\prec\!\!\prec} e^{x^{O(1)}}}$, show that $L f = g$ admits a distinguished solution, which is not necessarily grid-based, but whose support is always well-based and contained in a finitely generated group.
e) Show that (d) still holds if $A_{i,j} \in \mathbb{T}_{\underline{\prec\!\!\prec} e^{x^{O(1)}}}$.
f) Given a general $g \in \mathbb{T}$, show that $L f = g$ admits a well-based distinguished solution.
g) Give a bound for the cardinality of $\mathfrak{H}_L$.

Exercise 7.22. Let $\mathscr{O}$ be the space of partial grid-based operators $L: \mathbb{T} \rightharpoonup \mathbb{T}$, such that $\operatorname{dom} L$ is a space of finite codimension over C in $\mathbb{T}$. Two such operators are understood to be equal if they coincide on a space of finite codimension in $\mathbb{T}$.

a) Show that $\mathscr{O}$ is a $\mathbb{T}$-algebra under composition.

b) Show that each $L \in \mathbb{T}[\partial]^{\neq}$ induces a unique operator in $\mathscr{O}$ with $LL^{-1} = L^{-1}L = 1$.
c) Show that the skew fraction field $\mathbb{T}(\partial)$ of $\mathbb{T}[\partial]$ in $\mathscr{O}$ consists of operators $K^{-1}L$ with $K, L \in \mathbb{T}[\partial]$ and $K \neq 0$. Hint: show that for any $K, L \in \mathbb{T}[\partial]$ with $K \neq 0$, there exist $\tilde{K}, \tilde{L} \in \mathbb{T}[\partial]$ with $\tilde{K} \neq 0$ and $K\tilde{L} = L\tilde{K}$.

Exercise 7.23. Let $L \in \mathbb{E}[\partial]^{\neq}$, where $\mathbb{E} = C[[\mathfrak{E}]]$ denotes the field of exponential transseries.

a) If $L \asymp L_0$, then show that there exists a decomposition
$$L = c\mathfrak{m}(1 + K_1) \cdots (1 + K_n),$$
with $c\mathfrak{m} \in C\,\mathfrak{E}$, $K_1, \ldots, K_n \in \mathbb{T}(\partial)_{\preccurlyeq}$ and $\operatorname{supp} K_1 \lll \cdots \lll \operatorname{supp} K_n$.
b) If $c > 0$ and K_1 is sufficiently small, then show how to define $\log L$.
c) Given $\lambda \in C$, extend the definition of ∂^λ from exercise 7.7(c) to a definition of L^λ on a suitable strong subvector space of $\mathbb{T}$.

7.5 The deformed Newton polygon method

Let $L \in \mathbb{T}[\partial]^{\neq}$ be a linear differential operator and consider the problem of finding the solutions to the homogeneous equation $Lh = 0$. Modulo upward shiftings it suffices to consider the case when the coefficients of L can all be expanded w.r.t. a plane transbasis $\mathfrak{B}$. Furthermore, theorem 7.17 and its corollaries imply that it actually suffices to find the elements of $\mathfrak{H}_L$.

Now solving the equation $Lh = 0$ is equivalent to solving the equation $R_L(f) = 0$ for $f = h^\dagger$. As we will see in the next section, finding the dominant monomials of solutions is equivalent to solving the "Riccati equation modulo $o(1)$"

$$R_{L,+f,*}(0) = 0 \qquad (7.17)$$

for $f \in C[[\mathfrak{B}^C]]_{\succcurlyeq}$. It turns out that this equation is really a "deformation" of the algebraic equation

$$R_{L,\mathrm{alg}}(f) = 0. \qquad (7.18)$$

In this section, we will therefore show how to solve (7.17) using a deformed version of the Newton polygon method from chapter 3.

7.5.1 Asymptotic Riccati equations modulo $o(1)$

Let $\mathfrak{B}$ is a plane transbasis and $L \in C[[\mathfrak{B}^C]][\partial]^{\neq}$. We regard L as a linear differential operator on $C[[x^{\mathbb{N}}\,\mathfrak{B}^C]]$. Given $\mathfrak{v} \in \mathfrak{B}^C \cup \{\top\}$, consider the asymptotic versions

$$R_{L,+f,*}(0) = 0 \qquad (f \prec \mathfrak{v}) \qquad (7.19)$$

and

$$R_{L,\mathrm{alg}}(f) = 0 \qquad (f \prec \mathfrak{v}) \qquad (7.20)$$

of (7.17) resp. (7.18). We call (7.19) an *asymptotic Riccati equation modulo* $o(1)$. A solution $f \in C[[\mathfrak{B}^C]]_{\succ}$ of (7.19) is said to have *multiplicity* μ, if $R_{L,+f,\text{alg},i} \prec R_{L,+f}$ for all $i < \mu$ and $R_{L,+f,\text{alg},\mu} \asymp R_{L,+f}$.

Given $f \in C[[\mathfrak{B}^C]]_{\succ}$, we notice that for all k,

$$U_k(f) = f^k + O_f(f^{k-1}). \tag{7.21}$$

We say that $\mathfrak{m} \in (\mathfrak{B}^C)_{\succ}$ is a *starting monomial* of f relative to (7.19), if $\mathfrak{m}$ is a starting monomial of f relative to (7.20). Starting terms of solutions and their multiplicities are defined similarly. The *Newton degree* of (7.19) is defined to be the Newton degree of (7.20). The formula (7.21) yields the following analogue of proposition 3.4:

Proposition 7.23. *If* $f \in C[[\mathfrak{M}^C]]_{\succ}$ *is a solution to* (7.19), *then* τ_f *is a starting term of* f *relative to* (7.19).

Proof. Assume the contrary, so that there exists an index $i \in \{0, ..., r\}$ with $L_j f^j \prec L_i f^i$ for all $j \neq i$. But then

$$L_j U_j(f) \sim L_j f^j \prec L_i f^i \sim L_i U_i(f)$$

for all j. Hence

$$R_{L,+f,\text{alg},0} = R_L(f) \sim L_i f^i$$

and similarly

$$R_{L,+f,\text{alg},j} = R_{L^{(j)}}(f) \preccurlyeq L_i f^{i-j}$$

for all j. In other words,

$$R_{L,+f} \asymp R_{L,+f,\text{alg},0}$$

and $R_{L,+f,*}(0) = (L_i f^i)_* \neq 0$. □

7.5.2 Quasi-linear Riccati equations

We say that (7.19) is *quasi-linear* if its Newton degree is one (i.e. if (7.20) is quasi-linear). We have the following analogue of lemma 3.5:

Proposition 7.24. *If* (7.19) *is quasi-linear, then it admits a unique solution* $f \in C[[\mathfrak{B}^C]]_{\succ}$.

Proof. Let $\mathfrak{V} = \{\mathfrak{m} \in \mathfrak{B}^C : 1 \preccurlyeq \mathfrak{m} \prec \mathfrak{v}\}$ and consider the well-based operator

$$\begin{aligned} \Phi : C[[\mathfrak{V}]] &\longrightarrow C[[\mathfrak{V}]] \\ f &\longmapsto -\left(\frac{L_0 + L_2 U_2(f) + \cdots + L_r U_r(f)}{L_1}\right)_{\succ}. \end{aligned}$$

Since (7.19) is quasi-linear, we have

$$L_i \mathfrak{v}^i \preccurlyeq L_1 \mathfrak{v} \tag{7.22}$$

for all i and $L_0 \prec L_1\, \mathfrak{v}$. Moreover, on $\mathfrak{V} \subseteq \{\mathfrak{m} \in \mathfrak{B}^C\colon \mathfrak{m} \mathrel{\underline{\lll}} \mathfrak{v}\}$ we have $(\operatorname{supp} \partial) \preccurlyeq \mathfrak{v}^\dagger \lll \mathfrak{v}$. Since $U_i(f) - f^i$ is a differential polynomial of degree $< i$, we thus have

$$\operatorname{supp} U_i \preccurlyeq \mathfrak{v}^i, \tag{7.23}$$

when considering U_i as an operator on $C[[\mathfrak{V}]]$. Combining (7.22) and (7.23), we conclude that

$$\operatorname{supp} \Phi_1 \cup (\operatorname{supp} \Phi_2)\, \mathfrak{m} \cup \cdots \cup (\operatorname{supp} \Phi_r)\, \mathfrak{m}^{r-1} \prec 1$$

for all $\mathfrak{m} \in \mathfrak{B}^C$ with $1 \preccurlyeq \mathfrak{m} \prec \mathfrak{v}$. By theorem 6.14, it follows that the equation

$$\Phi(f) = f \tag{7.24}$$

admits a unique fixed point f in $C[[\mathfrak{V}]]$. We claim that this is also the unique solution to (7.19).

Let us first show that f is indeed a solution. From (7.24), we get

$$R_{L,+f,\mathrm{alg},0} = R_L(f) = o(L_1). \tag{7.25}$$

On the other hand, we have for $i \geqslant 1$:

$$\begin{aligned} R_{L,+f,\mathrm{alg},1} &= R_{L'}(f) \\ &= L_1 + O(L_2\, f) + \cdots + O(L_r\, f^{r-1}) \sim L_1 \qquad (7.26)\\ R_{L,+f,\mathrm{alg},i} &= R_{L^{(i)}}(f) \\ & \quad O(L_i) + \cdots + O(L_r\, f^{r-i}) \preccurlyeq L_1\, \mathfrak{v}^{1-i}. \qquad (7.27)\end{aligned}$$

In other words, $R_{L,+f} \asymp L_1$ and $R_{L,+f,*}(0) = 0$. Assume finally that $\tilde{f} \in C[[\mathfrak{V}]]$ is such that $1 \preccurlyeq \delta = \tilde{f} - f \prec \mathfrak{v}$. Then (7.25), (7.26) and (7.27) also imply that

$$R_{L,+\tilde{f},\mathrm{alg},0} = R_{L,+f}(\delta) \sim L_1\, \delta \succcurlyeq L_1 \sim R_{L,+\tilde{f},\mathrm{alg},1}.$$

In other words, $R_{L,+\tilde{f},*}(0) \neq 0$. □

7.5.3 Refinements

Given a *refinement*

$$f = \varphi + \tilde{f} \qquad (\tilde{f} \prec \tilde{\mathfrak{v}}), \tag{7.28}$$

where $1 \preccurlyeq \varphi \prec \mathfrak{v}$ and $\tilde{\mathfrak{v}} = \mathfrak{d}_\varphi$, the equation (7.19) becomes

$$R_{\tilde{L},+\tilde{f},*}(0) = 0 \qquad (\tilde{f} \prec \tilde{\mathfrak{v}}), \tag{7.29}$$

where $\tilde{L} = \mathrm{e}^{-\int \varphi} L_{\times \mathrm{e}^{\int \varphi}}$ satisfies $R_{\tilde{L}} = R_{L,+\varphi}$. We recall that the coefficients of the corresponding algebraic equation

$$R_{\tilde{L},\mathrm{alg}}(\tilde{f}) = 0 \qquad (\tilde{f} \prec \tilde{\mathfrak{v}}) \tag{7.30}$$

are given by

$$R_{\tilde{L},\mathrm{alg},i} = R_{L^{(i)}}(\varphi).$$

Let us show that the analogues of lemmas 3.6 and 3.7 hold.

Proposition 7.25. *Let $\varphi \in C[[\mathfrak{B}^C]]_{\succcurlyeq}$. Then the Newton degree of*

$$R_{L,+\varphi,+\tilde{f},*}(0)=0 \qquad (\tilde{f} \prec \varphi) \tag{7.31}$$

equals the multiplicity of τ_φ as a starting term of f relative to (7.19).

Proof. For a certain transmonomial $\mathfrak{n}$, the Newton polynomial relative to $\mathfrak{m}=\mathfrak{d}_\varphi$ is given by

$$N_{R_L,\mathfrak{m}}(c)=N_{R_{L,\text{alg}},\mathfrak{m}}(c)=L_{d,\mathfrak{n}/\mathfrak{m}^d}\,c^d+\cdots+L_{0,\mathfrak{n}}.$$

Then, similarly as in the proof of lemma 3.6, we have

$$\begin{aligned}\tilde{L}_i &= \frac{1}{i!}R_{L^{(i)}}(\varphi)\\ &= \frac{1}{i!}\sum_{k=i}^{n}\binom{k}{i}L_k\,(\varphi^{k-i}+O_\varphi(\varphi^{k-i-1}))\\ &= \frac{1}{i!}\sum_{k=i}^{n}\binom{k}{i}(L_{k,\mathfrak{n}\mathfrak{m}^{-k}}+o(1))\,\mathfrak{n}\,\mathfrak{m}^{-k}\,(c+o(1))^{k-i}\,\mathfrak{m}^{k-i}\\ &= \frac{1}{i!}N_{P,\mathfrak{m}}^{(i)}(c)\,\mathfrak{n}\,\mathfrak{m}^i+o(\mathfrak{n}\,\mathfrak{m}^i)\end{aligned}$$

for all i, and we conclude in the same way. □

Proposition 7.26. *Let d be the Newton degree of* (7.19). *If f admits a unique starting term τ of multiplicity d, then*

a) The equation

$$R_{L^{(d-1)},+\varphi,*}(0)=0 \qquad (\varphi \prec \mathfrak{v}) \tag{7.32}$$

is quasi-linear and has a unique solution with $\varphi=\tau+o(\tau)$.

b) Any refinement

$$\tilde{f}=\tilde{\varphi}+\tilde{\tilde{f}} \qquad (\tilde{\tilde{f}} \prec \tilde{\mathfrak{v}}) \tag{7.33}$$

transforms (7.32) *into an equation of Newton degree $<d$.*

Proof. Part (a) follows immediately from lemma 3.7(a) and the fact that $R_{L^{(d-1)},\text{alg}}=R_{L,\text{alg}}^{(d-1)}$. Now consider a refinement (7.33). As to (b), let $\tilde{\mathfrak{n}}$ be such that the the Newton polynomial associated to $\tilde{\mathfrak{m}}=\mathfrak{d}_{\tilde{\varphi}}$ is given by

$$N_{R_{\tilde{L}},\tilde{\mathfrak{m}}}(c)=N_{R_{\tilde{L},\text{alg}},\tilde{\mathfrak{m}}}(c)=\tilde{L}_{d,\tilde{\mathfrak{n}}/\tilde{\mathfrak{m}}^d}\,c^d+\cdots+\tilde{L}_{0,\tilde{\mathfrak{n}}}.$$

By the choice of φ, we have

$$\tilde{L}_{d-1}=R_{L^{(d-1)}}(\varphi)=R_{L^{(d-1)},+\varphi,\text{alg},0}\prec R_{L^{(d-1)},+\varphi,1}=R_{L^{(d)}}(\varphi)=\tilde{L}_d.$$

It follows that the term of degree $d-1$ in $N_{R_{\tilde{L}},\tilde{\mathfrak{m}}}(c)$ vanishes, so $N_{R_{\tilde{L}},\tilde{\mathfrak{m}}}$ cannot admit a root of multiplicity d. We conclude by proposition 7.25. □

7.5.4 An algorithm for finding all solutions

Putting together the results from the previous sections, we obtain the following analogue of `polynomial_solve`:

Algorithm `riccati_solve`
Input: An asymptotic Riccati equation (7.19) modulo $o(1)$.
Output: The set of solutions to (7.19) in $C[[\mathfrak{B}^C]]_{\succcurlyeq}$.

1. Compute the starting terms $c_1 \mathfrak{m}_1, \ldots, c_\nu \mathfrak{m}_\nu$ of f relative to (7.20).
2. If $\nu = 1$ and c_1 is a root of multiplicity d of $N_{P,\mathfrak{m}_1}$, then let φ be the unique solution to (7.32). Refine (7.28) and apply `riccati_solve` to (7.29). Return the so obtained solutions to (7.19).
3. For each $1 \leqslant i \leqslant \nu$, refine $f = c_i \mathfrak{m}_i + \tilde{f}$ $(\tilde{f} \prec \mathfrak{m}_i)$ and apply `riccati_solve` to the new equation in $\tilde{f}$. Collect and return the so obtained solutions to (7.19), together with 0, if $L_0 = 0$.

Proposition 7.27. *The algorithm* `riccati_solve` *terminates and returns all solutions to* (7.19) *in* $C[[\mathfrak{B}^C]]_{\succcurlyeq}$. □

Since C is only real closed, the equation (7.19) does not necessarily admit d starting terms when counting with multiplicities. Consequently, the equation may admit less than d solutions. Nevertheless, we do have:

Proposition 7.28. *If the Newton degree d of* (7.19) *is odd, then* (7.19) *admits at least one solution in* $C[[\mathfrak{B}^C]]_{\succcurlyeq}$.

Proof. If $d = 1$, then we apply the proposition 7.24. Otherwise, there always exists a starting monomial $\mathfrak{m}$, such that $\deg N_{R_L,\mathfrak{m}} - \operatorname{val} N_{R_L,\mathfrak{m}}$ is odd as well. Since C is real closed, it follows that their exists at least one starting term of the form $\tau = c\,\mathfrak{m}$ of odd multiplicity $\tilde{d}$. Modulo one application of proposition 7.26, we may assume that $\tilde{d} < d$, and the result follows by proposition 7.25 and induction over d. □

Example 7.29. Consider the linear differential operator

$$L = \mathrm{e}^{-2\mathrm{e}^x}\,\partial^3 - 2\,\mathrm{e}^{-\mathrm{e}^x}\,\partial^2 + \partial - 2\,\mathrm{e}^x,$$

with

$$R_{L,\mathrm{alg}} = \mathrm{e}^{-2\mathrm{e}^x}\,F^3 - 2\,\mathrm{e}^{-\mathrm{e}^x}\,F^2 + F - 2\,\mathrm{e}^x.$$

The starting terms for $R_L(f) = 0$ are $\tau = 2\,\mathrm{e}^x$ and $\tau = \mathrm{e}^{\mathrm{e}^x}$ (of multiplicity 2). The refinement $f = 2\,\mathrm{e}^x + \tilde{f}$ $(\tilde{f} \prec \mathrm{e}^x)$ leads to

$$R_{L,+2\mathrm{e}^x,\mathrm{alg}} = F + O(\mathrm{e}^{2x - \mathrm{e}^x}),$$

so $f = 2\,\mathrm{e}^x$ is a solution to (7.17). The other starting term $\tau = \mathrm{e}^{\mathrm{e}^x}$ leads to

$$R_{L,+\mathrm{e}^{\mathrm{e}^x},\mathrm{alg}} = \mathrm{e}^{-\mathrm{e}^x}\,F^3 + F^2 + 3\,\mathrm{e}^x\,F - \mathrm{e}^{\mathrm{e}^x + x} + \mathrm{e}^{2x} + \mathrm{e}^x,$$

and $R_{L,+\mathrm{e}^{\mathrm{e}^x},\mathrm{alg}}(\tilde{f})=0$ $(\tilde{f}\prec\mathrm{e}^{\mathrm{e}^x})$ admits two starting terms $\tilde{\tau}=\pm\,\mathrm{e}^{(\mathrm{e}^x+x)/2}$. After one further refinement, we obtain the following two additional solutions to (7.17):

$$\begin{aligned} f &= \mathrm{e}^{\mathrm{e}^x}+\mathrm{e}^{(\mathrm{e}^x+x)/2}-\frac{9}{4}\,\mathrm{e}^x-\frac{1}{4};\\ f &= \mathrm{e}^{\mathrm{e}^x}-\mathrm{e}^{(\mathrm{e}^x+x)/2}-\frac{9}{4}\,\mathrm{e}^x-\frac{1}{4}. \end{aligned}$$

7.6 Solving the homogeneous equation

Let $L\in C[[\mathfrak{B}^C]][\partial]^{\neq}$ be a linear differential operator on $C[[x^{\mathbb{N}}\,\mathfrak{B}^C]]$, where $\mathfrak{B}$ is a plane transbasis. Let $f_1,\ldots,f_s$ be the solutions to (7.17), as computed by `riccati_solve`, and $\mu_1,\ldots,\mu_s$ their multiplicities. We will denote

$$\mathfrak{H}_{L;\mathfrak{B}}^{*}=\{\mathrm{e}^{\int f_1},\ldots,x^{\mu_1-1}\,\mathrm{e}^{\int f_1},\ldots,\mathrm{e}^{\int f_s},\ldots,x^{\mu_s-1}\,\mathrm{e}^{\int f_s}\}.$$

The following proposition shows how to find the elements of $\mathfrak{H}_{L;\mathfrak{B}}$ when we consider L as an operator on $C[[\mathfrak{B}^C]]$:

Proposition 7.30. *We have*

$$\mathfrak{H}_{L;\mathfrak{B}}=\mathfrak{H}_{L;\mathfrak{B}}^{*}\cap x^{\mathbb{N}}\,\mathfrak{B}^C.$$

Proof. Let $x^i\,\mathfrak{m}\in x^{\mathbb{N}}\,\mathfrak{B}^C$ and consider the operator $K=\mathfrak{m}^{-1}\,L_{\times\mathfrak{m}}$. Then

$$\begin{aligned} x^i\,\mathfrak{m}\in\mathfrak{H}_L &\Leftrightarrow c(K)(x^i)=0\\ &\Leftrightarrow i<\min\{d\colon K_d\asymp K\}\\ &\Leftrightarrow i<\min\{d\colon R_{K,\mathrm{alg},d}\asymp R_K\}\\ &\Leftrightarrow i<\min\{d\colon R_{L,+\mathfrak{m}^\dagger,\mathrm{alg},d}\asymp R_{L,+\mathfrak{m}^\dagger}\} \end{aligned}$$

But $\min\{d\colon R_{L,+\mathfrak{m}^\dagger,\mathrm{alg},d}\asymp R_{L,+\mathfrak{m}^\dagger}\}$ is precisely the multiplicity of $\mathfrak{m}^\dagger\in C[[\mathfrak{B}^C]]_{\succcurlyeq}$ as a solution of (7.17). □

In order to find the elements of $\mathfrak{H}_L$ when we consider L as an operator on $\mathbb{T}$, we have to study the dependence of $\mathfrak{H}_{L;\mathfrak{B}}^{*}$ under extensions of $\mathfrak{B}$ and upward shifting. Now `riccati_solve` clearly returns the same solutions if we enlarge $\mathfrak{B}$. The proposition below ensures that we do not find essentially new solutions when shifting upwards. In the more general context of oscillating transseries, which will be developed in the next section, this proposition becomes superfluous (see remark 7.38).

Proposition 7.31. *Assume that*

$$\begin{aligned} \mathfrak{B} &= (\mathfrak{b}_1,\ldots,\mathfrak{b}_n)\\ \hat{\mathfrak{B}} &= \{\mathrm{e}^x,\mathfrak{b}_1\uparrow,\ldots,\mathfrak{b}_n\uparrow\}. \end{aligned}$$

Then

$$\mathfrak{H}^*_{L\uparrow;\hat{\mathfrak{B}}} = \mathfrak{H}^*_{L;\mathfrak{B}}\uparrow.$$

Proof. Assume that $g \in C[[\hat{\mathfrak{B}}^C]]_{\succcurlyeq}$ is a solution to

$$R_{L\uparrow,+g,*}(0) = 0 \tag{7.34}$$

of multiplicity l. Let $f = g_{\succ}\downarrow/x$, $\alpha = g_{\asymp}$ and let k be the multiplicity of f as a solution of (7.17). We have to prove that

$$l > 0 \Leftrightarrow \alpha \in \{0, ..., k-1\} \Rightarrow l = 1.$$

Let $\mathfrak{m}$ be $\preccurlyeq$-maximal in $\operatorname{supp} f \setminus \mathfrak{B}^C$ and set $\psi = \sum_{\mathfrak{n} \succ \mathfrak{m}} f_{\mathfrak{n}} \mathfrak{n}$. If such an $\mathfrak{m}$ does not exist, then set $\psi = f$. Then, modulo replacing L by $e^{\int \psi} L_{\times e^{-\int \psi}}$, we may assume without generality that either $\mathfrak{d}_f \notin \mathfrak{B}^C$ or $f = 0$.

Let us first consider the case when $\mathfrak{m} = \mathfrak{d}_f \notin \mathfrak{B}^C$. Since all starting monomials for $R_{L,\mathrm{alg}}(f) = 0$ are necessarily in $\mathfrak{B}^C$, there exists an i with $L_j \mathfrak{m}^j \prec L_i \mathfrak{m}^i$ for all $j \neq i$. It follows from (7.4) that

$$\begin{aligned}
(L\uparrow)_i (\mathfrak{m}\uparrow e^x)^i &= (L_i\uparrow + O(L_{i+1}\uparrow) + \cdots + O(L_r\uparrow))\, \mathfrak{m}\uparrow^i \\
&\asymp L_i\uparrow \mathfrak{m}\uparrow^i \\
(L\uparrow)_j (\mathfrak{m}\uparrow e^x)^j &= (O(L_j\uparrow) + \cdots + O(L_r\uparrow))\, \mathfrak{m}\uparrow^j \\
&\prec L_i\uparrow \mathfrak{m}\uparrow^i \asymp (L\uparrow)_i (\mathfrak{m}\uparrow e^x)^i \qquad (j \neq i).
\end{aligned}$$

In other words, $\mathfrak{d}_g = \mathfrak{m}\uparrow e^x$ is not a starting monomial for $R_{L\uparrow,\mathrm{alg}}(g) = 0$, so neither (7.17) nor (7.34) holds.

Let us now consider the case when $f = 0$ and observe that k is minimal with $L_{*,k} \neq 0$. If $k = 0$, then $R_{L\uparrow,*} = L_{0,*}$, so we neither have (7.17) nor (7.34). If $\alpha \notin \{0, ..., k-1\}$, then

$$\begin{aligned}
R_{L\uparrow,+\alpha}(0) &= e^{-\alpha x} (L\uparrow)(e^{\alpha x}) \\
&= (x^{-\alpha} L(x^\alpha))\uparrow \\
&\asymp L_k\uparrow e^{-kx} \\
&\asymp L\uparrow,
\end{aligned}$$

so g does not satisfy (7.34). Similarly, if $\alpha \in \{0, ..., k-1\}$, then $R_{L\uparrow,+\alpha}(0) \prec L_k\uparrow e^{-kx} \asymp L\uparrow$, which implies (7.34). Moreover, setting $K = e^{-\alpha x} L\uparrow_{\times e^{\alpha x}}$, we have

$$\begin{aligned}
R_{K\uparrow,+1}(0) &= R_{L\uparrow\uparrow,+\alpha e^x+1}(0) \\
&= (x^{-\alpha} \log^{-1} x\, L(x^\alpha \log x))\uparrow\uparrow \\
&\asymp L_k\uparrow\uparrow e^{-ke^x - x} \\
&\asymp K\uparrow e^{(l-1)x} \\
&\succcurlyeq K\uparrow
\end{aligned}$$

In other words, $R_{K\uparrow,+1,*}(0) \neq 0$, whence $l = 1$. □

Theorem 7.32. *Let $L \in \mathbb{T}[\partial]$ be a linear differential operator on $\mathbb{T}$ of order r, whose coefficients can be expanded w.r.t. a plane transbasis $\mathfrak{B}$. Assume that $f_1, \ldots, f_s$ are the solutions to* (7.17)*, with multiplicities $\mu_1, \ldots, \mu_s$. Then*

$$\mathfrak{H}_L = \{\mathrm{e}^{\int f_1}, \ldots, x^{\mu_1 - 1} \mathrm{e}^{\int f_1}, \ldots, \mathrm{e}^{\int f_s}, \ldots, x^{\mu_s - 1} \mathrm{e}^{\int f_s}\}; \tag{7.35}$$

$$H_L \subseteq C[[\mathfrak{B}^C]][x]_r \{\mathrm{e}^{\int f_1}, \ldots, \mathrm{e}^{\int f_s}\}. \tag{7.36}$$

Proof. Let $\mathfrak{E}$ denote the set of exponential transmonomials and let us first assume that $\mathfrak{H}_L \subseteq x^{\mathbb{N}} \mathfrak{E}$. Then there exists a supertransbasis $\hat{\mathfrak{B}}$ of $\mathfrak{B}$, with $\mathfrak{H}_L \subseteq x^{\mathbb{N}} \hat{\mathfrak{B}}^C \subseteq x^{\mathbb{N}} \mathfrak{E}$ and $\mathrm{e}^{\int f_1}, \ldots, \mathrm{e}^{\int f_s} \in \hat{\mathfrak{B}}^C$. Now `riccati_solve` returns the same solutions with respect to $\mathfrak{B}$ and $\hat{\mathfrak{B}}$. Therefore, proposition 7.30 yields

$$\mathfrak{H}_L = \mathfrak{H}_L \cap x^{\mathbb{N}} \hat{\mathfrak{B}}^C = \mathfrak{H}_{L;\hat{\mathfrak{B}}} = \mathfrak{H}^*_{L;\hat{\mathfrak{B}}} = \mathfrak{H}^*_{L;\mathfrak{B}}.$$

In general, we have $\mathfrak{H}_L \uparrow_l = \mathfrak{H}_{L\uparrow_l}$ for some $l \geqslant 0$. So applying the above argument to $L\uparrow_l$, combined with proposition 7.31, we again have (7.35). As to (7.36), assume that $\mathfrak{h} = x^j \mathrm{e}^{f_i} \in \mathfrak{H}_L$ and let $K = \mathrm{e}^{-\int f_i} L_{\times \mathrm{e}^{\int f_i}} \in C[[\mathfrak{B}^C]][\partial]$. Then

$$h^{\mathfrak{h}} = \mathfrak{h} - L^{-1} L \mathfrak{h} = (x^j - K^{-1} K x^j) \, \mathrm{e}^{\int f_i} \in C[[\mathfrak{B}^C]][x]_r \mathrm{e}^{\int f_i}.$$

The result now follows from the fact that the $h^{\mathfrak{h}}$ form a basis of H_L. □

Since the equation (7.17) may admit less than r solutions (see remark 7.27), we may have $\dim \mathfrak{H}_L < r$. Nevertheless, proposition 7.28 implies:

Corollary 7.33. *If $L \in \mathbb{T}[\partial]$ is a linear differential operator of odd order, then the equation $L h = 0$ admits at least one non-trivial solution in $\mathbb{T}$.* □

7.7 Oscillating transseries

Let $L \in \mathbb{T}[\partial]^{\neq}$ be a linear differential operator of order r. Since C is only real closed, the dimension of the solution space H_L of $L h = 0$ can be strictly less than r. In order to obtain a full solution space of dimension r, we have both to consider transseries with complex coefficients and the adjunction of oscillating transmonomials. In this section we will sketch how to do this.

7.7.1 Complex and oscillating transseries

Let $\mathfrak{T}$ be the set of transmonomials and consider the field

$$\tilde{\mathbb{T}} = \mathbb{T} \oplus \mathrm{i} \, \mathbb{T} \cong (C + \mathrm{i} C)[[\mathfrak{T}]] = \tilde{C}[[\mathfrak{T}]]$$

of transseries with complex coefficients. Then most results from the previous sections can be generalized in a straightforward way to linear differential operators $L \in \tilde{\mathbb{T}}[\partial]$. We leave it as an exercise for the reader to prove the following particular statements:

Proposition 7.34. *Let $L \in \tilde{\mathbb{T}}[\partial]^{\neq}$ be a linear differential operator on $\tilde{\mathbb{T}}$. Then L admits a distinguished right inverse L^{-1} and H_L admits a finite distinguished basis.*

Proposition 7.35. *Let $L \in \tilde{C}[[\mathfrak{B}^C]][\partial]^{\neq}$ be a linear differential operator, where $\mathfrak{B}$ is a plane transbasis, and $\mathfrak{v} \in \mathfrak{B}^C \cup \{\mathrm{T}\}$. If the Newton degree of* (7.19) *is d, then* (7.19) *admits d solutions, when counted with multiplicities.*

An *oscillating transseries* is an expression of the form

$$f = f_{;\psi_1} \mathrm{e}^{\mathrm{i}\psi_1} + \cdots + f_{;\psi_k} \mathrm{e}^{\mathrm{i}\psi_k}, \tag{7.37}$$

where $f_{;\psi_1}, \ldots, f_{;\psi_k} \in \tilde{\mathbb{T}}$ and $\psi_1, \ldots, \psi_k \in \mathbb{T}_{\succ}$. Such transseries can be differentiated in a natural way

$$f' = (f'_{;\psi_1} + \mathrm{i}\,\psi'_1)\,\mathrm{e}^{\mathrm{i}\psi_1} + \cdots + (f'_{;\psi_k} + \mathrm{i}\,\psi'_k)\,\mathrm{e}^{\mathrm{i}\psi_k}.$$

We denote by

$$\mathbb{O} = \bigoplus_{\psi \in \mathbb{T}_{\succ}} \tilde{\mathbb{T}}\,\mathrm{e}^{\mathrm{i}\psi}$$

the differential ring of all oscillating transseries. Given an oscillating transseries $f \in \mathbb{O}$, we call (7.37) the *spectral decomposition* of f. Notice that

$$\mathbb{O} \cong C[[\mathrm{e}^{\mathbb{T}_{\succ}}]],$$

where $\mathrm{e}^f \preccurlyeq \mathrm{e}^g$ if and only if $\Re f \leqslant \Re g$ and $\Im f = \Im g$.

7.7.2 Oscillating solutions to linear differential equations

Consider a linear differential operator $L \in \tilde{\mathbb{T}}[\partial]^{\neq}$. We have

$$L f = \sum_{\psi \in \mathbb{T}_{\succ}} (L_{;\psi}\, f_{;\psi})\,\mathrm{e}^{\mathrm{i}\psi},$$

where

$$L_{;\psi} := \mathrm{e}^{-\mathrm{i}\psi}\, L_{\times \mathrm{e}^{\mathrm{i}\psi}} \in \tilde{\mathbb{T}}[\partial],$$

since $(\mathrm{e}^{\mathrm{i}\psi})^{\dagger} \in \tilde{\mathbb{T}}$ for all $\psi \in \mathbb{T}_{\succ}$. In other words, L "acts by spectral components" and its trace T_L is determined by

$$\begin{aligned} \mathfrak{R}_L &= \bigcup_{\psi \in \mathbb{T}_{\succ}} \mathfrak{R}_{L_{;\psi}}\, \mathrm{e}^{\mathrm{i}\psi} \\ T_L(c\,\mathfrak{m}\,\mathrm{e}^{\mathrm{i}\psi}) &= T_{L_{;\psi}}(c\,\mathfrak{m})\,\mathrm{e}^{\mathrm{i}\psi}. \end{aligned}$$

Now let $g \in \mathbb{O}$ and consider the differential equation

$$L f = g. \tag{7.38}$$

This equation is equivalent to the system of all equations of the form

$$L_{;\psi}\, f_{;\psi} = g_{;\psi}. \tag{7.39}$$

By proposition 7.34, the operators $L_{;\psi}$ all admit distinguished right inverses. We call

$$f = L^{-1} g = \sum_{\psi \in \mathbb{T}_{\succ}} L_{;\psi}^{-1} g_{;\psi} \mathrm{e}^{\mathrm{i}\psi}$$

the *distinguished solution* of (7.38). The operator L^{-1}: $g \mapsto L^{-1} g$, which is strongly linear, is called the *distinguished right inverse* of L. The solutions to the homogeneous equation may be found as follows:

Theorem 7.36. *Let $L \in \tilde{\mathbb{T}}[\partial]$ be a linear differential operator on $\tilde{\mathbb{T}}$ of order r, whose coefficients can be expanded w.r.t. a plane transbasis $\mathfrak{B}$. Assume that $f_1, ..., f_s$ are the solutions to* (7.17)*, with multiplicities $\mu_1, ..., \mu_s$. Then*

$$\mathfrak{H}_L = \{\mathrm{e}^{\int f_1}, ..., x^{\mu_1 - 1} \mathrm{e}^{\int f_1}, ..., \mathrm{e}^{\int f_s}, ..., x^{\mu_s - 1} \mathrm{e}^{\int f_s}\}; \tag{7.40}$$

$$H_L \subseteq C[\![\mathfrak{B}^C]\!][x]_r \{\mathrm{e}^{\int f_1}, ..., \mathrm{e}^{\int f_s}\}. \tag{7.41}$$

Proof. Let $\mathfrak{h} = x^j \mathfrak{m}$, where $\mathfrak{m} = \mathrm{e}^{\int f_i}$, $1 \leqslant i \leqslant s$ and $0 \leqslant j < \mu_i$. Then $K = \mathfrak{m}^{-1} L_{\times \mathfrak{m}}$, considered as an operator on $\tilde{\mathbb{T}}$, satisfies

$$R_{K,\mathrm{alg},j} = R_{L,+f_i,\mathrm{alg},j} \prec R_{L,+f_i} = R_K.$$

Hence $K_j \prec K$, $x^j \in \mathfrak{H}_K$ and $\mathfrak{h} \in \mathfrak{H}_L$. Furthermore,

$$h^{\mathfrak{h}} = \mathfrak{h} - L^{-1} L \mathfrak{h} = (x^j - K^{-1} K x^j) \mathrm{e}^{\int f_i} \in C[\![\mathfrak{B}^C]\!][x]_r \mathrm{e}^{\int f_i}$$

is an element of H_L with dominant monomial $\mathfrak{h}$. By proposition 7.35, there are r such solutions $h^{\mathfrak{h}}$ and they are linearly independent, since they have distinct dominant monomials. Consequently, they form a basis of H_L, since $\dim H_L \leqslant r$. This proves (7.41). Since each element $\mathfrak{h} \in \mathfrak{H}_L$ induces an element $h^{\mathfrak{h}} = \mathfrak{h} - L^{-1} L \mathfrak{h}$ with dominant monomial $\mathfrak{h}$ in H_L, we also have (7.40). □

Corollary 7.37. *Let $L \in \tilde{\mathbb{T}}[\partial]$ be a linear differential operator on $\tilde{\mathbb{T}}$ of order r. Then $\dim H_L = r$.* □

Remark 7.38. Due to the fact that the dimension r of H_L is maximal in theorem 7.36, its proof is significantly shorter than the proof of theorem 7.32. In particular, we do not need the equivalent of proposition 7.31, which was essentially used to check that upward shifting does not introduce essentially new solutions.

Exercise 7.24. Assume that C is a subfield of K and consider a strongly linear operator L: $C[\![\mathfrak{M}]\!] \rightarrow C[\![\mathfrak{N}]\!]$. Show that L extends by strong linearity into a strongly linear operator $\tilde{L}$: $K[\![\mathfrak{M}]\!] \rightarrow K[\![\mathfrak{N}]\!]$. If L admits a strongly linear right inverse L^{-1}, then show that the same holds for $\tilde{L}$ and $(\tilde{L}^{-1})_{|C[\![\mathfrak{N}]\!]} = L^{-1}$.

Exercise 7.25. Let $L \in \tilde{C}[\![\mathfrak{B}^C]\!][\partial]^{\neq}$.

a) Let $\tau \succcurlyeq 1$ be a starting term for (7.19) and assume that φ is a solution of (7.20) with $\tau_\varphi = \tau$. Consider the refinement $f = \varphi + \tilde{f}$ $(\tilde{f} \prec \tau)$ and let $\tilde{P} = P_{+\varphi}$. Prove that $\tilde{P}_0 \preccurlyeq_\tau \tau^{-1} P_0$.

b) Prove that any sequence of refinements like in (a) is necessarily finite.
c) Design an alternative algorithm for solving (7.19).
d) Given a solution $f \in \mathbb{T}$ to (7.19), prove that there exists a $\hat{f}$ in the algebraic closure of $C\{L_0, ..., L_r\}$, such that $\hat{f} - f \prec 1$.

Exercise 7.26. Let $M \in \mathscr{M}_r(\tilde{\mathbb{T}})$ be an $r \times r$ matrix with coefficients in $\tilde{\mathbb{T}}$ and consider the equation

$$V' = M V' \tag{7.42}$$

for $V' \in \mathbb{O}^r$.

a) Show that the equation $L h = 0$ can be reduced to an equation of the form (7.42) and *vice versa*.
b) If $\int M \prec 1$, then show that

$$V = I + \int M + \int M \int M + \cdots$$

is a solution to (7.42).

c) Assume that M is a block matrix of the form

$$M = \begin{pmatrix} M_1 & M_2 \\ M_3 & M_4 \end{pmatrix},$$

where M_2, M_3, $M_4 \prec M_1$ and M_1 is invertible with $\mathfrak{d}(M_1^{-1}) = \mathfrak{d}(M_1)^{-1}$. Consider the change of variables

$$V = P \tilde{V} = \begin{pmatrix} I & E \\ 0 & I \end{pmatrix} \tilde{V},$$

which transforms M into

$$\begin{aligned} \tilde{M} &= P^{-1} M P - P^{-1} P' \\ &= \begin{pmatrix} M_1 - M_3 E & M_2 + M_1 E - E M_4 - E M_3 E - E' \\ M_3 & M_3 E + M_4 \end{pmatrix}. \end{aligned}$$

Show that

$$M_2 + M_1 E - E M_4 - E M_3 E - E' = 0$$

admits a unique infinitesimal solution E. Also show that the coefficient M_3 can be cleared in a similar way.

d) Show that the equation (7.42) can be put in the form from (c) modulo a constant change of variables $V = P \tilde{V}$ with $P \in \mathscr{M}_r(\tilde{C})$.
e) Give an algorithm for solving (7.42) when there exist r different dominant monomials of eigenvalues of M. What about the general case?
f) Check the analogue of exercise 7.25(d) in the present setting.

Exercise 7.27. Take $C = \mathbb{R}$ and let L be as in exercise 7.6, but with coefficients in $L_{i,j} \in \tilde{\mathbb{T}}_{\lll e^x}$.

a) Determine the maximal flat subspace of $\mathbb{O}$ on which L is defined.
b) Show that L admits a distinguished right-inverse on $\mathbb{O}_{\lll e^x}$. Can $\mathfrak{H}_L$ be infinite?
c) Same question for $\mathbb{O}_{\underline{\lll} e^x}$ instead of $\mathbb{O}_{\lll e^x}$.

7.8 Factorization of differential operators

7.8.1 Existence of factorizations

One important consequence of corollary 7.37, i.e. the existence a full basis of solutions of dimension r of H_R, is the possibility to factor the L as a product of linear operators:

Theorem 7.39. *Any linear differential operator $L \in \tilde{\mathbb{T}}[\partial]^{\neq}$ of order r admits a factorization*

$$L = L_r\,(\partial - a_1) \cdots (\partial - a_r)$$

with $a_1, \ldots, a_r \in \tilde{\mathbb{T}}[\partial]$.

Proof. We prove the theorem by induction over the order r. For $r = 0$ we have nothing to prove. If $r \geqslant 1$, then there exists a non-trivial solution $h \in \tilde{\mathbb{T}}^{\neq}$ to the equation $L h = 0$, by corollary 7.37. Now the division of L by $\partial - h^{\dagger}$ in the ring $\tilde{\mathbb{T}}[\partial]$ yields a relation

$$L = \tilde{L}\,(\partial - h^{\dagger}) + \rho,$$

for some $\rho \in \tilde{\mathbb{T}}$, and $L h = \rho h = 0$ implies $\rho = 0$. The theorem therefore follows by induction over r. □

Theorem 7.40. *Any linear differential operator $L \in \mathbb{T}[\partial]^{\neq}$ admits a factorization as a product of a transseries in $\mathbb{T}$ and operators*

$$\partial - a$$

with $a \in \mathbb{T}$, or

$$\begin{aligned} &\partial^2 - (2\,a + b^{\dagger})\,\partial + (a^2 + b^2 - a' + a\,b^{\dagger}) = \\ &\quad (\partial - (a - b\,\mathrm{i} + b^{\dagger}))\,(\partial - (a + b\,\mathrm{i})) \end{aligned}$$

with $a, b \in \mathbb{T}$.

Proof. We prove the theorem by induction over the order r of L. If $r = 0$ then we have nothing to do. If there exists a solution $h \in \mathbb{T}$ to $L h = 0$, then we conclude in a similar way as in theorem 7.39. Otherwise, there exists a solution $h^{\dagger} \in \tilde{\mathbb{T}}$ to the Riccati equation $R_L(h^{\dagger})$, such that $h^{\dagger} = a + b\,\mathrm{i}$ with $a, b \in \mathbb{T}$ and $b \neq 0$. Now division of L by $(\partial - (a - b\,\mathrm{i} + b^{\dagger}))\,(\partial - (a + b\,\mathrm{i}))$ in the ring $\mathbb{T}[\partial]$ yields

$$\begin{aligned} L &= \tilde{L}\,(\partial - (a - b\,\mathrm{i} + b^{\dagger}))\,(\partial - (a + b\,\mathrm{i})) + R \\ &= \tilde{L}\,(\partial - (a + b\,\mathrm{i} + b^{\dagger}))\,(\partial - (a - b\,\mathrm{i})) + R \end{aligned}$$

for some differential operator R of order < 2. Moreover, R is both a multiple of $\partial - (a + b\,\mathrm{i})$ and $\partial - (a - b\,\mathrm{i})$, when considered as an operator in $\tilde{\mathbb{T}}[\partial]$. But this is only possible if $R = 0$. We conclude by induction. □

7.8.2 Distinguished factorizations

We have seen in section 7.4 that the total ordering on the transmonomials allows us to isolate a distinguished basis of solutions to the equation $L\,h=0$. A natural question is whether such special bases of solutions induce special factorizations of L and vice versa.

We will call a series f *monic*, if f is regular and $c_f=1$. Similarly, a differential operator L of order r is said to be *monic* if $L_r=1$. A tuple of elements is said to be monic if each element is monic. Given a regular series f, the series $\operatorname{mon} f:=f/c_f$ is monic. In what follows we will consider bases of H_L as tuples $(h_1,\ldots,h_r)$. We will also represent factorizations $L=(\partial-f_1)\cdots(\partial-f_r)$ of monic differential operators by tuples $(f_1,\ldots,f_r)$.

Proposition 7.41. *Let $L\in\tilde{\mathbb{T}}[\partial]^{\neq}$ be a monic linear differential operator on $\mathbb{O}$ of order r. Then*

a) To any monic basis $(h_1,\ldots,h_r)$ of H_L, we may associate a factorization

$$\begin{aligned} L &= (\partial-f_1)\cdots(\partial-f_r),\\ f_i &= [(\partial-f_{i+1})\cdots(\partial-f_r)\,h_i]^{\dagger} \quad (i=r,\ldots,1), \end{aligned}$$

and we write $(f_1,\ldots,f_r)=\operatorname{fact}(h_1,\ldots,h_r)$.

b) To any factorization

$$L=(\partial-f_1)\cdots(\partial-f_r),$$

we may associate a monic basis $(h_1,\ldots,h_r)=\operatorname{sol}(f_1,\ldots,f_r)$ of H by

$$h_i=\operatorname{mon}\,[(\partial-f_{i+1})\cdots(\partial-f_r)]^{-1}\,\mathrm{e}^{\int f_i} \quad (i=r,\ldots,1).$$

We have $h_{i,\mathfrak{d}(h_j)}=0$ for all $i<j$.

c) For any factorization represented by $(f_1,\ldots,f_r)$ we have

$$\operatorname{fact}\operatorname{sol}(f_1,\ldots,f_r)=(f_1,\ldots,f_r).$$

d) If $(h_1,\ldots,h_r)$ is a monic basis of H_L such that $h_{i,\mathfrak{d}(h_j)}=0$ for all $i<j$, then

$$\operatorname{sol}\operatorname{fact}(h_1,\ldots,h_r)=(h_1,\ldots,h_r).$$

Proof. Assume that $(h_1,\ldots,h_r)$ is a monic basis of H_L and let us prove by induction that $(\partial-f_{i+1})\cdots(\partial-f_r)$ is a right factor of L for all $i=r,\ldots,0$. This is clear for $i=r$. Assume that

$$L=K\,(\partial-f_{i+1})\cdots(\partial-f_r)$$

for some $i\in\{1,\ldots,r\}$. Then

$$K\,(\partial-f_{i+1})\cdots(\partial-f_r)\,h_i=0$$

implies that $\partial-f_i$ is a right factor of K, in a similar way as in the proof of theorem 7.39. Hence (a) follows by induction.

As to (b), the h_i are clearly monic solutions of $L h=0$, and, more generally,

$$(\partial - f_{i+1}) \cdots (\partial - f_r) h_j = 0$$

for $j > i$. The distinguished property of $[(\partial - f_{i+1}) \cdots (\partial - f_r)]^{-1}$ therefore implies that $h_{i,\mathfrak{d}(h_j)} = 0$ for all $j > i$. This also guarantees the linear independence of the h_i. Indeed, assume that we have a relation

$$\lambda_1 h_1 + \cdots + \lambda_i h_i = 0.$$

Then

$$0 = (\lambda_1 h_1 + \cdots + \lambda_i h_i)_{\mathfrak{d}(h_i)} = \lambda_i,$$

and, repeating the argument, $\lambda_{i-1} = \cdots = \lambda_1 = 0$. This proves (b).

Now consider a factorization $L = (\partial - f_1) \cdots (\partial - f_r)$ and let

$$(\tilde{f}_1, \ldots, \tilde{f}_r) = \operatorname{fact} \operatorname{sol} (f_1, \ldots, f_r).$$

Given $i \in \{1, \ldots, r\}$ with $\tilde{f}_{i+1} = f_{i+1}, \ldots, \tilde{f}_r = f_r$, we get

$$\begin{aligned} \tilde{f}_i &= [(\partial - \tilde{f}_{i+1}) \cdots (\partial - \tilde{f}_r) \operatorname{mon} [(\partial - f_{i+1}) \cdots (\partial - f_r)]^{-1} e^{\int f_i}]^\dagger \\ &= [c^{-1} (\partial - f_{i+1}) \cdots (\partial - f_r) [(\partial - f_{i+1}) \cdots (\partial - f_r)]^{-1} e^{\int f_i}]^\dagger \\ &= (c^{-1} e^{\int f_i})^\dagger = f_i, \end{aligned}$$

where $c \in C^{\neq}$ is the dominant coefficient of

$$[(\partial - f_{i+1}) \cdots (\partial - f_r)]^{-1} e^{\int f_i}.$$

Applying the above argument for $i = r, \ldots, 1$, we obtain (c).

Let us finally consider a monic basis $(h_1, \ldots, h_r)$ of H_L such that $h_{i,\mathfrak{d}(h_j)} = 0$ for all $i < j$. Let

$$\begin{aligned} (f_1, \ldots, f_r) &= \operatorname{fact} (h_1, \ldots, h_r) \\ (\tilde{h}_1, \ldots, \tilde{h}_r) &= \operatorname{sol} (f_1, \ldots, f_r) \end{aligned}$$

Assume that $\tilde{h}_{i+1} = h_{i+1}, \ldots, \tilde{h}_r = h_r$ for some $i \in \{1, \ldots, r\}$ and let

$$K = (\partial - f_i) \cdots (\partial - f_r).$$

Then both $(h_i, \ldots, h_r)$ and $(\tilde{h}_i, h_{i+1}, \ldots, h_r)$ form monic bases for H_K and $h_{i,\mathfrak{d}(h_j)} = \tilde{h}_{i,\mathfrak{d}(h_j)} = 0$ for all $j > i$. It follows that $(\tilde{h}_i - h_i)_{\mathfrak{h}} = 0$ for all $\mathfrak{h} \in \mathfrak{H}_K$, whence $\tilde{h}_i = h_i$. Applying the argument for $i = r, \ldots, 1$, we obtain (d). □

The *distinguished basis* of H_L is the unique monic basis $(h_1, \ldots, h_r)$ such that $h_{i,\mathfrak{d}(h_j)} = 0$ for all $i < j$ and $h_1 \succ \cdots \succ h_r$. The corresponding factorization of L is called the *distinguished factorization*.

Exercise 7.28. Assume that $L \in \mathbb{T}[\partial]$ admits a factorization

$$L = (\partial - f_1) \cdots (\partial - f_r)$$

with $f_1, \ldots, f_r \in \mathbb{T}$. Then

a) Prove that there exists a unique such factorization with $f_1 \geqslant \cdots \geqslant f_r$.
b) Prove that this unique factorization is the distinguished factorization.

8

Algebraic differential equations

Let $\mathbb{T} = C[\![x]\!]$ be the field of grid-based transseries in x over a real closed field C and let $P \in \mathbb{T}\{F\}$ be a differential polynomial of order r. In this chapter, we show how to determine the transseries solutions of the equation

$$P(f) = 0.$$

More generally, given an initial segment $\mathfrak{V} \subseteq \mathfrak{T}$ of transmonomials, so that

$$\mathfrak{v} \in \mathfrak{V} \wedge \mathfrak{w} \preccurlyeq \mathfrak{v} \Rightarrow \mathfrak{w} \in \mathfrak{V},$$

we will study the asymptotic algebraic differential equation

$$P(f) = 0 \quad (f \in C[\![\mathfrak{V}]\!]). \tag{E}$$

Usually, we have $\mathfrak{V} = \mathfrak{T}$ or $\mathfrak{V} = \{\mathfrak{w} \in \mathfrak{T} : \mathfrak{w} \prec \mathfrak{v}\}$ for some $\mathfrak{v}$.

In order to solve (E), we will generalize the Newton polygon method from chapter 3 to the differential setting. This program gives rise to several difficulties. First of all, the starting monomials for differential equations cannot be read off directly from the Newton polygon. For instance, the equation $f' = e^{e^x}$ admits a starting monomial $e^{e^x - x}$ whereas the Newton polygon would suggest e^{e^x} instead. Also, it is no longer true that cancellations are necessarily due to terms of different degrees, as is seen for the equation $f' = f$, which admits e^x as a starting monomial.

In order to overcome this first difficulty, the idea is to find a criterion which tells us when a monomial $\mathfrak{m}$ is a starting monomial for the equation (E). The criterion we will use is the requirement that the differential Newton polynomial associated to $\mathfrak{m}$ admits a non-zero solution in the algebraic closure of C. Differential Newton polynomials are defined in section 8.3.1; modulo multiplicative conjugations, it will actually suffice to define them in the case when $\mathfrak{m} = 1$. In section 8.3.3, we will show how to compute starting monomials and terms. Actually, the starting monomials which correspond to cancellations between terms of different degrees can almost be read off from the Newton polygon. The other ones are computed using Riccati equations.

A second important difficulty with the differential Newton polygon method is that almost multiple solutions are harder to "unravel" using the differentiation technique from section 3.1.3. One obvious reason is that the quasi-linear equation obtained after differentiation is a differential equation with potentially multiple solutions. Another more pathological reason is illustrated by the example

$$f^2 + 2 f' + \frac{1}{x^2} + \frac{1}{x^2 \log^2 x} + \cdots + \frac{1}{x^2 \log^2 x \cdots \log_l^2 x} = 0. \tag{8.1}$$

Although the coefficient of f in this equation vanishes, the equation admits $\frac{1}{x}$ as a starting term of multiplicity 2. Indeed, setting $f = \frac{1}{x} \tilde{f}$, we get

$$\tilde{f}^2 + 2 \tilde{f}' - 2 \tilde{f} + 1 + \frac{1}{\log^2 x} + \cdots + \frac{1}{\log^2 x \cdots \log_l^2 x} = 0.$$

Differentiation yield the quasi-linear equation

$$2 \tilde{f} - 2 = 0,$$

but after the refinement $\tilde{f} = 1 + \tilde{\tilde{f}}$ ($\tilde{\tilde{f}} \prec 1$) and upward shifting, we obtain an equation

$$\tilde{\tilde{f}}^2 + 2 \tilde{\tilde{f}}' + \frac{1}{x^2} + \frac{1}{x^2 \log^2 x} + \cdots + \frac{1}{x^2 \log^2 x \cdots \log_{l-1}^2 x} = 0,$$

which has the same form as (8.1). This makes it hard to unravel almost multiple solutions in a constructive way. Nevertheless, as we will see in section 8.6, the strong finiteness properties of the supports of grid-based transseries will ensure the existence of a brute-force unravelling algorithm.

In section 8.7 we put all techniques of the chapter together in order to state an explicit (although theoretical) algorithm for the resolution of (E). In this algorithm, we will consider the computation of the distinguished solution to a quasi-linear equation as a basic operation. Quasi-linear equations are studied in detail in section 8.5.

In the last section, we prove a few structural results about the solutions of (E). We start by generalizing the notion of distinguished solutions to equations of Newton degree $d > 1$. We next prove that (E) admits at least one solution if d is odd. We will also prove a bound for the number of "new exponentials" which may occur in solutions to (E).

8.1 Decomposing differential polynomials

8.1.1 Serial decomposition

Let $P \in \mathbb{T}\{F\}$ be a differential polynomial over $\mathbb{T}$ of order r. In the previous chapter, we have already observed that we may interpret P as a series

$$P = \sum_{\mathfrak{m} \in \mathfrak{T}} P_{\mathfrak{m}} \mathfrak{m}, \tag{8.2}$$

where the coefficients are differential polynomials in $C\{F\}$. We call (8.2) the *serial decomposition* of P. As before, the embedding $\mathbb{T}\{F\} \hookrightarrow C\{F\}[[\mathfrak{T}]]$ induces definitions for the asymptotic relations $\preccurlyeq$, $\prec$, etc. and dominant monomials and coefficients of differential polynomials. We will denote by D_P the dominant coefficient of P.

8.1.2 Decomposition by degrees

The most natural decomposition of P is given by

$$P(f) = \sum_{\boldsymbol{i}} P_{\boldsymbol{i}} f^{\boldsymbol{i}}. \tag{8.3}$$

Here we use vector notation for tuples

$$\begin{aligned} \boldsymbol{i} &= (i_0, \ldots, i_r) \\ \boldsymbol{j} &= (j_0, \ldots, j_r) \end{aligned}$$

of integers:

$$\begin{aligned} \|\boldsymbol{i}\| &= i_0 + \cdots + i_r; \\ f^{\boldsymbol{i}} &= f^{i_0} (f')^{i_1} \cdots (f^{(r)})^{i_r}; \\ \boldsymbol{i} \leqslant \boldsymbol{j} &\Leftrightarrow i_0 \leqslant j_0 \wedge \cdots \wedge i_r \leqslant j_r; \\ \binom{\boldsymbol{j}}{\boldsymbol{i}} &= \binom{j_0}{i_0} \cdots \binom{j_r}{i_r}. \end{aligned}$$

We call (8.3) the *decomposition* of P *by degrees*. The i-th *homogeneous part* of P is defined by

$$P_i = \sum_{\|\boldsymbol{i}\| = i} P_{\boldsymbol{i}} f^{\boldsymbol{i}},$$

so that

$$P = \sum_i P_i. \tag{8.4}$$

We call (8.4) the *decomposition* of P *into homogeneous parts*. If $P \neq 0$, then the largest $d = \deg P$ with $P_d \neq 0$ is called the *degree* of P and the smallest $\nu = \operatorname{val} P$ with $P_\nu \neq 0$ the *differential valuation* of P.

8.1.3 Decomposition by orders

Another useful decomposition of P is its *decomposition by orders*:

$$P(f) = \sum_{\boldsymbol{\omega}} P_{[\boldsymbol{\omega}]} f^{[\boldsymbol{\omega}]} \tag{8.5}$$

In this notation, $\boldsymbol{\omega}$ runs through tuples $\boldsymbol{\omega} = (\omega_1, ..., \omega_l)$ of integers in $\{0, ..., r\}$ of length $l \leqslant \deg P$, and $P_{[\boldsymbol{\omega}]} = P_{[\omega_{\sigma(1)}, ..., \omega_{\sigma(l)}]}$ for all permutations of integers. We again use vector notation for such tuples

$$\begin{aligned}
|\boldsymbol{\omega}| &= l; \\
\|\boldsymbol{\omega}\| &= \omega_1 + \cdots + \omega_l; \\
f^{[\boldsymbol{\omega}]} &= f^{(\omega_1)} \ldots f^{(\omega_l)}; \\
\boldsymbol{\omega} \leqslant \boldsymbol{\tau} &\Leftrightarrow \omega_1 \leqslant \tau_1 \wedge \cdots \wedge \omega_l \leqslant \tau_l; \\
\binom{\boldsymbol{\tau}}{\boldsymbol{\omega}} &= \binom{\tau_1}{\omega_1} \cdots \binom{\tau_l}{\omega_l}.
\end{aligned}$$

For the last two definitions, we assume that $|\boldsymbol{\omega}| = |\boldsymbol{\tau}| = l$. We call $\|\boldsymbol{\omega}\|$ the *weight* of $\boldsymbol{\omega}$. The ω-th *isobaric part* of P is defined by

$$P_{[\omega]} = \sum_{\|\boldsymbol{\omega}\| = \omega} P_{[\boldsymbol{\omega}]} f^{[\boldsymbol{\omega}]},$$

so that

$$P = \sum_{\omega} P_{[\omega]}. \tag{8.6}$$

We call (8.6) the *decomposition* of P *into isobaric parts*. If $P \neq 0$, then the largest $\omega = \operatorname{wt} P$ with $P_{[\omega]} \neq 0$ is called the *weight* of P and the smallest $\omega = \operatorname{wv} P$ with $P_{[\omega]} \neq 0$ the *weighted differential valuation* of P.

8.1.4 Logarithmic decomposition

It is convenient to denote the successive logarithmic derivatives of f by

$$\begin{aligned}
f^{\dagger} &= f'/f; \\
f^{\langle i \rangle} &= f^{\dagger \cdots \dagger} \qquad (i \text{ times}).
\end{aligned}$$

Then each $f^{(i)}$ can be rewritten as a polynomial in $f, f^{\dagger}, ..., f^{\langle i \rangle}$:

$$\begin{aligned}
f &= f; \\
f' &= f^{\dagger} f; \\
f'' &= ((f^{\dagger})^2 + f^{\dagger\dagger} f^{\dagger}) f; \\
f''' &= ((f^{\dagger})^3 + 3 f^{\dagger\dagger} (f^{\dagger})^2 + (f^{\dagger\dagger})^2 f^{\dagger} + f^{\dagger\dagger\dagger} f^{\dagger\dagger} f^{\dagger}) f; \\
&\vdots
\end{aligned}$$

We define the *logarithmic decomposition* of P by

$$P(f) = \sum_{\boldsymbol{i} = (i_0, ..., i_r)} P_{\langle \boldsymbol{i} \rangle} f^{\langle \boldsymbol{i} \rangle}, \tag{8.7}$$

where

$$f^{\langle \boldsymbol{i} \rangle} = f^{i_0} (f^{\dagger})^{i_1} \cdots (f^{\langle r \rangle})^{i_r}.$$

Now consider the total lexicographical ordering $\leqslant^{\text{lex}}$ on $\mathbb{N}^{r+1}$, defined by

$$\begin{aligned} \boldsymbol{i} <^{\text{lex}} \boldsymbol{j} \iff & (i_0 < j_0) \vee \\ & (i_0 = j_0 \wedge i_1 < j_0) \vee \\ & \vdots \\ & (i_0 = j_0 \wedge \cdots \wedge i_{r-1} = j_{r-1} \wedge i_r < j_r). \end{aligned}$$

Assuming that $P \neq 0$, let $\boldsymbol{i}$ be maximal for $\leqslant^{\text{lex}}$ with $P_{\langle \boldsymbol{i} \rangle} \neq 0$. Then

$$P(f) \sim P_{\langle \boldsymbol{i} \rangle} f^{\langle \boldsymbol{i} \rangle} \tag{8.8}$$

for $f \to \infty_{\mathbb{T}}$ or $f \to -\infty_{\mathbb{T}}$.

8.2 Operations on differential polynomials

8.2.1 Additive conjugation

Given a differential polynomial $P \in \mathbb{T}\{F\}$ and a transseries $h \in \mathbb{T}$, the *additive conjugation* of P with h is the unique differential polynomial $P_{+h} \in \mathbb{T}\{F\}$, such that

$$P_{+h}(f) = P(h + f)$$

for all $f \in \mathbb{T}$. The coefficients of P_{+h} are explicitly given by

$$P_{+h,\boldsymbol{i}} = \sum_{\boldsymbol{j} \geqslant \boldsymbol{i}} \binom{\boldsymbol{j}}{\boldsymbol{i}} h^{\boldsymbol{j}-\boldsymbol{i}} P_{\boldsymbol{j}}. \tag{8.9}$$

Notice that for all $i \in \mathbb{N}$, we have

$$\left(\frac{\partial P}{\partial F^{(i)}}\right)_{+\varphi} = \frac{\partial P_{+\varphi}}{\partial F^{(i)}}.$$

Proposition 8.1. *If $h = c + \varepsilon$ with $c \in C$ and $\varepsilon \prec 1$, then*

$$\begin{aligned} P_{+h} &\asymp P \\ D_{P_{+h}} &= D_{P,+c} \end{aligned}$$

Proof. The relation (8.9) both yields $P_{+h} \preccurlyeq P$ and

$$P = P_{+h,-h} \preccurlyeq P_{+h},$$

so $P_{+h} \asymp P$. Furthermore,

$$P_{+h,\boldsymbol{i}} = P_{\boldsymbol{i}} + \sum_{\boldsymbol{j} \geqslant \boldsymbol{i}} (c^{\boldsymbol{j}-\boldsymbol{i}} + o(1))\, P_{\boldsymbol{j}} = P_{+c,\boldsymbol{i}} + o(P)$$

for all $\boldsymbol{i}$, whence the second relation. □

8.2.2 Multiplicative conjugation

The *multiplicative conjugation* of a differential polynomial $P \in \mathbb{T}\{F\}$ with a transseries $h \in \mathbb{T}$ is the unique differential polynomial $P_{\times h} \in \mathbb{T}\{F\}$, such that

$$P_{\times h}(f) = P(h\,f)$$

for all $f \in \mathbb{T}$. The coefficients of $P_{\times h}$ are given by

$$P_{\times h,[\omega]} = \sum_{\tau \geqslant \omega} \binom{\tau}{\omega} h^{[\tau-\omega]} P_{[\tau]}. \tag{8.10}$$

Proposition 8.2.

a) If $h \succ\!\!\succ x$, then for all i,

$$P_{i,\times h} \asymp_h h^i\,P_i.$$

b) If $h \succ\!\!\succ x$, then

$$P_{\times h} \asymp^{*}_{h} P.$$

c) If P and $h > 0$ are exponential, then

$$P_{\times h} \asymp^{*}_{\log h} h\,P.$$

Proof. If $h \succ\!\!\succ x$, then the equation (8.10) implies $P_{i,\times h} \preccurlyeq_h h^i\;P_i$ and $P_i \preccurlyeq_h h^{-i}\;P_{i,\times h}$, whence ($a$). Part ($b$) follows directly from (a), and (c) is proved in a similar way. □

8.2.3 Upward and downward shifting

The *upward* and *downward shiftings* of a differential polynomial P are the unique differential polynomials $P\uparrow$ resp. $P\downarrow$ in $\mathbb{T}\{F\}$ such that

$$\begin{aligned} P\uparrow(f\uparrow) &= P(f)\uparrow \\ P\downarrow(f\downarrow) &= P(f)\downarrow \end{aligned}$$

for all $f \in \mathbb{T}$. The non-linear generalizations of the formulas (7.4) and (7.5) for the coefficients of $P\uparrow$ and $P\downarrow$ are

$$(P\uparrow)_{[\omega]} = \sum_{\tau \geqslant \omega} s_{\tau,\omega}\,\mathrm{e}^{-\|\tau\|x}\,(P_{[\tau]}\uparrow) \tag{8.11}$$

$$(P\downarrow)_{[\omega]} = \sum_{\tau \geqslant \omega} S_{\tau,\omega}\,x^{\|\omega\|}\,(P_{[\tau]}\downarrow), \tag{8.12}$$

where the $s_{\tau,\omega}$ are generalized Stirling numbers of the first kind

$$\begin{aligned} s_{\tau,\omega} &= s_{\tau_1,\omega_1} \cdots s_{\tau_l,\omega_l} \\ (f(\log x))^{(j)} &= \sum_{i=0}^{j} s_{j,i}\,x^{-j} f^{(i)}(\log x) \end{aligned}$$

and the $S_{\boldsymbol{\tau},\boldsymbol{\omega}}$ are generalized Stirling numbers of the second kind

$$S_{\boldsymbol{\tau},\boldsymbol{\omega}} = S_{\tau_1,\omega_1} \cdots S_{\tau_l,\omega_l}$$
$$(f(\mathrm{e}^x))^{(j)} = \sum_{i=0}^{j} S_{j,i}\, \mathrm{e}^{ix} f^{(i)}(\mathrm{e}^x).$$

Proposition 8.3. *We have*

$$P\uparrow \asymp^*_{\mathrm{e}^x} \mathfrak{d}_P\uparrow.$$

Proof. We get $P\uparrow \preccurlyeq^*_{\mathrm{e}^x} \mathfrak{d}_P\uparrow$ from (8.11) and $P = P\uparrow\downarrow \preccurlyeq^*_x \mathfrak{d}_{P\uparrow}\downarrow$ from (8.12). □

Proposition 8.4. *If $P \in \mathbb{T}\{F\}$ is exponential, then*

$$D_{P\uparrow} = D_{D_P\uparrow}.$$

Proof. Since $P = (D_P + o_{\mathrm{e}^x}(1))\, \mathfrak{d}_P$, the equation (8.11) yields

$$P\uparrow = (D_P\uparrow + o_{\mathrm{e}^{\mathrm{e}^x}}(1))\, (\mathfrak{d}_P\uparrow)$$

and $\operatorname{supp} D_P\uparrow \subseteq \{\mathrm{e}^{-\mathbb{N}x}\} \asymp_{\mathrm{e}^{\mathrm{e}^x}} 1$. This clearly implies the relation. □

Exercise 8.1. Let $g \in \mathbb{T}^{>,\succ}$ and $P \in \mathbb{T}\{F\}$.

a) Show that there exists a unique $P_{\circ g} \in \mathbb{T}\{F\}$ with

$$P_{\circ g}(f \circ g) = P(f) \circ g$$

for all $f \in \mathbb{T}$.

b) Give an explicit formula for $P_{\circ g,[\boldsymbol{\omega}]}$ for all $\boldsymbol{\omega}$.

c) Show that $\cdot_{\circ g}$ is a differential ring homomorphism:

$$\begin{aligned} (\mathbb{T}\{F\}, \partial) &\longrightarrow (\mathbb{T}\{F\}, (g')^{-1}\partial) \\ P &\longmapsto P_{\circ g} \end{aligned}$$

Exercise 8.2. Let $P \in \mathbb{T}\{F_1, ..., F_k\}$ and $Q_1, ..., Q_k \in \mathbb{T}\{F_1, ..., F_l\}$.

a) Let $P \circ (Q_1, ..., Q_k) \in \mathbb{T}\{F_1, ..., F_l\}$ be the result of the substitution of Q_i for each F_i in P. Show that $P \mapsto P \circ (Q_1, ..., Q_l)$ is a morphism of differential rings.

b) Reinterpret additive and multiplicative conjugation using composition like above.

c) Show that $\mathbb{T}[\partial]$ is isomorphic to $(\mathbb{T}\{F\}_{\mathrm{lin}}, +, \circ)$, where

$$\mathbb{T}\{F\}_{\mathrm{lin}} = \mathbb{T}\, F \oplus \mathbb{T}\, F' \oplus \cdots.$$

Exercise 8.3. Let $P = \sum_i P_i F^i \in \mathbb{T}_{\prec\!\!\prec \mathrm{e}^x}[[F, F', ...]]$.

a) If (P_i) forms a grid-based family, then show that $P(f)$ is well-defined for all $f \in \mathbb{T}^{\preccurlyeq}_{\prec\!\!\prec \mathrm{e}^x}$.

b) For two operators P and Q like in (a), with $Q \preccurlyeq 1$, show that $P \circ Q$ is well-defined.

c) Generalize (b) to operators in several variables and to more general subspaces of the form $C[[\mathfrak{V}]]$ of $\mathbb{T}_{\prec\!\!\prec \mathrm{e}^x}$.

8.3 The differential Newton polygon method

8.3.1 Differential Newton polynomials

Recall from the introduction that, in order to generalize the Newton polygon method to the differential setting, it is convenient to first define the differential Newton polynomial associated to a monomial $\mathfrak{m}$. We will start with the case when $\mathfrak{m}=1$ and rely on the following key observations:

Lemma 8.5. *Let $P\in C\{F\}$ be isobaric, of weight ν and assume that $D_{P\uparrow}=P$. Then $P\in C[F]\,(F')^{\nu}$.*

Proof. For all isobaric $H\in C\{F\}$ of weight ν, let us denote

$$H^{*}=\sum_{j} H_{(j,\nu,0,\ldots,0)}\,F^{j}\,(F')^{\nu}.$$

Then $Q=P-P^{*}$ satisfies $D_{Q\uparrow}=Q$ and $Q^{*}=0$. Furthermore, (8.11) yields

$$Q\uparrow=\mathrm{e}^{-\nu x}\,Q.$$

Consequently, if $Q(f)=0$ for some $f\in\mathbb{T}$, then

$$Q(f\uparrow)=\mathrm{e}^{\nu x}\,(Q\uparrow)(f\uparrow)=\mathrm{e}^{\nu x}\,(Q(f)\uparrow)=0.$$

Since $Q^{*}=0$ implies $Q(x)=0$, it follows by induction that $Q(\exp_i x)=0$ for any iterated exponential of x. From (8.8), we conclude that $Q=0$ and $P\in C[F]\,(F')^{\nu}$. □

Theorem 8.6. *Let P be a differential polynomial with exponential coefficients. Then there exists a polynomial $Q\in C[F]$ and an integer ν, such that for all $l\geqslant \operatorname{wt} P$, we have $D_{P\uparrow_l}=Q\,(F')^{\nu}$.*

Proof. By formula (8.11), we have $D_P\uparrow\asymp\mathrm{e}^{-(\operatorname{wv} D_P)x}$ and

$$D_{P\uparrow}(F)=\sum_{\|\omega\|=\operatorname{wv} D_P}\left(\sum_{\tau\geqslant\omega} s_{\tau,\omega}\,D_{P,[\tau]}\right)F^{[\omega]}. \qquad (8.13)$$

Consequently,

$$\operatorname{wt} D_P\geqslant\operatorname{wv} D_P=\operatorname{wt} D_{P\uparrow}\geqslant\operatorname{wv} D_{P\uparrow}=\operatorname{wt} D_{P\uparrow\uparrow}\geqslant\cdots.$$

Hence, for some $l\leqslant\operatorname{wt} P$, we have $\operatorname{wt} D_{P\uparrow_{l+1}}=\operatorname{wv} D_{P\uparrow_{l+1}}=\operatorname{wt} D_{P\uparrow_l}$. Now (8.13) applied on $P\uparrow_l$ instead of P yields $D_{P\uparrow_{l+1}}=D_{P\uparrow_l}$. Proposition 8.4 therefore gives

$$D_{P\uparrow_l}=D_{P\uparrow_{l+1}}=D_{D_{P\uparrow_l}\uparrow}=D_{D_{P\uparrow_{l+1}}\uparrow}=D_{P\uparrow_{l+2}}=\cdots.$$

We conclude by applying lemma 8.5 with $D_{P\uparrow_l}$ for P. □

Given an arbitrary differential polynomial P, the above theorem implies that there exists a polynomial $Q\in C[F]$ and an integer ν, such that $D_{P\uparrow_l}=Q\,(F')^{\nu}$ for all sufficiently large l. We call

$$N_P=Q\,(F')^{\nu}$$

the *differential Newton polynomial* for P. More generally, if $\mathfrak{m}$ is an arbitrary monomial, then we call $N_{P_{\times\mathfrak{m}}}$ the *differential Newton polynomial for P associated to* $\mathfrak{m}$. If P is exponential and $N_P=D_P$, then we say that P is *transparent*. Notice that a transseries is transparent if and only if it is exponential.

8.3.2 Properties of differential Newton polynomials

Proposition 8.7.

a) $N_{P\uparrow}=N_P$ *for all* P.
b) *If* $c\in C$ *and* $\varepsilon\prec 1$*, then* $N_{P_{+c+\varepsilon}}=N_{P,+c}$.
c) *If* $\mathfrak{m}\prec\mathfrak{n}$*, then* $\operatorname{val}N_{P_{\times\mathfrak{m}}}\leqslant\deg N_{P_{\times\mathfrak{m}}}\leqslant\operatorname{val}N_{P_{\times\mathfrak{n}}}\leqslant\deg N_{P_{\times\mathfrak{n}}}$.

Proof. Assertion (a) is trivial, by construction.

In (b), modulo a sufficient number of upward shiftings, we may assume without loss of generality that P, $P_{+c+\varepsilon}$ and ε are transparent. Dividing P by $\mathfrak{d}_P$, we may also assume that $P\asymp 1$. Then (8.9) implies

$$P_{+c+\varepsilon}=D_{P,+c+\varepsilon}+o_{\mathrm{e}^x}(1)=D_{P,+c}+o_{\mathrm{e}^x}(1),$$

so that $N_{P_{+c+\varepsilon}}=D_{P_{+c+\varepsilon}}=D_{P,+c}=N_{P,+c}$.

As to (c), it clearly suffices to consider the case when $\mathfrak{m}\prec 1$ and $\mathfrak{n}=1$. After a finite number of upward shiftings, we may also assume that P and $P_{\times\mathfrak{m}}$ are transparent and $\mathfrak{m}\succ\!\!\succ x$. Let $d=\operatorname{val}P$. Then for all $i>d$ we have $P_i\preccurlyeq P_d$, whence

$$P_{\times\mathfrak{m},d}=P_{d,\times\mathfrak{m}}\asymp_{\mathfrak{m}}\mathfrak{m}^d\,P_d\succ\mathfrak{m}^i\,P_i\asymp_{\mathfrak{m}}P_{i,\times\mathfrak{m}}=P_{\times\mathfrak{m},i},$$

by proposition 8.2(a). This implies $\deg D_{\times\mathfrak{m}}\leqslant d$, as desired. □

Proposition 8.8. *Let* $P\in\mathbb{T}\{F\}^{\neq}$, $\mathfrak{m}\succcurlyeq\!\!\succcurlyeq\mathrm{e}^x$ *and* $T=\sum_{\mathfrak{u}\asymp_{\mathfrak{m}}P}P_{\mathfrak{u}}\,\mathfrak{u}$. *Then we have* $N_{P_{\times\mathfrak{n}}}=N_{T_{\times\mathfrak{n}}}$ *for all* $\mathfrak{n}\prec\!\!\prec\mathfrak{m}$.

Proof. Since $\mathfrak{m}\succcurlyeq\!\!\succcurlyeq\mathrm{e}^x$, we first notice that

$$T\uparrow=\sum_{\mathfrak{u}\asymp_{\mathfrak{m}\uparrow}P}P\uparrow_{\mathfrak{u}}\,\mathfrak{u}.$$

Hence, modulo division by $\mathfrak{d}_P$ and a sufficient number of upward shiftings, we may assume without loss of generality that $P\asymp 1$, that P and $\mathfrak{n}$ are exponential, that $N_{P_{\times\mathfrak{n}}}=D_{P_{\times\mathfrak{n}}}$, and $N_{T_{\times\mathfrak{n}}}=D_{T_{\times\mathfrak{n}}}$. Then

$$(P-T)_{\times\mathfrak{n}}\asymp_{\mathfrak{n}}(P-T)\,\mathfrak{n}\prec_{\mathfrak{m}}\mathfrak{n}$$

and $P_{\times\mathfrak{n}}\asymp_{\mathfrak{n}}\mathfrak{n}$, whence $P_{\times\mathfrak{n}}=T_{\times\mathfrak{n}}+o_{\mathfrak{m}}(P_{\times\mathfrak{n}})$. We conclude that $N_{P_{\times\mathfrak{n}}}=D_{P_{\times\mathfrak{n}}}=D_{T_{\times\mathfrak{n}}}=N_{T_{\times\mathfrak{n}}}$. □

8.3.3 Starting terms

We call $\mathfrak{m} \in \mathfrak{V}$ a *starting monomial*, if $N_{P_{\times \mathfrak{m}}}$ admits a non-zero root c in the algebraic closure C^{alg} of C. This is the case if and only if $N_{P_{\times \mathfrak{m}}} \notin C F^{\mathbb{N}}$. We say that $\mathfrak{m}$ is *algebraic* if $N_{P_{\times \mathfrak{m}}}$ is non-homogeneous, and *differential* if $N_{P_{\times \mathfrak{m}}} \notin C[F]$. A starting monomial, which is both algebraic and differential, is said to be *mixed*.

Example 8.9. Let $\mathfrak{m}$ be a starting monomials for $P(f) = 0$, where $P = LF$ and $L \in \mathbb{T}[\partial]$. Then $N_{L_{\times \mathfrak{m}}\uparrow_l} = D_{L_{\times \mathfrak{m}}\uparrow_l} \in C\, F'$ for all sufficiently large l. By proposition 7.6, it follows that $\mathfrak{m}\uparrow_l \in \mathfrak{H}_{L\uparrow_l}$ for all sufficiently large l, whence $\mathfrak{m} \in \mathfrak{H}_L$. Similarly, if $\mathfrak{m}$ is not a starting monomial, then $N_{L_{\times \mathfrak{m}}\uparrow_l} = D_{L_{\times \mathfrak{m}}\uparrow_l} \in C F$ for all sufficiently large l, and $\mathfrak{m} \notin \mathfrak{H}_L$.

Assuming that we have determined a starting monomial $\mathfrak{m}$ for (E), let $c \in C^{\text{alg}}$ be a non-zero root of $N_{P_{\times \mathfrak{m}}}$. If $c \in C$, then we call $c\mathfrak{m}$ a *starting term* for (E). If $N_{P_{\times \mathfrak{m}}} = Q\,(F')^{\nu}$ with $Q \in C[F]$ and $Q(c) = 0$, then $c\mathfrak{m}$ is said to be an *algebraic starting term*. If $\nu \neq 0$, then we say that $c\mathfrak{m}$ is a *differential starting term*. The *multiplicity* of c (and of $c\mathfrak{m}$) is the differential valuation of $N_{P_{\mathfrak{m}},+c}$. Notice that the definition of the multiplicity extends to the case when $c = 0$.

Proposition 8.10. *Assume that f is a non-zero transseries solution to* (E). *Then τ_f is a starting term.*

Proof. Assume that $\tau_f = c\mathfrak{m}$ is not a starting term. Modulo normalization, we may assume without loss of generality that P is transparent and $\mathfrak{m} = \mathfrak{d}_P = 1$. Then

$$P(f) = N_P(f) + o_{e^x}(1) = N_P(c) + o_{e^x}(1) \neq 0,$$

since $N_P(c) \neq 0$. □

The *Newton degree* of (E) is defined to be the maximum $d = \deg_{\mathfrak{V}} P$ of val P and the largest possible degree of $N_{P_{\times \mathfrak{m}}}$ for monomials $\mathfrak{m} \in \mathfrak{V}$. The above proposition shows that equations of Newton degree zero do not admit solutions.

Proposition 8.11. *If $\varphi \in C[\![\mathfrak{V}]\!]$, then*

$$\deg_{\mathfrak{V}} P_{+\varphi} = \deg_{\mathfrak{V}} P.$$

Proof. Consider a monomial $\mathfrak{m} \in \mathfrak{V}$ with $\mathfrak{m} \succcurlyeq \varphi$. Modulo a multiplicative conjugation with $\mathfrak{m}$ we may assume without loss of generality that $\mathfrak{m} = 1$, so that $\varphi = c + \varepsilon$ with $c \in C$ and $\varepsilon \prec 1$. Modulo upward shifting, we may also assume that P, $P_{+\varphi}$ and φ are transparent. Then $\deg N_{P_{+\varphi}} = \deg N_{P,+c} = \deg N_P$, by proposition 8.7($b$). □

Geometrically speaking, we may consider the Newton degree as "the multiplicity of zero as a root of P modulo $\mathfrak{V}$". More generally, given an initial segment $\mathfrak{W} \subseteq \mathfrak{V}$, we say that $\varphi \in C[\![\mathfrak{V}]\!]$ is a *solution* to (E) *modulo* $\mathfrak{W}$, if the Newton degree of

$$P_{+\varphi}(\tilde{f}) = 0 \qquad (\tilde{f} \in C[\![\mathfrak{W}]\!]) \tag{8.14}$$

is strictly positive. The *multiplicity* of such a solution is defined to be the Newton degree of (8.14). If $\psi \in \varphi + C[\![\mathfrak{W}]\!]$, then the multiplicities of φ and ψ as solutions of (E) modulo $\mathfrak{W}$ coincide, by proposition 8.11. In particular, if φ is a solution of (E) modulo $\mathfrak{W}$, then so is $\psi = \varphi_{\mathfrak{V} \setminus \mathfrak{W}} = \sum_{\mathfrak{m} \in \mathfrak{V} \setminus \mathfrak{W}} \varphi_{\mathfrak{m}} \mathfrak{m}$. We call ψ a *normalized solution*, because it is the unique solution in $\varphi + C[\![\mathfrak{W}]\!]$ such that $\psi_{\mathfrak{m}} = 0$ for all $\mathfrak{m} \in \mathfrak{W}$.

8.3.4 Refinements

Given a starting term $\tau = c\,\mathfrak{m}$ for (E), we will generalize the technique of refinements in order to compute the remaining terms. In its most general form, a *refinement* for (E) is a change of variables together with an asymptotic constraint

$$f = \varphi + \tilde{f} \qquad (\tilde{f} \in C[\![\tilde{\mathfrak{V}}]\!]), \tag{R}$$

where $\varphi \in C[\![\mathfrak{V}]\!]$ and $\tilde{\mathfrak{V}} \subseteq \mathfrak{V}$ is an initial segment of transmonomials. Such a refinement transforms (E) into

$$\tilde{P}(\tilde{f}) = P_{+\varphi}(\tilde{f}) = 0 \qquad (\tilde{f} \in C[\![\tilde{\mathfrak{V}}]\!]). \tag{RE}$$

Usually, we take $\tilde{\mathfrak{V}} = \{\mathfrak{w} \in \mathfrak{T} : \mathfrak{w} \prec \varphi\}$, in which case (RE) becomes

$$\tilde{P}(\tilde{f}) = 0 \qquad (\tilde{f} \prec \varphi). \tag{8.15}$$

In particular, we may take $\varphi = c\mathfrak{m}$, but, as in section 3.3.2, it is useful to allow for more general φ in presence of almost multiple solutions.

Consider a refinement (R) and a second refinement

$$\tilde{f} = \tilde{\varphi} + \tilde{\tilde{f}} \qquad (\tilde{\tilde{f}} \in C[\![\tilde{\tilde{\mathfrak{V}}}]\!]) \tag{RR}$$

with $\tilde{\varphi} \in C[\![\tilde{\mathfrak{V}}]\!]$ and $\tilde{\tilde{\mathfrak{V}}} \subseteq \tilde{\mathfrak{V}}$. Then we may compose (R) and (RR) so as to yield another refinement

$$f = \varphi + \tilde{\varphi} + \tilde{\tilde{f}} \qquad (\tilde{\tilde{f}} \in C[\![\tilde{\tilde{\mathfrak{V}}}]\!]). \tag{8.16}$$

Refinements of the form (8.16) are said to be *finer* as (R).

Proposition 8.12. *Consider a refinement* (R) *with* $\varphi \in C[\![\mathfrak{V}]\!]$. *Then the Newton degree of* (RE) *is bounded by the Newton degree of* (E).

Proof. By the definition of Newton degree, the result is clear if $\varphi = 0$. In general, we may decompose the refinement in a refinement with $\tilde{\mathfrak{V}} = \mathfrak{V}$ and a refinement with $\varphi = 0$. We conclude by proposition 8.11. □

Proposition 8.13. *Let* $\varphi \in C[[\mathfrak{V}]]$ *and* $\mathfrak{m} \succcurlyeq \varphi$. *Then the Newton degree of*

$$\tilde{P}(\tilde{f}) = P_{+\varphi}(\tilde{f}) = 0 \qquad (\tilde{f} \prec \mathfrak{m})$$

is equal to the multiplicity $\tilde{d}$ *of* $c = \varphi_{\mathfrak{m}}$ *as a root of* $N_{P_{\times \mathfrak{m}}}$.

Proof. Let us first show that $\deg N_{\tilde{P}_{\times \mathfrak{n}}} \leqslant \tilde{d}$ for any monomial $\mathfrak{n} \prec \mathfrak{m}$. Modulo multiplicative conjugation and upward shifting, we may assume without loss of generality that $\mathfrak{m} = 1$ and that P, $\tilde{P}_{\times \mathfrak{n}}$, $\mathfrak{n}$ and φ are transparent. The differential valuation of $N_{P,+c} = D_{\tilde{P}}$ being $\tilde{d}$, we have in particular $\tilde{P}_{\tilde{d}} \asymp \tilde{P}$. Hence,

$$\tilde{P}_{\times \mathfrak{n},i} \asymp_{\mathfrak{n}} \tilde{P}_i \mathfrak{n}^i \prec_{\mathfrak{n}} \tilde{P}_{\tilde{d}} \mathfrak{n}^{\tilde{d}} \asymp_{\mathfrak{n}} \tilde{P}_{\times \mathfrak{n},\tilde{d}}$$

for all $i > \tilde{d}$. We infer that $\deg N_{\tilde{P}_{\times \mathfrak{n}}} \leqslant \tilde{d}$.

At a second stage, we have to show that $\deg N_{\tilde{P}_{\times \mathfrak{n}}} \geqslant \tilde{d}$. Without loss of generality, we may again assume that $\mathfrak{m} = 1$, and that P and φ are transparent. The differential valuation of $N_{P,+c} = D_{\tilde{P}}$ being $\tilde{d}$, we have $\tilde{P}_i \prec \tilde{P}$ for all $i < \tilde{d}$. Taking $\mathfrak{n} = x^{-1}$, we thus get

$$\tilde{P}_{\times \mathfrak{n},i} \asymp_{\mathrm{e}^x} \tilde{P}_i \prec_{\mathrm{e}^x} \tilde{P} \asymp \tilde{P}_{\tilde{d}} \asymp_{\mathrm{e}^x} \tilde{P}_{\times \mathfrak{n},\tilde{d}}$$

for all $i < \tilde{d}$. We conclude that $\deg N_{\tilde{P}_{\times \mathfrak{n}}} \geqslant \tilde{d}$. □

Exercise 8.4. If $N_P = D_P \in C[F](F')^k$, then show that

a) $D_{P\uparrow} = D_P$.
b) $P\uparrow \asymp \mathfrak{d}_{P\uparrow} \mathrm{e}^{-kx}$.

Exercise 8.5. If $P = LF + g$, with $L \in \mathbb{T}[\partial]$ and $g \in \mathbb{T}^{\neq}$, then show that $T_L^{-1}(\tau_g)$ is the unique algebraic starting term for $P(f) = 0$.

Exercise 8.6.

a) Give a definition for the composition

$$f = \varphi + \tilde{f} \qquad (\tilde{f} \in C[[\tilde{\mathfrak{V}}]])$$

of an infinite sequence of refinements

$$\begin{aligned} f = f_0 &= \varphi_1 + f_1 \quad (f_1 \in C[[\mathfrak{V}_1]]) \\ f_1 &= \varphi_2 + f_2 \quad (f_2 \in C[[\mathfrak{V}_2]]) \\ &\vdots \end{aligned}$$

b) What can be said about the Newton degree of (RE)?

Exercise 8.7. Let $P, Q \in \mathbb{T}\{F\}$ and let $\mathfrak{V} \subseteq \mathfrak{T}$ be an initial segment.

a) Show that $\deg_{\mathfrak{V}} PQ = \deg_{\mathfrak{V}} P + \deg_{\mathfrak{V}} Q$.
b) What can be said about $\deg_{\mathfrak{V}} (P + Q)$?

c) If $\deg_{\mathfrak{V}} P > 0$ and $A_0, \ldots, A_n \in \mathbb{T}$, then show that

$$\deg_{\mathfrak{V}} (A_0 P + \cdots + A_n P^{(n)}) > 0.$$

Hint: first reduce to the case when $\mathfrak{V} = \{\mathfrak{v} \in \mathfrak{T}: \mathfrak{v} \prec 1\}$. Next, considering $P = 0, \ldots, P^{(n)} = 0$ as *algebraic* equations in $F, \ldots, F^{(r+n)}$, show that there exists a common solution $F = \phi_0, \ldots, F^{(r+n)} = \phi_{r+n}$ with $\phi_i \prec 1$ for all i (i.e. we do not require that $\phi_{i+1} = \phi_i'$ for $i < r + n - 1$).

Exercise 8.8. Improve the bound $i \geqslant \mathrm{wt}\, P$ in theorem 8.6 for P of degree $\leqslant 3$.

Exercise 8.9. Show that r upward shiftings may indeed be needed in theorem 8.6.

Exercise 8.10. Let $P \in C\{F'\}$ and let Λ be such that

$$\Lambda' = \frac{1}{x \log x \log_2 x \cdots}.$$

a) Show that

$$\mathfrak{d}_P^{\mathrm{as}} = \mathfrak{d}_{P(\Lambda)} = x^{-i_0} (\log x)^{-i_1} (\log_2 x)^{-i_3} \cdots,$$

with $i_0 \geqslant i_1 \geqslant \cdots \geqslant 1$.

b) Let $C\{F'\}_{d,w}$ be the subset of $C\{F'\}$ of homogeneous and isobaric polynomials of degree d and weight w. For $P \in C\{F'\}_{d,w}$, show that

$$\mathfrak{d}_P^{\mathrm{as}} = x^{-w} (\log x)^{-i_1} (\log_2 x)^{-i_2} \cdots$$

and $\lim_{k \to \infty} i_k = d$.

c) If l is such that $N_P = D_{P\uparrow_l}$, then show that

$$\mathfrak{d}_{P\uparrow_l} = (\exp_l x)^{-i_0} \cdots (\exp x)^{-i_{l-1}}.$$

d) Show that $N_P = D_{P\uparrow_l}$ if and only if $i_l = i_{l+1} = \cdots$.

8.4 Finding the starting monomials

8.4.1 Algebraic starting monomials

The algebraic starting monomials correspond to the slopes of the Newton polygon in the non-differential setting. However, they can not be determined directly from the dominant monomials of the P_i, because of the introductory example $f' = \mathrm{e}^{\mathrm{e}^x}$ and because there may be some cancellation of terms in the different homogeneous parts during multiplicative conjugations. Instead, the algebraic starting monomials are determined by successive approximation:

Proposition 8.14. *Let $i < j$ be such that $P_i \neq 0$ and $P_j \neq 0$.*

a) If P is exponential, then there exists a unique exponential monomial $\mathfrak{m}$, such that $P_{i, \times \mathfrak{m}} \asymp P_{j, \times \mathfrak{m}}$.

b) Denoting by $\mathfrak{m}_P$ the monomial $\mathfrak{m}$ in (a), there exists an integer $k \leqslant \mathrm{wt}\, P$, such that for all $l \geqslant k$ we have $\mathfrak{m}_{P\uparrow_l} = \mathfrak{m}_{P\uparrow_k} \uparrow_{l-k}$.

c) *There exists a unique monomial* $\mathfrak{m}$*, such that* $N_{(P_i+P_j)\times\mathfrak{m}}$ *is non-homogeneous.*

Proof. In (a), let $\mathfrak{B}=(\mathfrak{b}_1,\ldots,\mathfrak{b}_n)$ be a plane transbasis for the coefficients of P. We prove the existence of $\mathfrak{m}$ by induction over the least k, such that $\mathfrak{d}(P_i)/\mathfrak{d}(P_j)=\mathfrak{b}_1^{\alpha_1}\cdots\mathfrak{b}_k^{\alpha_k}$ for some $\alpha_1,\ldots,\alpha_k$. If $k=0$, then we have $\mathfrak{m}=1$. Otherwise, let $Q=P_{\times\mathfrak{n}}$ with $\mathfrak{n}=\mathfrak{b}_k^{\alpha_k/(j-i)}$. Then

$$Q_i\asymp_{\mathfrak{b}_k}P_i\,\mathfrak{n}^i\asymp_{\mathfrak{b}_k}P_j\,\mathfrak{n}^j\asymp_{\mathfrak{b}_k}Q_j,$$

so that $\mathfrak{d}(Q_i)/\mathfrak{d}(Q_j)=\mathfrak{b}_1^{\beta_1}\cdots\mathfrak{b}_l^{\beta_l}$ for some $l<k$ and $\beta_1,\ldots,\beta_l$. By the induction hypothesis, there exists a exponential monomial $\mathfrak{w}$, such that $Q_{i,\times\mathfrak{w}}\asymp Q_{j,\mathfrak{w}}$. Hence we may take $\mathfrak{m}=\mathfrak{n}\,\mathfrak{w}$. As to the uniqueness of $\mathfrak{m}$, assume that $\mathfrak{n}=\mathfrak{m}\,\mathfrak{b}_1^{\alpha_1}\cdots\mathfrak{b}_k^{\alpha_k}$ with $\alpha_k\neq 0$. Then

$$P_{i,\times\mathfrak{n}}\asymp_{\mathfrak{b}_k}P_{i,\times\mathfrak{m}}\,\mathfrak{b}_k^{i\alpha_k}\not\asymp_{\mathfrak{b}_k}P_{j,\mathfrak{m}}\,\mathfrak{b}_k^{j\alpha_k}\asymp_{\mathfrak{b}_k}P_{j,\times\mathfrak{n}}.$$

This proves (a).

The above argument also shows that $\mathfrak{m}_{P\uparrow}=\mathfrak{m}_P\!\uparrow\mathrm{e}^{\alpha x}$ for some $\alpha\in\mathbb{Q}$, since

$$P_{i,\times\mathfrak{m}}\!\uparrow\mathrm{e}^{(\mathrm{wv}\,P_{i,\times\mathfrak{m}})x}\asymp P_{j,\times\mathfrak{m}}\!\uparrow\mathrm{e}^{(\mathrm{wv}\,P_{j,\times\mathfrak{m}})x}.$$

Now, with the notations from theorem 8.6, we have shown that $\mathrm{wt}\,D_{P_i\uparrow}\leqslant\mathrm{wt}\,D_{P_i}$ and that equality occurs if and only if $D_{P_i}=F^{i-\mathrm{wt}\,D_{P_i}}(F')^{\mathrm{wt}\,D_{P_i}}$. Because of (8.10), we also notice that $\mathrm{wt}\,D_{P_i,\times\mathrm{e}^{\alpha x}}=\mathrm{wt}\,D_{P_i}$ for all $\alpha\in\mathbb{Q}$. It follows that

$$\mathrm{wt}\,D_{P_i,\times\mathfrak{m}_P}\geqslant\mathrm{wt}\,D_{P_i\uparrow,\times\mathfrak{m}_{P\uparrow}}\geqslant\cdots$$

and similarly for P_j instead of P_i. We finally observe that $\mathrm{wt}\,D_{P_i,\times\mathfrak{m}_P}=\mathrm{wt}\,D_{P_i\uparrow,\times\mathfrak{m}_{P\uparrow}}$ and $\mathrm{wt}\,D_{P_j,\times\mathfrak{m}_P}=\mathrm{wt}\,D_{P_j\uparrow,\times\mathfrak{m}_{P\uparrow}}$ imply that $\mathfrak{m}_{P\uparrow}=\mathfrak{m}_P\!\uparrow$, since

$$\mathrm{wt}\,D_{(F^\alpha(F')^\beta)_{\times\mathrm{e}^{\gamma x}}}=0\neq\beta=\mathrm{wt}\,D_{F^\alpha(F')^\beta}$$

whenever $\beta\neq 0$ and $\gamma\neq 0$. Consequently, $\mathrm{wt}\,D_{P_i\uparrow_l,\times\mathfrak{m}_{P\uparrow_l}}$ and $\mathrm{wt}\,D_{P_j\uparrow_l,\times\mathfrak{m}_{P\uparrow_l}}$ stabilize for $l\geqslant k$ with $k\leqslant\mathrm{wt}\,P$. For this k, we have (b).

With the notations from (b), $\mathfrak{m}_{P\uparrow_k}\!\downarrow_k$ is actually the unique monomial $\mathfrak{m}$ such that

$$D_{(P_i+P_j)\times\mathfrak{m}\uparrow_l}=D_{P_i,\times\mathfrak{m}\uparrow_k}+D_{P_j,\times\mathfrak{m}\uparrow_k}$$

is non-homogeneous for all sufficiently large l. Now $N_{(P_i+P_j)\times\mathfrak{m}}=D_{(P_i+P_j)\times\mathfrak{m}\uparrow_l}$ for sufficiently large l. This proves (c) for exponential differential polynomials P, and also for general differential polynomials, after sufficiently many upward shiftings. □

The unique monomial $\mathfrak{m}=\mathfrak{e}_{P,i,j}$ from part (c) of the above proposition is called the (i,j)-*equalizer* for P. An algebraic starting monomial is necessarily an equalizer. Consequently, there are only a finite number of algebraic starting monomials and they can be found as described in the proof of proposition 8.14.

Remark 8.15. From the proof of proposition 8.14, it follows that if P can be expanded w.r.t. a plane transbasis $\mathfrak{B} = (\mathfrak{b}_1, ..., \mathfrak{b}_n)$, then all equalizers for P belong to $(\log_{\text{wt} P}^C \mathfrak{b}_1) \cdots (\log^C \mathfrak{b}_1) \mathfrak{B}^C$.

8.4.2 Differential starting monomials

In order to find the differential starting monomials, it suffices to consider the homogeneous parts P_i of P, since $N_{P_{\times \mathfrak{m}},i} = N_{P_{i,\times \mathfrak{m}}}$, if $F'|N_{P_{\times \mathfrak{m}}}$ and $N_{P_{\times \mathfrak{m}},i} \neq 0$. Now, using (7.6), we may rewrite

$$P_i(f) = R_{P_i}(f^\dagger)\, f^i,$$

where R_{P_i} is a differential polynomial of order $\leqslant r-1$ in $f^\dagger$. We call R_{P_i} the *differential Riccati polynomial* associated to P_i.

For a linear differential operator L with exponential coefficients, we have seen in the previous chapter that finding the starting terms for the equation $L h = 0$ is equivalent to solving $R_L(f^\dagger) = 0$ modulo $o(1)$. Let us now show that finding the starting monomials for the equation $P_i(f) = 0$ is equivalent to solving $R_{P_i}(f^\dagger) = 0$ modulo $o(\frac{1}{x \log x \log \log x \cdots})$. In the exponential case, this is equivalent to solving the equation $R_{P_i}(f^\dagger) = 0$ modulo $o(1)$.

Proposition 8.16. *The monomial* $\mathfrak{m} \prec \mathfrak{v}$ *is a starting monomial of* f *w.r.t.*

$$P_i(f) = 0 \tag{8.17}$$

if and only if the equation

$$R_{P_i,+\mathfrak{m}^\dagger}(f^\dagger) = 0 \qquad \left(f^\dagger \prec \frac{1}{x \log x \log \log x \cdots}\right) \tag{8.18}$$

has strictly positive Newton degree.

Proof. We first notice that $R_{(P\uparrow)_i} = (R_{P_i}\uparrow)_{\times \mathrm{e}^{-x}}$ for all P and i. We claim that the equivalence of the proposition holds for P and $\mathfrak{m}$ if and only if it holds for $P\uparrow$ and $\mathfrak{m}\uparrow$. Indeed, $\mathfrak{m}$ is starting monomial w.r.t. (8.17), if and only if $\mathfrak{m}$ is a starting monomial w.r.t.

$$P_i\uparrow(f\uparrow) = 0 \tag{8.19}$$

and (8.18) has strictly positive Newton degree if and only if

$$R_{P_i,+\mathfrak{m}^\dagger}\uparrow(f^\dagger\uparrow) = 0 \qquad \left(f^\dagger\uparrow \prec \frac{1}{\mathrm{e}^x\, x \log x \cdots}\right) \tag{8.20}$$

has strictly positive Newton degree. Now the latter is the case if and only if

$$(R_{P_i,+\mathfrak{m}^\dagger}\uparrow)_{\times \mathrm{e}^{-x}}(f\uparrow^\dagger) = 0 \qquad \left(f\uparrow^\dagger \prec \frac{1}{x \log x \log \log x \cdots}\right)$$

has strictly positive Newton degree. But

$$(R_{P_i,+\mathfrak{m}^\dagger}\uparrow)_{\times \mathrm{e}^{-x}} = (R_{P_i}\uparrow)_{+\mathfrak{m}^\dagger\uparrow, \times \mathrm{e}^{-x}} = (R_{P_i}\uparrow)_{\times \mathrm{e}^{-x}, +\mathfrak{m}\uparrow^\dagger} = R_{(P\uparrow)_i, +\mathfrak{m}\uparrow^\dagger}.$$

This proves our claim.

Now assume that $\mathfrak{m}$ is a starting monomial w.r.t. (8.17). In view of our claim, we may assume without loss of generality that $P_{i,\times\mathfrak{m}}$ and $\mathfrak{m}$ are transparent. Since P_i is homogeneous, we have $D_{P_{i,\times\mathfrak{m}}} = \alpha\, F^{i-j}\,(F')^j$ for some $\alpha \in C^{\neq}$ and $j>0$, and

$$D_{R_{P_i,+\mathfrak{m}^\dagger}} = \alpha\, F^j.$$

Since $R_{P_i,+\mathfrak{m}^\dagger}$ is exponential, it follows that $N_{R_{P_i,+\mathfrak{m}^\dagger,\times x^{-2}}}$ has degree j, so that the Newton degree of (8.18) is at least $j>0$. Similarly, if $\mathfrak{m}$ is not a starting monomial w.r.t. (8.17), then $D_{P_{i,\times\mathfrak{m}}} = \alpha\, F^i$ and

$$D_{R_{P_i,+\mathfrak{m}^\dagger}} = \alpha$$

for some $\alpha \in C^{\neq}$. Consequently, $N_{R_{P_i,+\mathfrak{m}^\dagger,\times\mathfrak{n}}} = \alpha$ for any infinitesimal monomial $\mathfrak{n}$, and the Newton degree of (8.18) vanishes. □

8.4.3 On the shape of the differential Newton polygon

Proposition 8.17. *Let d be the Newton degree of* (E). *Then the algebraic starting monomials are equalizers of the form*

$$\mathfrak{e}_{P,i_0,i_1} \prec \mathfrak{e}_{P,i_1,i_2} \prec \cdots \prec \mathfrak{e}_{P,i_{l-1},i_l},$$

where $i_0 = \operatorname{val} P < i_1 < \cdots < i_{l-1} < i_l = d$.

Proof. Let us prove the proposition by induction over $d - \operatorname{val} P$. If $d = \operatorname{val} P$, then there is nothing to prove, so assume that $d > \operatorname{val} P$. Let $i < d$ be such that $\mathfrak{m} = \mathfrak{e}_{P,i,d}$ is maximal for $\preccurlyeq$. Modulo a multiplicative conjugation with $\mathfrak{m}$ and upward shifting, we may assume without loss of generality that $\mathfrak{m} = 1$ and that P is transparent.

We claim that 1 is a starting monomial for (E). Indeed, let $\mathfrak{n} \in \mathfrak{V}$ be such that $d = \deg N_{P_{\times\mathfrak{n}}}$. By proposition 8.7($c$), we already have $1 \preccurlyeq \mathfrak{n} \in \mathfrak{V}$, since otherwise

$$d = \operatorname{val} N_{P_{\times\mathfrak{n}}} = \operatorname{val} N_{(P_i+P_d)_{\times\mathfrak{n}}} \leqslant \operatorname{val} N_{P_i+P_d} = i.$$

Now assume for contradiction that 1 is not a starting monomial for (E), so that $P \succ P_i \asymp P_d$, and let j be such that $P \asymp P_j$. We must have $j < d$, since proposition 8.7(c) implies

$$\deg N_P \leqslant \deg N_{P_{\times\mathfrak{n}}} = d.$$

Now consider the equalizer $\mathfrak{v} = \mathfrak{e}_{P,j,d} \preccurlyeq 1$. After sufficiently many upward shiftings, we may assume without loss of generality that $P_{\times\mathfrak{v}}$ and $\mathfrak{v}$ are transparent. But then

$$P_{\times\mathfrak{v},j} \asymp_{\mathfrak{v}} \mathfrak{v}^j P_j \succ_{\mathfrak{v}} \mathfrak{v}^d P_d \asymp_{\mathfrak{v}} P_{\times\mathfrak{v},d},$$

which contradicts the fact that $P_{\times\mathfrak{v},j} \asymp P_{\times\mathfrak{v},d}$.

Having proved our claim, let $k = \operatorname{val} N_P$ and $N_P = Q\,(F')^\nu$. Since P is exponential, we have $P = N_P + o_{e^x}(1)$, whence

$$P_{\times x^{-1}\uparrow\uparrow} = ((-1)^\nu Q_{k-\nu} + o_{e^x}(1))\, F^k\, e^{-(k+\nu)e^x}.$$

In other words, $N_{P_{\times x^{-1}}} = (-1)^\nu Q_{k-\nu} F^k$. It follows that the equation

$$P(f) = 0 \qquad (f \prec 1)$$

has Newton degree k. We conclude by applying the induction hypothesis to this equation. □

Proposition 8.18. *Assume that $\mathfrak{m}$ is a non-algebraic starting monomial for (E). Then, with the notations from proposition 8.17, there exists a unique $p \in \{0, \ldots, l\}$ such that*

$$\operatorname{val} N_{P_{\times \mathfrak{m}}} = \deg N_{P_{\times \mathfrak{m}}} = i_p.$$

Moreover, $p > 0 \Rightarrow \mathfrak{e}_{P, i_{p-1}, i_p} \prec \mathfrak{m}$ and $p < l \Rightarrow \mathfrak{m} \prec \mathfrak{e}_{P, i_p, i_{p+1}}$.

Proof. By proposition 8.7(c),

$$p = \min\{q : \mathfrak{m} \prec \mathfrak{e}_{P, i_q, i_{q+1}} \vee q = l\} = \max\{q : \mathfrak{e}_{P, i_{q-1}, i_q} \prec \mathfrak{m} \vee q = 0\}$$

fulfills the requirements. □

Exercise 8.11. Compute the starting terms for

$$e^{-e^x} f^3 + f'' f - (f')^2 + x^4 e^{-3x} f''' + e^{-e^x} = 0.$$

Exercise 8.12. Let $P \in \mathbb{E}\{F\}^{\neq}$ be a differential polynomial with exponential coefficients and assume that $x^{\alpha_0} \cdots \log_l^{\alpha_l} x$ with $\alpha_l \neq 0$ is a starting monomial for $P(f) = 0$. Then prove that $l \leqslant \operatorname{wt} P$. Hint: if P is homogeneous, then show that

$$\operatorname{wt} D_P > \operatorname{wt} D_{P_{\times x^{\alpha_0}\uparrow}} > \cdots > \operatorname{wt} D_{P_{\times x^{\alpha_0} \cdots \log_l^{\alpha_l} x \uparrow_l}}.$$

Exercise 8.13. Let K be a differential field and $f \in K$, $P \in K\{F\}$. If $P(f) = 0$, then show that there exists a homogeneous $H \in K\{F\}$ of degree $\leqslant \operatorname{wt} P + \deg P$, such that $H(e^{\int f}) = 0$.

Exercise 8.14. Prove that there are exactly $d - \operatorname{val} P$ algebraic starting terms in $C^{\text{alg}}\mathfrak{T}$ for an equation (E) of Newton degree d.

Exercise 8.15. Let $\mathbb{T}\{F\}_d$ denote the space of homogeneous $P \in \mathbb{T}\{F\}$ of degree d. Given $P \in \mathbb{T}\{F\}_2$, let $\varphi(P) \in \mathbb{T}[F, F^\dagger]$ be the result of substituting $F^{\dagger\dagger} = F^{\dagger\dagger\dagger} = \cdots = 0$ in the logarithmic decomposition of R_P.

a) Show that $\varphi(P) \in \mathbb{T}[F, F']$, when rewriting $F^\dagger = F'/F$.
b) Show that $\varphi: \mathbb{T}\{F\}_2 \to \mathbb{T}[F, F']$ is an isomorphism.
c) What about higher degrees?

8.5 Quasi-linear equations

8.5.1 Distinguished solutions

The equation (E) is said to be *quasi-linear* if its Newton degree is one. A solution f to a quasi-linear equation is said to be *distinguished* if we have $f_{\mathfrak{d}(\tilde{f}-f)} = 0$ for all other solutions $\tilde{f}$ to (E). Distinguished solutions are unique: if f and $\tilde{f}$ are distinct distinguished solutions, then we would have $f_{\mathfrak{d}(\tilde{f}-f)} = \tilde{f}_{\mathfrak{d}(f-\tilde{f})} = 0$, whence $(f-\tilde{f})_{\mathfrak{d}(f-\tilde{f})} = 0$, which is absurd.

Lemma 8.19. *Assume that the equation* (E) *is quasi-linear and that the coefficients of* P *can be expanded w.r.t. a plane transbasis* $\mathfrak{B} = (\mathfrak{b}_1, \ldots, \mathfrak{b}_n)$. *Assume also that* $P \asymp 1$, $P_0 \prec_{\mathfrak{b}_n} 1$, *and let*

$$\mathfrak{I} = \{\mathfrak{m} \in (\log \mathfrak{b}_1)^{\mathbb{N}} \mathfrak{B}^C : \mathfrak{m} \prec_{\mathfrak{b}_n} 1\}.$$

Then, considering $L = -P_{1, \asymp_{\mathfrak{b}_n} 1}$ *and* $R = P - P_0 + L$ *as operators on* $C[[\mathfrak{I}]]$, *the equation* (E) *admits a distinguished solution* f *given by*

$$f = L^{-1} (\mathrm{Id} - R L^{-1})^{-1} P_0. \tag{8.21}$$

Proof. Since $C[x][[\mathfrak{b}_1; \ldots; \mathfrak{b}_{n-1}]]\, \mathfrak{b}_n^{\alpha}$ is stable under L and L^{-1} for each $\alpha \in C$, the operator $R L^{-1}$ is strictly extensive on $C[[\mathfrak{I}]]$ and supp $R L^{-1}$ is grid-based. By theorem 6.15, the operator $\mathrm{Id} - R L^{-1}$ therefore admits an inverse

$$(\mathrm{Id} - R L^{-1})^{-1} = \mathrm{Id} + R L + (R L^{-1})^2 + \cdots.$$

This shows that f is well-defined. In order to show that f is the distinguished solution, assume that $\tilde{f}$ is another solution and let $\mathfrak{d} = \mathfrak{d}_{\tilde{f}-f}$. If $\mathfrak{d} \asymp_{\mathfrak{b}_n} 1$, then we clearly have $f_{\mathfrak{d}} = 0$, since $f \prec_{\mathfrak{b}_n} 1$. If $\mathfrak{d} \prec_{\mathfrak{b}_n} 1$, then let

$$\delta = \sum_{\mathfrak{m} \asymp_{\mathfrak{b}_n} \mathfrak{d}} (\tilde{f} - f)_{\mathfrak{m}}\, \mathfrak{m}.$$

Since $P(\tilde{f}) - P(f) = 0$, we have $L\delta = 0$, so that $\mathfrak{d} = \mathfrak{d}_{\delta}$ is the dominant monomial of a solution to the equation $L h = 0$. Hence $f_{\mathfrak{d}} = 0$, since $f \in \operatorname{im} L^{-1}$. □

Lemma 8.20. *Consider a quasi-linear equation* (E) *whose coefficients can be expanded w.r.t. a plane transbasis* $\mathfrak{B} = (\mathfrak{b}_1, \ldots, \mathfrak{b}_n)$. *Assume that* $P \asymp P_0 \asymp P_1$ *and* $N_P = D_P$. *Then* (E) *admits a distinguished solution*

$$f \in C[[\log_{n-1} x; \ldots; x; \mathfrak{b}_1; \ldots; \mathfrak{b}_n]].$$

Proof. Modulo division of the equation by $\mathfrak{d}_P$, we may assume without loss of generality that $P \asymp 1$. We prove the result by induction over n. If $n = 0$, then

$$P = D_P = N_P = \alpha + \beta F$$

for some $\alpha \in C$ and $\beta \in C^{\neq}$. Hence $f = -\frac{\beta}{\alpha}$ is the distinguished solution to $P(f)=0$. Assume now that $n>0$. By the induction hypothesis, there exists a distinguished solution to the quasi-linear equation

$$P_{\asymp_{\mathfrak{b}_n} 1}(\varphi)=0 \qquad (\varphi \prec C[[\mathfrak{V}]]) \tag{8.22}$$

with $\varphi \in C[[\log_{n-2} x; \ldots; x; \mathfrak{b}_1; \ldots; \mathfrak{b}_{n-1}]]$. By lemma 8.19, the equation

$$P_{+\varphi \uparrow_{n-1}}(\psi)=0 \quad (\psi \prec C[[\mathfrak{V}]] \uparrow_{n-1})$$

admits a distinguished solution

$$\psi \in C[x][[\exp x, \ldots, \exp_{n-2} x; \mathfrak{b}_1 \uparrow_{n-1}; \ldots; \mathfrak{b}_n \uparrow_{n-1}]]$$

with $\psi \prec_{\mathfrak{b}_n} 1$. Then the distinguished properties of φ and ψ imply that $f = \varphi + \psi \downarrow_{n-1}$ is the distinguished solution to (E). □

Theorem 8.21. *Assume that the equation* (E) *is quasi-linear. Then it admits a distinguished transseries solution f. Moreover, if the coefficients of P can be expanded w.r.t. a plane transbasis $\mathfrak{B} = (\mathfrak{b}_1, \ldots, \mathfrak{b}_n)$, then*

$$f \in C[[\log_n x; \ldots; x; \mathfrak{b}_1; \ldots; \mathfrak{b}_n]].$$

Proof. If $P_0 = 0$, then 0 is the trivial distinguished solution of (E). Assume therefore that $P_0 \neq 0$. Modulo some upward shiftings we may assume without loss of generality that the coefficients of P and the transbasis $\mathfrak{B}$ are exponential. Modulo a multiplicative conjugation and using proposition 8.14(a), we may also assume that $P_0 \asymp P_1$. Now consider the $(0,1)$-equalizer $\mathfrak{e} = \mathfrak{e}_{P,0,1}$ for P, which is also the only algebraic starting monomial. If

$$D_{P_0+P_1} = \alpha + \beta_\nu F^{(\nu)} + \cdots + \beta_l F^{(l)}$$

with $\beta_\nu \neq 0$, then $\mathfrak{e} = x^\nu$ and

$$D_{P \uparrow_{\times \mathfrak{e} \uparrow}} = \alpha + \beta_\nu \nu^\nu F.$$

In other words, after one more upward shifting and a multiplicative conjugation with $\mathfrak{e}\uparrow$, we may also assume that $N_P = D_P$. We conclude by lemma 8.20. □

8.5.2 General solutions

Lemma 8.22. *Consider a quasi-linear equation* (E) *with exponential coefficients and a solution f which is not exponential. Let $\mathfrak{l}$ be the largest monomial in* supp f *which is not exponential. Then $\mathfrak{l} = x^k \mathfrak{l}^\sharp$ for some $k \in \mathbb{N}$ and an exponential monomial $\mathfrak{l}^\sharp \in \mathfrak{H}_{P_{+f},1}$.*

Proof. Consider the exponential transseries $\varphi = \sum_{\mathfrak{m} \succ \mathfrak{l}} f_\mathfrak{m} \mathfrak{m}$. Then

$$P_{+\varphi}(\tilde{f})=0 \quad (\tilde{f} \in C[[\mathfrak{V}]])$$

admits $\tilde{f} = f - \varphi$ as a solution, so it is quasi-linear and $\mathfrak{l}$ is a starting monomial. Consequently, $\mathfrak{l}$ is also a starting monomial for the equation $L\tilde{f} = -P_{+\varphi,0}$, where $L = P_{+\varphi,1}$. It follows that $\mathfrak{l} = x^k\, \mathfrak{l}^\sharp$ for some exponential monomial $\mathfrak{l}^\sharp \in \mathfrak{H}_L$.

Let us show that $\mathfrak{l}^\sharp \in \mathfrak{H}_{\tilde{L}}$, where $\tilde{L} = P_{+f,1}$. Modulo an additive conjugation with φ, a multiplicative conjugation with $\mathfrak{l}^\sharp$, and division of the equation by $\mathfrak{d}_P$, we may assume without loss of generality that $\varphi = 0$, $\mathfrak{l}^\sharp = 1$ and $P \asymp 1$. Since the equation $P\uparrow_{\times \mathrm{e}^{kx}}(\tilde{f})$ $(\tilde{f} \preccurlyeq 1)$ is quasi-linear, we have

$$P\uparrow_{\times \mathrm{e}^{kx}} = P_0\uparrow + L_k\uparrow \partial^k\uparrow_{\times \mathrm{e}^{kx}} + o_{\mathrm{e}^x}(1).$$

It follows that

$$P_{+f}\uparrow_{\times \mathrm{e}^{kx}} = P\uparrow_{\times \mathrm{e}^{kx}, +f\uparrow \mathrm{e}^{kx}} = L_k\uparrow \partial^k\uparrow_{\times \mathrm{e}^{kx}} + o_{\mathrm{e}^x}(1),$$

whence

$$P_{+f}\uparrow = \partial^k\uparrow + o_{\mathrm{e}^x}(\mathrm{e}^{-kx}).$$

In other words, 1 is a starting monomial for the equation

$$(\tilde{L}\uparrow)(h) = 0.$$

We conclude that $1 \in \mathfrak{H}_{\tilde{L}\uparrow}$ and $1 \in \mathfrak{H}_{\tilde{L}}$. □

Theorem 8.23. *Let f be a solution to a quasi-linear equation* (E)*. If the depths of the coefficients of P are bounded by d, then the depth of f is bounded by $d + r$.*

Proof. For each i, such that the depth of f is $> d + i$, let $\mathfrak{l}_i$ be the minimal element in the support of f of depth $> d + i$. By the previous lemma, we have $\mathfrak{l}_i\uparrow_{d+i} \in \mathfrak{H}_{P_{+f,1}\uparrow_{d+i}}$, whence $\mathfrak{l}_i \in \mathfrak{H}_{P_{+f,1}}$. Therefore, $\mathfrak{l}_i\uparrow_{d+i} \in x^{\mathbb{N}}\, \mathfrak{E}$, where $\mathfrak{E}$ denotes the set of exponential transmonomials. The result now follows from the fact that $\operatorname{card} \mathfrak{H}_{P_{+f,1}} = \dim H_{P_{+f,1}} \leqslant r$. □

Corollary 8.24. *If the coefficients of P can be expanded w.r.t. a plane transbasis $(\mathfrak{b}_1, \ldots, \mathfrak{b}_n)$, then the distinguished solution to* (E) *belongs to $C[[\log_{r-1} x; \ldots; x; \mathfrak{b}_1; \ldots; \mathfrak{b}_n]]$.*

Theorem 8.25. *Let f be a solution to a quasi-linear equation* (E)*. Then f may be written in a unique way as*

$$f = f^* + h_1 + \cdots + h_s,$$

where f^* is the distinguished solution to (E)*, $s \leqslant r$, and*

$$h_1 \succ \ldots \succ h_s \in \mathbb{T}^{\neq}$$

are such that each $h_i - \tau(h_i)$ is the distinguished solution to the equation

$$P_{+f^* + h_1 + \cdots + h_{i-1} + \tau(h_i)}(\varphi) = 0 \quad (\varphi \prec C[[\mathfrak{W}]]).$$

Proof. Consider the sequence $h_1, h_2, \ldots$ with $h_i = \tau_i + \delta_i$ for all i, where

$$\tau_i = \tau(f - h_1 - \cdots - h_{i-1})$$

and δ_i is the distinguished solution to

$$P_{+f^*+h_1+\cdots+h_{i-1}+\tau_i}(\varphi) = 0 \quad (\varphi \prec C[\![\mathfrak{V}]\!]).$$

Since the equation

$$P_{+f^*+h_1+\cdots+h_{i-1}+\tau_i}(\varphi) = 0 \quad (\varphi \prec \tau_i)$$

is quasi-linear (it admits $f - h_1 - \cdots - h_{i-1} - \tau_i$ as a solution), δ_i is also the distinguished solution to this latter equation, whence $\delta_i \prec \tau_i$. By induction, it follows that $h_1 \succ h_2 \succ \cdots$.

Let us now prove that the sequence $h_1, h_2, \ldots$ has length at most r. Assume the contrary and consider

$$\tilde{P} = P_{+f^*+h_1+\cdots+h_{r+1}}.$$

Then

$$\tilde{P}(-h_i - \cdots - h_{r+1}) = 0$$

for all $i \in \{1, \ldots, r+1\}$, so $\mathfrak{d}_{h_1}, \ldots, \mathfrak{d}_{h_{r+1}}$ are starting monomials for

$$\tilde{P}(\tilde{f}) = 0 \quad (\tilde{f} \prec C[\![\mathfrak{V}]\!]).$$

Since this equation is quasi-linear and $\tilde{P}_0 = 0$, it follows that $\mathfrak{d}_{h_1}, \ldots, \mathfrak{d}_{h_{r+1}}$ are also starting monomials for the linear differential equation

$$\tilde{L}\tilde{f} = \tilde{P}_1(\tilde{f}) = 0.$$

In other words, $\{\mathfrak{d}_{h_1}, \ldots, \mathfrak{d}_{h_{r+1}}\} \subseteq \mathfrak{H}_{\tilde{L}}$. But then

$$r + 1 \leqslant \operatorname{card} \mathfrak{H}_{\tilde{L}} = \dim H_{\tilde{L}} \leqslant r. \qquad \square$$

Exercise 8.16. If f is the distinguished solution to a quasi-linear equation (E) and $\varphi \trianglelefteq f$ a truncation of f, then show that $\tilde{f} = f - \varphi$ is the distinguished solution to

$$P_{+\varphi}(\tilde{f}) = 0 \quad (\tilde{f} \in C[\![\mathfrak{V}]\!]).$$

Exercise 8.17. Assume that (E) is quasi-linear, with distinguished solution f. Show that the equation $P_{\times\mathfrak{m}}(g) = 0$ $(g \in \mathfrak{m}^{-1}\,\mathfrak{V})$ is also quasi-linear, with distinguished solution $g = \mathfrak{m}^{-1} f$. And if $\mathfrak{m}$ is replaced by a transseries?

Exercise 8.18. Show that $f \in C[\![\log_{\operatorname{expo}(\mathfrak{b}_n)-1} x; \ldots; x; \mathfrak{b}_1; \ldots; \mathfrak{b}_n]\!]$ in theorem 8.21.

Exercise 8.19. Show that the dependence of f on $\log_{d+r-1} x$ is polynomial in theorem 8.23.

Exercise 8.20. Give an example of a quasi-linear equation (E) such that the set

$$\{\mathfrak{d}_{g-f}: P(f) = P(g) = 0 \wedge f, g \in C[[\mathfrak{V}]] \wedge g \neq f\}$$

is infinite.

Exercise 8.21. Can you give an example for which

$$f \in C[[\log_{r-1} x; \ldots; x; \mathfrak{b}_1; \ldots; \mathfrak{b}_n]] \setminus C[[\log_{r-2} x; \ldots; x; \mathfrak{b}_1; \ldots; \mathfrak{b}_n]]$$

in corollary 8.24?

8.6 Unravelling almost multiple solutions

As pointed out in the introduction, "unravelling" almost multiple solutions is a more difficult task than in the algebraic setting. As our ultimate goal, a *total unravelling* is a refinement

$$f = \varphi + \tilde{f} \qquad (\tilde{f} \prec \tilde{\mathfrak{v}}), \tag{8.23}$$

such that $\deg_{\preccurlyeq \tilde{\mathfrak{v}}} P = d$ and $\deg_{\prec \tilde{\mathfrak{v}}} P < d$. Unfortunately, total unravellings can not be read off immediately from the equation or its derivatives. Nevertheless, we will show how to "approximate" total unravellings by so called partial unravellings which are constructed by repeatedly solving suitable quasi-linear equations.

8.6.1 Partial unravellings

In order to effectively construct a total unravelling, consider a starting monomial $\mathfrak{m}$ such that $N_{P_{\times \mathfrak{m}}}$ admits a root of multiplicity d. Assume that $l \in \mathbb{Z}$ is sufficiently large so that $P_{\times \mathfrak{m}}$ is exponential and

$$N_{P_{\times \mathfrak{m}} \uparrow_l} = D_{P_{\times \mathfrak{m}} \uparrow_l} = a\, (F - c)^{d-k}\, (F')^k$$

for some $a, c \in C^{\neq}$ and k. Let

$$Q = \begin{cases} \left(\dfrac{\partial^{d-1} P_{\times \mathfrak{m}} \uparrow_l}{(\partial F)^{d-1-k}\, (\partial F')^k} \right)_{\times \mathfrak{m}^{-1}} \downarrow_l & \text{if } k < d \\[2ex] \left(\dfrac{\partial^{d-1} P_{\times \mathfrak{m}} \uparrow_l}{(\partial F')^{d-1}} \right)_{\times \mathfrak{m}^{-1}} \downarrow_l & \text{if } k = d \end{cases} \tag{8.24}$$

and consider a refinement (R) such that

AU1. The Newton degree of (RE) equals d.
AU2. $Q(\varphi) = 0$ and $\mathfrak{d}_\varphi = \mathfrak{m}$.
AU3. We have $\tilde{\mathfrak{V}} = \{\mathfrak{v} \in \mathfrak{T} : \mathfrak{v} \prec \mathfrak{h}\}$ for $\mathfrak{h} = \mathfrak{m}$ or some starting monomial $\mathfrak{h}$ for

$$\tilde{Q}(h) = Q_{+\varphi}(h) = 0 \qquad (h \in C[[\mathfrak{V}]]).$$

Then we call (R) an *atomic unravelling*.

Proposition 8.26. *Let $\mathscr{S}$ be a set of atomic unravellings for* (E)*. Then $\mathscr{S}$ admits a finest element.*

Proof. Assume for contradiction that there exists an infinite sequence

$$\begin{aligned} f &= \varphi_0 + f_1 & (f_1 \prec \mathfrak{v}_1) \\ f &= \varphi_0 + \varphi_1 + f_2 & (f_2 \prec \mathfrak{v}_2) \\ &\vdots \end{aligned}$$

of finer and finer atomic unravellings in $\mathscr{S}$, so that

$$\varphi_i \prec \mathfrak{v}_i \preccurlyeq \varphi_{i-1}$$

for all $i > 0$. Setting

$$\psi = \varphi_0 + \cdots + \varphi_{r+1},$$

it follows for all $i > 0$ that

$$Q_{+\psi}(-\varphi_i - \cdots - \varphi_{r+1}) = 0.$$

Consequently, $\mathfrak{d}_{\varphi_i}$ is a starting monomial for $Q_{+\psi,1}(h) = 0$ and $i \in \{1, ..., r+1\}$. But this is impossible, since $\operatorname{card} \mathfrak{H}_{Q_{+\psi,1}} \leqslant r$. □

Given an atomic unravelling (R) followed by a second refinement (RR) such that the Newton degree of

$$\tilde{\tilde{P}}(\tilde{\tilde{f}}) = \tilde{P}_{+\tilde{\varphi}}(\tilde{\tilde{f}}) = 0 \qquad (\tilde{\tilde{f}} \in C[[\tilde{\tilde{\mathfrak{V}}}]])$$

equals d, we say that (RR) is *compatible* with (R) if $\tilde{\varphi} \neq 0$, $\tilde{\tilde{\mathfrak{V}}} \prec \tilde{\varphi}$ and $\mathfrak{d}_{\tilde{\varphi}}$ is not a starting monomial for

$$\tilde{Q}(h) = 0 \qquad (h \in C[[\tilde{\mathfrak{V}}]]). \tag{8.25}$$

If the second refinement (RR) is not compatible with (R), then we may construct a finer atomic unravelling

$$f = \varphi + \psi + \tilde{f} \qquad (\tilde{f} \prec \psi)$$

such that $\tau(\psi) = \tau(\tilde{\varphi})$. Indeed, it suffices to take $\psi = \tau(\tilde{\varphi}) + h$, where h is the distinguished solution to the equation

$$Q_{\varphi + \tau(\tilde{\varphi})}(h) = 0 \qquad (h \prec \tilde{\varphi}).$$

In other words, during the construction of solutions of (E) we "follow" the solutions to $Q(h) = 0$ as long as possible whenever the Newton degree remains d.

A *partial unravelling* is the composition of a finite number l of compatible atomic unravellings. We call l the length of the partial unravelling. By convention, the identity refinement

$$f = \tilde{f} \ \ (\tilde{f} \in C[[\mathfrak{V}]])$$

is a partial unravelling of length 0. We have shown the following:

Proposition 8.27. *Assume that* (E) *has Newton degree d. Given a partial unravelling* (R) *and a starting term $\tilde{\tau}$ for* (RE) *of multiplicity d, there exists a finer partial unravelling*

$$f = \varphi + \tilde{\varphi} + \tilde{\tilde{f}} \qquad (\tilde{\tilde{f}} \prec \tilde{\tau})$$

with $\tilde{\varphi} \sim \tilde{\tau}$. □

8.6.2 Logarithmic slow-down of the unravelling process

The introductory example (8.1) shows that an atomic unravelling does not necessarily yield a total unravelling. Nevertheless, when applying a succession of compatible atomic unravellings, the following proposition shows that the corresponding monomials $\mathfrak{m}$ change by factors which decrease logarithmically.

Theorem 8.28. *Consider an atomic unravelling* (R)*, followed by a compatible refinement* (RR)*. Then, denoting $\tilde{\mathfrak{m}} = \mathfrak{d}_{\tilde{\varphi}}$, there exists an $\tilde{\tilde{\mathfrak{m}}} \in \tilde{\tilde{\mathfrak{V}}}$ with*

$$\frac{\tilde{\mathfrak{m}}}{\tilde{\tilde{\mathfrak{m}}}} \preccurlyeq\!\!\!\prec \log \frac{\mathfrak{m}}{\tilde{\mathfrak{m}}}.$$

Proof. Modulo some upward or downward shiftings, we may assume without loss of generality that $l = 0$ in (8.24), so that $P_{\times \mathfrak{m}}$ is exponential. Modulo a multiplicative conjugation with $\mathfrak{m}$ and division of P by $\mathfrak{d}_P$, we may also assume that $\mathfrak{m} = 1$ and that $P \asymp 1$. By proposition 8.1 it follows that $\tilde{P} \asymp \tilde{\tilde{P}} \asymp Q \asymp \tilde{Q} \asymp 1$.

Let us first show that $\tilde{\mathfrak{m}} \not\succcurlyeq\!\!\!\succ \mathrm{e}^x$. Assuming the contrary, we have either $\varphi - c \prec\!\!\!\prec \mathrm{e}^x$ or $\varphi - c \not\succcurlyeq\!\!\!\succ \mathrm{e}^x$, where $c = c_\varphi$. In the first case, $\mathfrak{v} = \mathfrak{d}_{\varphi - c} \prec\!\!\!\prec \mathrm{e}^x$ is a starting monomial for

$$Q_{+c}(\tilde{f}) = 0 \qquad (\tilde{f} \prec 1),$$

and $D_{Q_{+c}} \in C\, F^{\mathbb{N}}\, (F')^{\mathbb{N}}$. Since Q_{+c} is exponential, it follows that $N_{Q_{+c}} = D_{Q_{+c}}$, as well as $N_{Q_{+c}, \times \mathfrak{v}} = N_{D_{Q_{+c}}, \times \mathfrak{v}}$, by proposition 8.8. So $\mathfrak{v}$ is also a starting monomial for the equation $N_{Q_{+c}}(h) = 0$ $(h \prec 1)$. But this is impossible, since $N_{Q_{+c}} \in C\, F^{\mathbb{N}}\, (F')^{\mathbb{N}}$. In the second case, $\mathfrak{v} = \tilde{\mathfrak{m}} \prec\!\!\!\prec \mathrm{e}^x$ is a starting monomial for

$$P_{+c}(\tilde{f}) = 0 \qquad (\tilde{f} \prec 1).$$

Again P_{+c} is exponential and $D_{P_{+c}} \in C\, F^{\mathbb{N}}\, (F')^{\mathbb{N}}$, so we obtain a contradiction in a similar way as above.

Since $\tilde{\mathfrak{m}}$ is not a starting monomial for (8.25), we have

$$\tilde{Q}(\tilde{\varphi})\!\uparrow_p = \tilde{Q}_{\times \tilde{\mathfrak{m}}}(\tilde{\varphi}/\tilde{\mathfrak{m}})\!\uparrow_p \asymp \mathfrak{d}(\tilde{Q}_{\times \tilde{\mathfrak{m}}}\!\uparrow_p)$$

for a sufficiently large $p \in \mathbb{N}$ such that $\mathfrak{m}\uparrow_p$, $\tilde{Q}\uparrow_p$ and $\tilde{\varphi}\uparrow_p$ are exponential and $D_{\tilde{Q}_{\times \tilde{\mathfrak{m}}}\uparrow_p} = N_{\tilde{Q}_{\times \tilde{\mathfrak{m}}}\uparrow_p}$. Using proposition 8.3 and the fact that $\tilde{\mathfrak{m}} \succcurlyeq\!\!\succ e^x$, it follows that

$$\tilde{Q}(\tilde{\varphi}) \asymp^*_{\log \tilde{\mathfrak{m}}} \mathfrak{d}(\tilde{Q}_{\times \tilde{\mathfrak{m}}}).$$

On the other hand,

$$\frac{\mathfrak{d}(\tilde{Q}_{\times \tilde{\mathfrak{m}}})}{\mathfrak{d}(\tilde{Q})\, \tilde{\mathfrak{m}}} \preccurlyeq\!\!\prec \tilde{\mathfrak{m}}^\dagger \preccurlyeq\!\!\prec \log \tilde{\mathfrak{m}},$$

whence

$$\tilde{Q}(\tilde{\varphi}) \succcurlyeq^*_{\log \tilde{\mathfrak{m}}} \mathfrak{d}(\tilde{Q})\, \tilde{\mathfrak{m}} = \tilde{\mathfrak{m}}.$$

We conclude that

$$\tilde{\tilde{P}}_{d-1} \succcurlyeq^*_{\log \tilde{\mathfrak{m}}} \tilde{\mathfrak{m}},$$

since $\tilde{Q}(\tilde{\varphi})$ is the coefficient of $F^{d-1-k}(F')^k$ in $\tilde{\tilde{P}}$ for some k.

Now let $\mathfrak{n}$ be a monomial with $\mathfrak{n} \prec^*_{\log \tilde{\mathfrak{m}}} \tilde{\mathfrak{m}}$, so that $\mathfrak{n} \prec^*_{\log \mathfrak{n}} \tilde{\mathfrak{m}}$ and $\tilde{\tilde{P}}_{d-1} \succ^*_{\log \mathfrak{n}} \mathfrak{n}$. Then, proposition 8.2 implies

$$\tilde{\tilde{P}}_{\times \mathfrak{n}, d-1} \asymp^*_{\log \mathfrak{n}} \tilde{\tilde{P}}_{d-1} \mathfrak{n}^{d-1} \succ^*_{\log \mathfrak{n}} \mathfrak{n}^d \asymp^*_{\log \mathfrak{n}} \tilde{\tilde{P}}_{\times \mathfrak{n}, d},$$

From propositions 8.3 and 8.8, it therefore follows that the degree of $N_{\tilde{\tilde{P}}_{\times \mathfrak{n}}}$ cannot exceed $d-1$. We conclude that there exists an $\tilde{\tilde{\mathfrak{m}}} = \mathfrak{n} \in \tilde{\tilde{\mathfrak{V}}}$ with

$$\tilde{\tilde{\mathfrak{m}}} \succcurlyeq^*_{\log \tilde{\mathfrak{m}}} \tilde{\mathfrak{m}},$$

since (8.26) has Newton degree d. □

8.6.3 On the stagnation of the depth

This section deals with two important consequences of proposition 8.28. Roughly speaking, after one atomic unravelling, the terms of degree $> d$ do no longer play a role in the unravelling process. If $\tilde{P}$ is exponential, and modulo the hypothesis that $\tilde{P}_d(h) = 0$ only admits exponential starting monomials, it will follow that the process only involves monomials in $x^{\mathbb{N}}\, \mathfrak{E}$, where $\mathfrak{E}$ denotes the set of exponential transmonomials.

Lemma 8.29. *Consider an equation* (E) *of Newton degree d and assume that* $P_0, \ldots, P_{d-1} \in C[x][[\mathfrak{E}]]$ *and* $P_d \in C[[\mathfrak{E}]]$. *Then any non-differential starting term of multiplicity d is in* $C^{\neq}\, x^{\mathbb{N}}\, \mathfrak{E}$.

Proof. Let $c\,\mathfrak{m}$ be a non-differential starting term of multiplicity d, so that $N_{P_{\times \mathfrak{m}}} = a\,(F - c)^d$ for some $a \in C$. Then $\mathfrak{m}$ is the (i, j)-equalizer for all $0 \leqslant i < j \leqslant d$. In particular, $c\,\mathfrak{m}$ is a starting term for the linear equation $P_0 + P_1(f) = 0$. Hence, $\mathfrak{m} \in x^{\mathbb{N}}\, \mathfrak{E}$, by proposition 7.8 and the incomplete transbasis theorem. □

Theorem 8.30. *Consider an atomic unravelling* (R) *for an equation* (E) *of Newton degree d, followed by a compatible refinement* (RR) *such that* $\tilde{\varphi}_{\tilde{\tilde{\mathfrak{V}}}}=0$. *Assume that P and* φ *are exponential and that* $\tilde{P}_d(h)=0$ *admits only exponential starting monomials. Then* $\tilde{\varphi}\in C[x][\![\mathfrak{E}]\!]$.

Proof. If $\tilde{\varphi}=0$, then we have nothing to prove, so assume that $\tilde{\varphi}\neq 0$. By **U1** and lemma 8.29, it follows that $\mathfrak{d}_{\tilde{\varphi}}\in x^{\mathbb{N}}\,\mathfrak{E}$. Modulo a multiplicative conjugation with an element in $\mathfrak{E}$ and the division of $\tilde{P}$ by $\mathfrak{d}_{\tilde{P}}$, we may therefore assume without loss of generality that $\tilde{\mathfrak{m}}\in x^{\mathbb{N}}$ and $\tilde{P}\asymp\tilde{Q}\asymp 1$. Notice that $\mathfrak{m}/\tilde{\mathfrak{m}}\mathrel{\underline{\succ\!\!\succ}}\mathrm{e}^x$ since $\mathfrak{m}\succ\tilde{\mathfrak{m}}$ and $\mathfrak{m}$ is exponential.

By theorem 8.28, our assumption $\tilde{\varphi}_{\tilde{\tilde{\mathfrak{V}}}}=0$ implies

$$\frac{\tilde{\mathfrak{m}}}{\mathfrak{v}}\mathrel{\underline{\prec\!\!\prec}}\log\frac{\mathfrak{m}}{\tilde{\mathfrak{m}}}$$

for all $\mathfrak{v}\in\operatorname{supp}\tilde{\varphi}$. Since $\mathfrak{m}\succ 1$ is exponential, this relation simplifies to

$$\mathfrak{v}\mathrel{\underline{\prec\!\!\prec}}\log\mathfrak{m}.$$

Now assume that $\tilde{\varphi}\notin C[x][\![\mathfrak{E}]\!]$, let $\mathfrak{n}\in\operatorname{supp}\tilde{\varphi}$ be maximal with $\tilde{\varphi}\notin x^{\mathbb{N}}\,\mathfrak{E}$, and let $\psi=\sum_{\mathfrak{u}\succ\mathfrak{n}}\tilde{\varphi}_{\mathfrak{u}}\,\mathfrak{u}$. Since $\tau_{\varphi}\asymp\mathfrak{m}$ is a starting term for (E) of multiplicity d, we have $P_{\times\mathfrak{m},i}\preccurlyeq P_{\times\mathfrak{m},d}$ for all $i>d$. It follows that $\tilde{P}_{\times\mathfrak{m},i}\preccurlyeq\tilde{P}_{\times\mathfrak{m},d}$, $\tilde{P}_i\preccurlyeq_{\mathfrak{m}}\tilde{P}_d/\mathfrak{m}$ and $\tilde{P}_{+\psi,i}\preccurlyeq_{\mathfrak{m}}\tilde{P}_{+\psi,d}/\mathfrak{m}$ for all $i>d$. Now consider

$$T=\sum_{\mathfrak{u}\prec\!\!\prec\mathfrak{m}}\tilde{P}_{+\psi,\mathfrak{u}}\,\mathfrak{u}.$$

By what precedes, we have $\deg T=d$. Furthermore, $T_0,\ldots,T_{d-1}\in C[x][\![\mathfrak{E}]\!]$ and $T_d\in C[\![\mathfrak{E}]\!]$. By proposition 8.8, $\mathfrak{n}$ is a starting monomial for

$$T(g)=0\qquad(g\prec\operatorname{supp}\psi).$$

Moreover, $\mathfrak{n}$ is a differential starting monomial, by lemma 8.29. Since

$$T_d=\sum_{\mathfrak{u}\prec\!\!\prec\mathfrak{m}}\tilde{P}_{d,\mathfrak{u}}\,\mathfrak{u},$$

proposition 8.8 also implies that $\mathfrak{n}$ is a starting monomial for $\tilde{P}_d(h)=0$. Our assumptions thus result in the contradiction that $\mathfrak{m}\in\mathfrak{E}$. □

8.6.4 Bounding the depths of solutions

If we can bound the number of upward shiftings which are necessary for satisfying the conditions of proposition 8.30, then the combination of propositions 8.28 and 8.30 implies that any sequence of compatible atomic unravellings is necessarily finite. Now the problem of finding such a bound is a problem of order $r-1$, by proposition 8.16. Using induction, we obtain the following theorem:

Theorem 8.31. *Consider an equation* (E) *of Newton degree d and weight w, with exponential coefficients. If $f \in \mathbb{T}$ is a normalized solution to* (E) *modulo an initial segment $\mathfrak{W} \subsetneq \mathfrak{V}$, then f has depth $\leqslant B_{r,d,w}$, where $B_{0,d,w} = 0$ and $B_{r,d,w} = 2\,d\,(4\,w)^{r-1}$ if $r > 0$.*

Proof. We prove the theorem by a double recursion over r and d. If $r = 0$, then the theorem follows from corollary 3.9. In the case when $d = 0$ we also have nothing to prove, since there are no solutions. So assume that $r > 0$, $d > 0$ and that we have proved the theorem for all strictly smaller r or for the same r and all strictly smaller d. We may also assume that $f \neq 0$, since the theorem is clearly satisfied when $f = 0$.

Let $\mathfrak{m} \in \mathfrak{W} \setminus \mathfrak{V}$ be the dominant monomial of f. If f is algebraic, then proposition 8.14 implies that its depth is bounded by w. If $\mathfrak{m}$ is differential, then $r > 0$ and $\mathfrak{m}^\dagger$ is a root of R_{P_i} modulo $o(\frac{1}{x \log x \log\log x})$ for some i. Hence, its depth is bounded by $A_{r-1,w} = B_{r-1,w,w-1} \geqslant w$, because of the induction hypothesis. Modulo $A_{r-1,w}$ upward shiftings and a multiplicative conjugation with $\mathfrak{m}$, we may thus reduce the general case to the case when $\mathfrak{m} = 1$ and $N_P = D_P$. It remains to be shown that f has depth $\leqslant B_{r,d,w} - A_{r-1,w}$.

If $c = c_f$ is a root of multiplicity $< d$ of N_P, then the Newton degree of

$$P_{+c}(\tilde{f}) = 0 \;\; (\tilde{f} \prec \mathfrak{m})$$

is $< d$ by proposition 8.13 and $f - c$ is a root of this equation modulo $\mathfrak{W}$. The induction hypothesis now implies that $f - c$ has depth $\leqslant B_{r,d-1,w} \leqslant B_{r,d,w} - A_{r-1,w}$.

Assume now that c is a root of multiplicity d of N_P. Consider a finest atomic unraveling (R) for which $\tilde{f} = f - \varphi \in C[[\tilde{\mathfrak{V}}]]$. Then $\varphi\uparrow_r$ and $\tilde{P}\uparrow_r$ are exponential, by theorem 8.23. Let $\tilde{\varphi} \trianglelefteq \tilde{f}$ be the longest truncation of $\tilde{f}$, such that the Newton degree of

$$\tilde{\tilde{P}}(\tilde{\tilde{f}}) = \tilde{P}_{+\tilde{\varphi}}(\tilde{\tilde{f}}) = P_{+\varphi+\tilde{\varphi}}(\tilde{\tilde{f}}) = 0 \qquad (\tilde{\tilde{f}} \in C[[\mathfrak{V}]] \wedge \tilde{\tilde{f}} \prec \operatorname{supp} \tilde{\varphi})$$

is equal to d. By the induction hypothesis, $\tilde{P}_d\uparrow_{r+A_{r-1,w}}$ only admits exponential solutions. Now theorem 8.30 implies that $\tilde{\varphi}$ has depth $\leqslant r + A_{r-1,w} + 1$. If $\tilde{\tilde{f}} = \tilde{f} - \tilde{\varphi} = 0$, then we are done. Otherwise, $\tau_{\tilde{\tilde{f}}}$ is a starting term of multiplicity $< d$ for $\tilde{\tilde{P}}$, by the definition of $\tilde{\varphi}$. By what precedes, we conclude that $\tilde{\tilde{f}}$ has depth $\leqslant r + A_{r-1,w} + 1 + A_{r-1,w} + B_{r,d-1,w} \leqslant B_{r,d,w} - A_{r-1,w}$. □

Corollary 8.32. *Consider an equation* (E) *of Newton degree d and a non-empty set $\mathscr{S}$ of partial unravellings for* (E). *Then $\mathscr{S}$ admits a finest element.*

Proof. Let us first assume for contradiction that there exists an infinite sequence of compatible atomic unravellings

$$\begin{aligned} f = f_0 &= \varphi_0 + f_1 \qquad (f_1 \prec \mathfrak{v}_1) \\ f_1 &= \varphi_1 + f_2 \qquad (f_2 \prec \mathfrak{v}_2) \\ &\vdots \end{aligned}$$

Modulo a finite number of upward shiftings, it follows from theorem 8.31 that we may assume without loss of generality that the coefficients of $P_{+\varphi_0}$ are exponential and that $P_{+\varphi_0,d}$ only admits exponential solutions. Then theorem 8.30 implies that $\varphi_i \in C[x][[\mathfrak{E}]]$ for all $i \geqslant 1$. From theorem 8.28 it also follows that $\frac{\varphi_{i+1}}{\varphi_{i+2}} \precneqq \log \frac{\varphi_i}{\varphi_{i+1}}$ for all $i \geqslant 1$. But this is impossible.

Now pick a partial unravelling (R) in $\mathscr{S}$ of maximal length. Then any finer partial unravelling in $\mathscr{S}$ is obtained by replacing the last atomic unravelling which composes (R) by a finer one. The result now follows from proposition 8.26. □

Exercise 8.22. In theorem 8.30, show that whenever $\mathfrak{m}$ is a starting monomial for $P_d(h) = 0$ of the form $(\log_d x)^{\alpha_d} \cdots x^{\alpha_0} \mathfrak{m}^\sharp$ with $\mathfrak{m}^\sharp \in \mathfrak{E}$ and $\alpha_d \neq 0$, then $d \leqslant \operatorname{wt} P - 1$.

Exercise 8.23. Improve the bound in theorem 8.31 in the case when $r = 1$.

Exercise 8.24. Show how to obtain a total unravelling (8.23) *a posteriori*, by computing Q w.r.t. the monomial $\tilde{\mathfrak{v}}$ instead of $\mathfrak{m}$.

8.7 Algorithmic resolution

In this section, we will give explicit, but theoretical algorithms for solving (E). In order to deal with integration constants, we will allow for computations with infinite sets of transseries. In practice, one rather needs to compute with finite sets of "parameterized transseries". However, the development of such a theory (see [vdH97, vdH01a]) falls outside the scope of the present book.

8.7.1 Computing starting terms

Theorem 8.6 implies that we may compute the Newton polynomial of a differential polynomial $P \in \mathbb{T}\{F\}^{\neq}$ using the algorithm below. Recall that a monomial $\mathfrak{m}$ is a starting monomial if and only if $N_{P_{\times \mathfrak{m}}} \notin C F^{\mathbb{N}}$.

Algorithm N_P
Input: $P \in \mathbb{T}\{F\}^{\neq}$.
Output: The differential Newton polynomial N_P of P.

1. If P is not exponential or $D_P \notin C[F]\,(F')^{\mathbb{N}}$, then return $N_{P\uparrow}$.
2. Return D_P.

The algebraic starting monomials can be found by computing all equalizers and keeping only those which are starting monomials. The equalizers are computed using the method from the proof of proposition 8.14.

Algorithm $\mathfrak{e}_{P,i,j}$
Input: $P \in \mathbb{T}\{F\}^{\neq}$ and integers i, j with $P_i \neq 0$ and $P_j \neq 0$.
Output: The (i, j)-equalizer $\mathfrak{e}_{P,i,j}$ for P.

1. If P is not exponential or $D_{P_i+P_j} \notin C[F]\,(F')^{\mathbb{N}}$, then return $\mathfrak{e}_{P\uparrow,i,j}\downarrow$.
2. If $\mathfrak{d}(P_i) = \mathfrak{d}(P_j)$ then return 1.
3. Let $\mathfrak{m} := \sqrt[j-i]{\mathfrak{d}(P_i)/\mathfrak{d}(P_j)}$ and return $\mathfrak{m}\,\mathfrak{e}_{P_{\times\mathfrak{m}},i,j}$.

Algorithm `alg_st_mon`$(P, \mathfrak{V})$
Input: $P \in \mathbb{T}\{F\}^{\neq}$ and an initial segment $\mathfrak{V} \subseteq \mathfrak{T}$.
Output: The set of algebraic starting monomials for (E).

1. Compute $\mathfrak{M} := \{\mathfrak{e}_{P,i,j}: i < j \leqslant \deg P \wedge P_i \neq 0 \wedge P_j \neq 0\} \cap \mathfrak{V}$.
2. Return $\{\mathfrak{m} \in \mathfrak{M}: N_{P_{\times\mathfrak{m}}} \notin C F^{\mathbb{N}}\}$.

In fact, using proposition 8.17, it is possible to optimize the algorithm so that only a linear number of equalizers needs to be computed. This proposition also provides us with an efficient way to compute the Newton degree.

Algorithm `Newton_degree`$(P, \mathfrak{V})$
Input: $P \in \mathbb{T}\{F\}^{\neq}$ and an initial segment $\mathfrak{V} \subseteq \mathfrak{T}$.
Output: The Newton degree of (E).

1. Compute $\mathfrak{M} :=$ `alg_st_mon`$(P, \mathfrak{V})$.
2. Return $\max\,\{\deg N_{P_{\times\mathfrak{m}}}: \mathfrak{m} \in \mathfrak{M}\} \cup \{\operatorname{val} P\}$.

The algorithm for finding the differential starting terms is based on proposition 8.16 and a recursive application of the algorithm `ade_solve` (which will be specified below) in order to solve the Riccati equations modulo $o(\frac{1}{x \log x \log \log x \cdots})$.

Algorithm `diff_st_mon`$(P, \mathfrak{V})$
Input: $P \in \mathbb{T}\{F\}^{\neq}$ and an initial segment $\mathfrak{V} \subseteq \mathfrak{T}$.
Output: The set of differential starting monomials for (E).

1. If P is homogeneous, then
   Let $G :=$ `ade_solve`$(R_P, \mathfrak{T}, \{\mathfrak{m} \in \mathfrak{T}: \int \mathfrak{m} \prec 1\})$
   Return $\{e^{\int g}: g \in G\} \cap \mathfrak{V}$.
2. Let $\mathfrak{M}_i :=$ `diff_st_mon`$(P_i, \mathfrak{V})$ for each $i \leqslant \deg P$ with $P_i \neq 0$.
3. Return $\{\mathfrak{m} \in \mathfrak{M}_i: i \leqslant \deg P \wedge P_i \neq 0 \wedge N_{P_{\times\mathfrak{m}}} \notin C F^{\mathbb{N}}\}$.

Having computed the sets of algebraic and differential starting monomials, it suffices to compute the roots of the corresponding Newton polynomials in order to find the starting terms.

Algorithm `st_term`$(P, \mathfrak{V})$
Input: $P \in \mathbb{T}\{F\}^{\neq}$ and an initial segment $\mathfrak{V} \subseteq \mathfrak{T}$.
Output: The set of starting terms for (E).

1. Let $\mathfrak{D} := \texttt{alg\_st\_mon}(P, \mathfrak{V}) \cup \texttt{diff\_st\_mon}(P, \mathfrak{V})$.
2. Return $\{c\mathfrak{m} \in C^{\neq}\mathfrak{D} : N_{P_{\times\mathfrak{m}}}(c) = 0\}$.

8.7.2 Solving the differential equation

Let us now show how to find all solutions to (E) and, more generally, all normalized solutions of (E) modulo an initial segment $\mathfrak{W} \subseteq \mathfrak{V}$. First of all, 0 is a solution if and only if the Newton degree of $P(f) = 0$ $(f \in C[\![\mathfrak{W}]\!])$ is > 0. In order to find the other solutions, we first compute all starting terms τ in $\mathfrak{V} \setminus \mathfrak{W}$. For each such τ, we next apply the subalgorithm `ade_solve_sub` in order to find the set of solutions which starting term τ.

Algorithm `ade_solve`$(P, \mathfrak{V}, \mathfrak{W})$
Input: $P \in \mathbb{T}\{F\}^{\neq}$ and initial segments $\mathfrak{W} \subseteq \mathfrak{V} \subseteq \mathfrak{T}$.
Output: The set of normalized solutions to (E) modulo $\mathfrak{W}$.

1. Compute $\mathrm{T} := \texttt{st\_term}(P, \mathfrak{V}) \setminus C\mathfrak{W}$.
2. Let $S := \bigcup_{\tau \in \mathrm{T}} \texttt{ade\_solve\_sub}(P, \tau, \mathfrak{V}, \mathfrak{W})$.
3. If $\texttt{Newton\_degree}(P, \mathfrak{W}) > 0$ then $S := S \cup \{0\}$.
4. Return S.

Let d be the Newton degree of (E). In order to find the normalized solutions with starting terms τ of multiplicity $< d$, we may simply use the refinement

$$f = \tau + \tilde{f} \qquad (\tilde{f} \prec \tau)$$

and recursively solve

$$P_{+\tau}(\tilde{f}) = 0 \qquad (\tilde{f} \prec \tau).$$

The other starting terms require the unravelling theory from section 8.6: we start by computing the quasi-linear differentiated equation

$$Q(f) = 0 \ (f \in C[\![\mathfrak{V}]\!]), \tag{8.26}$$

with Q as in (8.24) and we will "follow" solutions to this equation as long as possible using the subalgorithm `unravel`.

Algorithm `ade_solve_sub`$(P, \tau, \mathfrak{V}, \mathfrak{W})$
Input: $P \in \mathbb{T}\{F\}^{\neq}$, initial segments $\mathfrak{W} \subseteq \mathfrak{V} \subseteq \mathfrak{T}$ and a starting term $\tau = c\mathfrak{m} \in C^{\neq}(\mathfrak{V} \setminus \mathfrak{W})$ for (E).
Output: The set of normalized solutions to (E) modulo $\mathfrak{W}$ with dominant term τ.

1. Let $\mu := \operatorname{val} N_{P_{\times\mathfrak{m}}, +c}$ and $d := \texttt{Newton\_degree}(P, \mathfrak{V})$.
2. If $\mu < d$, then return $\tau + \texttt{ade\_solve}(P_{+\tau}, \{\mathfrak{n} \in \mathfrak{T} : \mathfrak{n} \prec \mathfrak{m}\}, \mathfrak{W})$.

3. Compute Q using (8.24), with minimal l, and let $\varphi = \tau + h$, where h is the distinguished solution to

$$Q_{+\tau}(h) = 0 \qquad (h \prec \tau). \tag{8.27}$$

4. Return $\varphi_{\mathfrak{V} \setminus \mathfrak{W}} + \texttt{unravel}(P_{+\varphi}, Q_{+\varphi}, \{\mathfrak{n} \in \mathfrak{T} : \mathfrak{n} \prec \mathfrak{m}\}, \mathfrak{W})$.

The algorithm `unravel` is analogous to `ade_solve`, except that we now compute the solutions with a given starting term using the subalgorithm `unravel_sub` instead of `ade_solve_sub`.

Algorithm $\texttt{unravel}(P, Q, \mathfrak{V}, \mathfrak{W})$
Input: $P, Q \in \mathbb{T}\{F\}^{\neq}$ and initial segments $\mathfrak{W} \subseteq \mathfrak{V} \subseteq \mathfrak{T}$.
Output: The set of normalized solutions to (E) modulo $\mathfrak{W}$ with dominant term τ.

1. Compute $\mathrm{T} := \texttt{st\_term}(P, \mathfrak{V}) \setminus C\,\mathfrak{W}$.
2. Let $S := \bigcup_{\tau \in \mathrm{T}} \texttt{unravel\_sub}(P, Q, \tau, \mathfrak{V}, \mathfrak{W})$.
3. If $\texttt{Newton\_degree}(P, \mathfrak{W}) > 0$, then $S := S \cup \{0\}$.
4. Return S.

In `unravel_sub`, we follow the solutions to (8.26) as far as possible. More precisely, let Q be as in (8.24). Then the successive values of Q for calls to `unravel` and `unravel_sub` are of the form $Q_{+h_1}, \ldots Q_{+h_1 + \cdots + h_l}$, where $h_1 \succ \cdots \succ h_l$ satisfy $Q(h_1 + \cdots + h_i) = 0$ for each $i \in \{1, \ldots, l\}$. At the end, the refinement

$$f = h_1 + \cdots + h_l + \tilde{f} \qquad (\tilde{f} \prec h_l) \tag{8.28}$$

is an atomic unravelling for the original equation. Moreover, at the recursive call of `ade_solve_sub`, the next refinement will be compatible with (8.28).

Algorithm $\texttt{unravel\_sub}(P, Q, \tau, \mathfrak{V}, \mathfrak{W})$
Input: P, $Q \in \mathbb{T}\{F\}^{\neq}$, initial segments $\mathfrak{W} \subseteq \mathfrak{V} \subseteq \mathfrak{T}$ and a starting term $\tau = c\mathfrak{m} \in C^{\neq}(\mathfrak{V} \setminus \mathfrak{W})$ for (E).
Output: The set of normalized solutions to (E) modulo $\mathfrak{W}$ with dominant term τ.

1. If $N_{Q_{\times \mathfrak{m}}}(c) \neq 0$, then return $\texttt{ade\_solve\_sub}(P, \tau, \mathfrak{V}, \mathfrak{W})$.
2. Let $\varphi = \tau + h$, where h is the distinguished solution to (8.27).
3. Return $\varphi_{\mathfrak{V} \setminus \mathfrak{W}} + \texttt{unravel}(P_{+\varphi}, Q_{+\varphi}, \{\mathfrak{n} \in \mathfrak{T} : \mathfrak{n} \prec \mathfrak{m}\}, \mathfrak{W})$.

The termination of our algorithms are verified by considering the three possible loops. In successive calls of `solve` and `solve_sub` we are clearly done, since the Newton degree strictly decreases. As to successive calls of `unravel` and `unravel_sub`, we have $l \leqslant r$ in (8.28), by theorem 8.25. Finally, any global loop via `solve_sub` and `unravel`, during which the Newton degree d remains constant, corresponds to a sequence of compatible atomic unravellings. But such sequences are necessarily finite, by theorems 8.25, 8.30 and 8.31.

Exercise 8.25. Assume that $P \in C⟦x⟧\{F\}$ and that we search for zeros of (E) in the set of well-based transseries of finite exponential and logarithmic depths $C[[[x]]]$.

a) Given $Q \in C[[[x]]]\{F\}$, show there exists an l with $D_{Q\uparrow_l} \in C[F]\,(F')^{\mathbb{N}}$. Give a definition for the differential Newton polynomial N_Q of Q. Generalize proposition 8.10.
b) Given $i<j$ with $P_i \neq 0$ and $P_j \neq 0$, prove that there is at most one well-based transmonomial $\mathfrak{m}$ such that $N_{(P_i+P_j)_{\times \mathfrak{m}}}$ is non-homogeneous.
c) Show that proposition 8.16 still holds for well-based transmonomials.
d) Show that the set of solutions to (E) in $C⟦x⟧$ as computed by `ade_solve` coincides with the set of solutions to (E) in $C[[[x]]]$.
e) Show that $\zeta(x)$,
$$\varphi(x) = \frac{1}{x} + \frac{1}{x^{\pi}} + \frac{1}{x^{\pi^2}} + \cdots$$
and
$$\psi(x) = \frac{1}{x} + \frac{1}{\mathrm{e}^{\log^2 x}} + \frac{1}{\mathrm{e}^{\log^4 x}} + \cdots$$
do not satisfy an algebraic differential equation with coefficients in $\mathbb{T}$.
f) Does $\varphi(x)$ satisfy an algebraic differential equation with coefficients in $\mathbb{T}\{\zeta(x)\}$? And does $\psi(x)$ satisfy an algebraic differential equation with coefficients in $\mathbb{T}\{\zeta(x), \varphi(x)\}$?

8.8 Structure theorems

8.8.1 Distinguished unravellers

Theorem 8.33. *Let (E) be an equation of Newton degree $d > 1$. Then there exists a unique $\varphi \in C⟦\mathfrak{V}⟧$ which is longest for $\trianglelefteq$ with the properties that*

a) $\deg_{\tilde{\mathfrak{V}}} P_{+\varphi} = d$, for $\tilde{\mathfrak{V}} = \{\mathfrak{m} \in \mathfrak{V} : \mathfrak{m} \prec \operatorname{supp} \varphi\}$.
b) For any $\mathfrak{m} \in \operatorname{supp} \varphi$, the term $\varphi_{\mathfrak{m}} \mathfrak{m}$ is an algebraic starting term for

$$P_{+\varphi_{\succ \mathfrak{m}}}(\tilde{f}) = 0 \qquad (\tilde{f} \preccurlyeq \mathfrak{m}). \tag{8.29}$$

Proof. Consider the set $\mathscr{S}$ of all partial unravellings

$$f = \xi + \tilde{f} \qquad (\tilde{f} \in C⟦\tilde{\mathfrak{V}}⟧), \tag{8.30}$$

such that $\varphi = \xi_{\mathfrak{V} \setminus \tilde{\mathfrak{V}}}$ satisfies (a) and (b). Since $\mathscr{S}$ contains the identity refinement, we may choose (8.30) to be finest in $\mathscr{S}$, by corollary 8.32. We claim that φ is maximal for $\trianglelefteq$, such that (a) and (b) are satisfied.

Indeed, assume for contradiction that some $\psi \vartriangleright \varphi$ also satisfies (a) and (b). Then $c\,\mathfrak{m} = \tau(\psi - \varphi)$ is the unique algebraic starting term for (8.29) and it has multiplicity d. By proposition 8.27, there exists a partial unravelling

$$f = \xi + \tilde{\xi} + \tilde{\tilde{f}} \qquad (\tilde{\tilde{f}} \prec \tilde{\xi}),$$

which is finer than (8.30), and such that $\tilde{\xi} \sim c\,\mathfrak{m}$. By what precedes, $\tilde{\varphi} = (\xi + \tilde{\xi})_{\succcurlyeq \tilde{\xi}} = \varphi + c\,\tau$ satisfies (a). Moreover, $\tilde{\varphi}$ satisfies (b), since $\psi \trianglerighteq \tilde{\varphi}$ does. This contradicts the maximality of (8.30).

Let us now prove the uniqueness of φ. Assume for contradiction that $\psi \neq \varphi$ with $\psi \ntrianglelefteq \varphi$ and $\varphi \ntrianglelefteq \psi$ also satisfies (a) and (b). Let $\delta = \psi - \varphi$ and $\xi = \sum_{\mathfrak{m} \succ \delta} \varphi_{\mathfrak{m}}\,\mathfrak{m}$. Then

$$P_{+\xi}(\tilde{f}) = 0 \qquad (\tilde{f} \prec \operatorname{supp} \xi)$$

admits both $\tau(\varphi - \xi)$ and $\tau(\psi - \xi)$ as algebraic starting terms of multiplicity d. But this is impossible. □

The transseries φ from the theorem is called the *distinguished unraveller* for (E). It has the property that for any algebraic starting term $\tilde{\tau}$ for

$$P_{+\varphi}(\tilde{f}) = 0 \qquad (\tilde{f} \preccurlyeq \operatorname{supp} \varphi), \tag{8.31}$$

the refinement

$$f = \varphi + \tilde{\tau} + \tilde{\tilde{f}} \qquad (\tilde{\tilde{f}} \prec \tilde{\tau})$$

is a total unravelling.

Remark 8.34. It is easily checked that theorem 8.33 also holds for $d = 1$, and that φ coincides with the distinguished solution of (E) in this case.

Recall that $\mathfrak{L}$ stands for the group of logarithmic monomials.

Proposition 8.35. *Let φ be as in theorem 8.33 and assume that $P \in C[[\mathfrak{B}^C]]\{F\}^{\neq}$ for a plane transbasis $\mathfrak{B} = (\mathfrak{b}_1, \ldots, \mathfrak{b}_n)$. Then $\varphi \in C[[\mathfrak{L}\,\mathfrak{B}^C]]$.*

Proof. Assume the contrary, let $\mathfrak{m} \in \operatorname{supp} \varphi$ be maximal, such that $\mathfrak{m} \notin C[[\mathfrak{L}\,\mathfrak{B}^C]]$, and let $\psi = \sum_{\mathfrak{n} \succ \mathfrak{m}} \varphi_{\mathfrak{n}}\,\mathfrak{n}$. Modulo a finite number of upward shiftings, we may assume without loss of generality that P and ψ are exponential. But then $\mathfrak{m} = \mathfrak{d}_{\varphi - \psi}$ is an algebraic starting monomial for

$$P_{+\psi}(\tilde{f}) \qquad (\tilde{f} \prec \operatorname{supp} \psi).$$

By remark 8.15, we conclude that $\mathfrak{m} \in C[[\mathfrak{L}\,\mathfrak{B}^C]]$. □

8.8.2 Distinguished solutions and their existence

A solution $\varphi \in \mathbb{T}$ to (E) is said to be *distinguished*, if for all $\mathfrak{m} \in \operatorname{supp} \varphi$, the term $\varphi_{\mathfrak{m}}\,\mathfrak{m}$ is an algebraic starting term for the equation

$$P_{+\varphi}(\tilde{f}) = 0 \qquad (\tilde{f} \preccurlyeq \mathfrak{m}).$$

If d is odd, then there exists at least one distinguished solution.

Theorem 8.36. *Any equation* (E) *of odd Newton degree admits at least one distinguished solution in* $\mathbb{T}$. *Moreover, if the coefficients of* P *can be expanded w.r.t. a plane transbasis* $\mathfrak{B} = (\mathfrak{b}_1, ..., \mathfrak{b}_n)$, *then any such solution is in* $C[[\mathfrak{L}\,\mathfrak{B}^C]]$.

Proof. We prove the theorem by induction over d. For $d=1$, the result follows from corollary 8.24. So let $d > 1$ and assume that the theorem holds for all smaller d.

Now proposition 8.17 implies that there exists at least one starting monomial and equalizer $\mathfrak{e} \in \mathfrak{L}\,\mathfrak{B}^C$ such that $\deg N_{P_{\times \mathfrak{e}}} - \operatorname{val} N_{P_{\times \mathfrak{e}}}$ is odd. It follows that $P = A\,(F')^\nu$ for some $A \in C[F]$ of odd degree. Since C is real closed, it follows that A admits a root c of odd multiplicity $\tilde{d}$.

If $\tilde{d} < d$, then proposition 8.13 and the induction hypothesis imply that

$$\tilde{P}(\tilde{f}) = P_{+c\mathfrak{e}}(\tilde{f}) = 0 \qquad (\tilde{f} \prec \mathfrak{e}) \tag{8.32}$$

admits a distinguished solution $\tilde{f} = C[[\mathfrak{L}\,\mathfrak{B}^C]]$, whence

$$f = c\,\mathfrak{e} + \tilde{f} \in C[[\mathfrak{L}\,\mathfrak{B}^C]]$$

is a distinguished solution to (E). Inversely, if $f \neq 0$ is a distinguished solution to (E) whose dominant term $c\,\mathfrak{e}$ has multiplicity $\tilde{d} < d$, then $\mathfrak{e}$ is necessarily an equalizer, and

$$\tilde{f} = f - c\,\mathfrak{e} \in C[[\mathfrak{L}\,\mathfrak{B}^C]]$$

a distinguished solution to (8.32), whence $f \in C[[\mathfrak{L}^C]]$.

If $\tilde{d} = d$, then let φ be the distinguished unraveller for (E), so that the equation

$$\tilde{P}(\tilde{f}) = P_{+\varphi}(\tilde{f}) = 0 \qquad (\tilde{f} \prec \operatorname{supp} \varphi) \tag{8.33}$$

does not admit an algebraic starting term of multiplicity d. Modulo some upward shiftings and by what precedes, it follows that (8.33) admits a distinguished solution $\tilde{f} \in C[[\mathfrak{L}\,\mathfrak{B}^C]]$. We conclude that

$$f = \varphi + \tilde{f} \in C[[\mathfrak{L}\,\mathfrak{B}^C]]$$

is a distinguished solution to (E). Inversely, we have $\varphi \trianglelefteq f$ for any distinguished solution f of (E), and $\tilde{f} = f - \varphi$ is a distinguished solution to (8.33), whence $f \in C[[\mathfrak{L}\,\mathfrak{B}^C]]$. □

8.8.3 On the intrusion of new exponentials

In this chapter, we have shown how to solve (E) directly as an equation in F, ..., $F^{(r)}$. A more advanced method for solving (E) is to use *integral refinements*

$$f = e^{\int \varphi + \tilde{f}} \qquad (\tilde{f} \in C[[\tilde{\mathfrak{V}}]])$$

in addition to usual refinements. This gives a better control over the number of exponentials and integration constants introduced in the resolution process, because $e^{\int \varphi + \tilde{f}}$ is often "strongly transcendental" over the field generated by the coefficients of P, so that the equation rewritten in $\tilde{f}$ has lower order. A full exposition of these techniques is outside the scope of this book, but the proof of the following theorem will illustrate some of the involved ideas to the reader.

Theorem 8.37. *Consider $P \in C[[\mathfrak{B}^C]]\{F\}^{\neq}$ of order r for some plane transbasis $\mathfrak{B}$. Then for each exponential solution $f \in \mathbb{T}$ to* (E), *there exists a transbasis $\hat{\mathfrak{B}}$ for f with* $\operatorname{card} \hat{\mathfrak{B}} \setminus \mathfrak{B} \leqslant r$.

Proof. Let us construct sequences $f_0, \ldots, f_l \in \mathbb{T}$, $\varphi_0, \ldots, \varphi_l \in \mathbb{T}$ and $\mathfrak{x}_1, \ldots, \mathfrak{x}_l \in \mathfrak{T}$ such that

1. $\mathfrak{X}_i = \mathfrak{B} \cup \{\mathfrak{x}_1, \ldots, \mathfrak{x}_i\}$ is totally ordered for $\prec\!\!\prec$.
2. $\varphi_i \in C[[\mathfrak{X}_i^C]]$ for each $i = \{0, \ldots, l\}$ (where we understand that $\mathfrak{X}_0 = \mathfrak{B}$).

We take $f_0 = f$. Given $i \geqslant 0$, let φ_i be the longest truncation of f_i, such that $\varphi_i \in C[[\mathfrak{X}_i^C]]$. If $\varphi_i = f_i$, then the sequence is complete. Otherwise, we let

$$\begin{aligned} \mathfrak{x}_{i+1} &= \mathfrak{d}(f_i - \varphi_i) \\ f_{i+1} &= (f_i - \varphi_i)^{\dagger}. \end{aligned}$$

If $\tilde{\mathfrak{B}}$ is an arbitrary transbasis for f, then

$$C[[\mathfrak{X}_0^C]] \subsetneq \cdots \subsetneq C[[\mathfrak{X}_l^C]] \subseteq C[[\tilde{\mathfrak{B}}^C]],$$

so that the construction finishes for $l \leqslant \operatorname{card} \tilde{\mathfrak{B}} \setminus \mathfrak{B}$. Setting $\hat{\mathfrak{B}} = \mathfrak{X}_l$, we also observe that $\log \mathfrak{x}_i \trianglelefteq \int f_i \in C[[\hat{\mathfrak{B}}]]$ for all $i \in \{1, \ldots, l\}$. It follows that $\hat{\mathfrak{B}}$ is a transbasis for f.

Let us now consider another sequence $\mathfrak{y}_1, \ldots, \mathfrak{y}_l$ with

$$\mathfrak{y}_{i+1} = \frac{f_i - \varphi_i}{c(f_i - \varphi_i)} \sim \mathfrak{x}_{i+1},$$

so that

$$f_{i+1} = \mathfrak{y}_{i+1}^{\dagger}.$$

Denoting $\mathfrak{Y}_i = \mathfrak{B} \cup \{\mathfrak{y}_1, \ldots, \mathfrak{y}_i\}$ for all $i \in \{1, \ldots, l\}$, we notice that $C[[\mathfrak{Y}_l^C]]$ is isomorphic to $C[[\mathfrak{X}_l^C]]$. Now for all $i \in \{1, \ldots, l-1\}$, we have

$$\mathfrak{y}_i' = \mathfrak{y}_i \mathfrak{y}_i^{\dagger} = \mathfrak{y}_i f_i = \mathfrak{y}_i (\varphi_i + c(f_i - \varphi_i) \mathfrak{y}_{i+1}) \in C[[\mathfrak{Y}_{i+1}^C]].$$

By strong linearity, it follows that for all $g \in C[[\mathfrak{Y}_i^C]]$ and $i \in \{0, \ldots, l-1\}$, we have $g' \in C[[\mathfrak{Y}_{i+1}^C]]$. Moreover, if

$$g \in C[[\mathfrak{Y}_{i-1}^C]] \oplus C[[\mathfrak{Y}_{i-1}^C]]^{\neq} \mathfrak{y}_i,$$

then the above formula also yields

$$g' \in C[[\mathfrak{Y}_i^C]] \oplus C[[\mathfrak{Y}_i^C]]^{\neq} \mathfrak{y}_{i+1}.$$

In particular,

$$f^{(i)} \in C[[\mathfrak{Y}_i^C]] \oplus C[[\mathfrak{Y}_i^C]]^{\neq} \mathfrak{y}_{i+1},$$

for all $i \in \{0, ..., l-1\}$.

Now assume for contradiction that $l > r$ and let $f^{(r)} = g + h\,\mathfrak{y}_{r+1}$ with g, $h \in C[[\mathfrak{Y}_r^C]]$. Then substitution of $f^{(i)} \in C[[\mathfrak{Y}_r^C]]$ for $F^{(i)}$ in P for all $i < r$ and $g + h\,F$ for $F^{(r)}$ yields a non-zero polynomial $S \in C[[\mathfrak{Y}_r^C]][F]$, which admits $\mathfrak{y}_{r+1} \notin C[[\mathfrak{Y}_r^C]]$ as a root. But this contradicts the fact that $C[[\mathfrak{Y}_r^C]]$ is real closed. We conclude that $l \leqslant r$, whence $\hat{\mathfrak{B}}$ is a transbasis for f with $\operatorname{card} \hat{\mathfrak{B}} \setminus \mathfrak{B} = l \leqslant r$. □

Corollary 8.38. *Consider $P \in C[[\mathfrak{B}^C]]\{F\}$ of order r for some transbasis $\mathfrak{B}$. Then for each solution $f \in \mathbb{T}$ to* (E)*, there exists a transbasis $\hat{\mathfrak{B}}$ for f with* $\operatorname{card} \hat{\mathfrak{B}} \setminus (\mathfrak{B} \cup \exp_{\mathbb{Z}} x) \leqslant r$.

Exercise 8.26. Give an alternative algorithm for the resolution of (E), where, after the computation of a starting term τ, we perform the refinement

$$f = \tau + \varphi + \tilde{f} \;\; (\tilde{f} \prec \tau),$$

where φ is the distinguished unraveller for $P_{+\tau}(\tilde{f}) = 0$ $(\tilde{f} \prec \tau)$.

Exercise 8.27. If, in the algorithms of section 8.7, we let `st_term` only return the algebraic starting terms, then show that the algorithm `ade_solve` will return the set of all distinguished solutions.

Exercise 8.28. Show that there exist at most $d = \deg_{\mathfrak{V}} P$ distinguished solutions to (E).

Exercise 8.29. If f is a distinguished solution to (E) and $\varphi \trianglelefteq f$, then show that $\tilde{f} - \varphi$ is a distinguished solution to $P_{+\varphi}(\tilde{f}) = 0$ $(\tilde{f} \prec \operatorname{supp} \varphi)$.

Exercise 8.30. Improve theorem 8.31 and show that we can take $B_{r,d,w} = 2\,r\,d\,w$. Hint: use exercise 8.22 in combination with the proof technique from theorem 8.37.

9

The intermediate value theorem

The main aim of this chapter is to prove the intermediate value theorem: given a differential polynomial $P \in \mathbb{T}\{F\}$ over the transseries and $f < g \in \mathbb{T}$ with $P(f)\,P(g) < 0$, there exists an $h \in \mathbb{T}$ with $f < h < g$ and $P(h) = 0$. In particular, any differential polynomial $P \in \mathbb{T}\{F\}$ of odd degree admits a zero in $\mathbb{T}$.

The intermediate value theorem is interesting from several points of view. First of all, it gives a simple sufficient condition for the existence of zeros of differential polynomials. This is complementary to the theory from the previous section, in which we gave a theoretical algorithm to compute all solutions, but no simple criterion for the existence of a solution (except for theorem 8.33).

Secondly, the intermediate value theorem has a strong geometric appeal. When considering differential polynomials as functions on $\mathbb{T}$, a natural question is to determine their geometric behaviour and in particular to localize their zeros. Another question would be to find the extremal and inflexion points. It is already known that extremal values are not necessarily attained. For instance, the differential polynomial

$$P(f) = f^2 + 2\,f'$$

admits its minimal "value"

$$-\frac{1}{x^2} - \frac{1}{x^2 \log^2 x} - \frac{1}{x^2 \log^2 x \log_2^2 x \cdots} - \cdots$$

"in"

$$f = \frac{1}{x} + \frac{1}{x \log x} + \frac{1}{x \log x \log_2 x} + \cdots.$$

In the future, we plan to classify all such non-standard "cuts" which occur as local extrema of differential polynomials. In particular, we expect that a cut occurs as a local minimum if and only of it occurs as a local maximum for another differential polynomial.

Finally, the intermediate value theorem is a starting point for the further development of the model theory for ordered differential algebra. Indeed, the field of transseries is a good candidate for an existentially closed model of this theory, i.e. a "real differentially algebraically closed field". Such fields are necessarily closed under the resolution of first order linear differential equations and they satisfy the intermediate value theorem. It remains to be investigated which additional properties should be satisfied and the geometric aspects of real differential polynomials may serve as a source of inspiration.

In order to prove the intermediate value theorem, the bulk of this chapter is devoted to a detailed geometric study of the "transline" $\mathbb{T}$ and differentially polynomial functions on it. Since the field of transseries is highly non-archimedean, it contains lots of cuts. Such cuts may have several origins: incompleteness of the constant field (if $C \neq \mathbb{R}$), the grid-based serial nature of $\mathbb{T}$, and exponentiation. In sections 9.1, 9.2, 9.3 and 9.4 we study these different types of cuts and prove a classification theorem.

Although the classification of cuts gives us a better insight in the geometry of the transline, the representation we use is not very convenient with respect to differentiation. In section 9.5, we therefore introduce another way to represent cuts using integral nested sequences of the form

$$\bar{f} = \varphi_0 + \epsilon_0 \, \mathrm{e}^{\int \varphi_1 + \epsilon_1 \mathrm{e}^{\int \cdot^{\cdot^{\cdot} \varphi_{k-1} + \epsilon_{k-1} \mathrm{e}^{\int \bar{f}_k}}}}.$$

This representation makes it possible to characterize the behaviour of differential polynomials in so called "integral neighbourhoods" of cuts, as we will see in section 9.6. In the last section, we combine the local properties of differential polynomials near cuts with the Newton polygon method from chapter 8, and prove the intermediate value theorem. We essentially use a generalization of the well-known dichotomic method for finding roots.

9.1 Compactification of total orderings

9.1.1 The interval topology on total orderings

Any totally ordered set E has a natural topology, called the *interval topology*, whose open sets are arbitrary unions of open intervals. We recall that an *interval* is a subset I of E, such that for each $x < y < z$ with $x, z \in I$, we have $y \in I$. An interval $I \subseteq E$ is said to be *open*, if for each $x \in I$ we have: x is minimal resp. maximal in I, if and only if x is minimal resp. maximal in E.

A set $U \subseteq E$ is *open* if every point in U is contained in an open interval $I \subseteq U$. Arbitrary unions of open sets are clearly open. The intersection of two open intervals I and J is again open: if x is minimal or maximal in $I \cap J$, then it is in particular minimal resp. maximal in I or J, whence x is minimal resp. maximal in E. It follows that the intersection of two open sets is also open, so the open sets of E form a topology.

We observe that an increasing union of open intervals is again an open interval. Hence, given an open set U and $x \in U$, there exists a maximal open interval $M_x \subseteq U$ with $x \in U$. It follows that each open set U admits a unique decomposition

$$U = \coprod \{M_x : x \in U\} \tag{9.1}$$

as the disjoint union of its maximal open subintervals.

Proposition 9.1. *A totally ordered set E with the interval topology is Hausdorff if and only if for each $x < y \in E$ there exists a $z \in E$, with $x < z < y$.*

Proof. Assume that E is Hausdorff and let $x < y \in E$. There exist open subsets $U \ni x$ and $V \ni y$ with $U \cap V = \varnothing$. Without loss of generality, we may assume that we have replaced U and V by subintervals which contain x resp. y. Since x is not maximal in E and U is open, there exists an $x' \in U$ with $x' > x$. We must also have $x' < y$: otherwise $y \in U$ whence $y \in U \cap V = \varnothing$, since U is an interval.

Conversely, assume that for all $x < y \in E$ there exists a $z \in E$, with $x < z < y$. Then given $x \neq y \in E$, and assuming by symmetry that $x < y$, there exists a $z \in E$, with $x < z < y$. Then $(\leftarrow, z) = \{u \in E : u < z\}$ and $(z, \rightarrow) = \{u \in E : u > z\}$ are disjoint intervals with $x \in (\leftarrow, z)$ and $y \in (z, \rightarrow)$. Moreover, for any $u \in (\leftarrow, z)$ there exists a $v \in E$ with $u < v < z$, and u is minimal in $(\leftarrow, z)$ if and only if it is minimal in E. Hence $(\leftarrow, z)$ is open, and similarly for $(z, \rightarrow)$. □

Example 9.2. Any totally ordered field E is Hausdorff.

9.1.2 Dedekind cuts

Given a totally ordered set E, let $\overline{E}$ denote the set of its open initial segments without maximal elements, ordered by inclusion. We have a natural increasing mapping

$$\begin{aligned} \iota: E &\longrightarrow \overline{E} \\ x &\longmapsto \operatorname{interior}(\leftarrow, x). \end{aligned}$$

Elements in $\overline{E} \setminus \iota(E)$ are called *cuts*. If E is Hausdorff, then we have already seen that $(\leftarrow, x)$ is open for all $x \in E$, so ι yields a natural inclusion of E into $\overline{E}$.

The elements $\bot_{\overline{E}} = \varnothing$ and $\top_{\overline{E}} = \bigcup \overline{E}$ are minimal and maximal in $\overline{E}$. If E admits no maximal element, then $\top_{\overline{E}} = E$. More generally, any non-empty subset of $\overline{E}$ admits an infimum and a supremum:

Proposition 9.3. *Any non-empty subset of $\overline{E}$ admits a supremum and an infimum in $\overline{E}$.*

Proof. Let $\overline{S} \neq \varnothing$ be a subset of $\overline{E}$ and consider the open initial segment without a maximal element

$$\overline{u} = \bigcup \overline{S}.$$

We claim that $\overline{u} = \sup \overline{S}$. By construction, $\overline{v} \leqslant \overline{u}$ for all $\overline{v} \in \overline{S}$. Conversely, if $\overline{v} \in \overline{E}$ satisfies $\overline{v} < \overline{u}$, then we may pick $x \in \overline{u} \setminus \overline{v}$. Now let $\overline{w} \in \overline{S}$ be such that $x \in \overline{w}$. Then $\overline{v} \subseteq \overline{w}$, whence $\overline{v} \leqslant \overline{w} \in \overline{S}$. In a similar way, it can be shown that the interior of $\bigcap \overline{S}$ equals the infimum of $\overline{S}$. □

Proposition 9.4. *Let I be an interval of a Hausdorff total ordering E. Then there exists unique $f \leqslant g \in \overline{E}$ such that I has one and only one of the following forms:*

a) $I = (f, g) \cap E$.
b) $I = [f, g) \cap E$ and $f \in E$.
c) $I = (f, g] \cap E$ and $g \in E$.
d) $I = [f, g] \cap E$ and $f, g \in E$.

Proof. Let $f = \inf I$ and $g = \sup I$. Then clearly

$$(\leftarrow, f) \cap E = (g, \rightarrow) \cap E = \varnothing$$

and $(f, g) \cap E \subseteq I$. Consequently,

$$I \subseteq [f, g] \cap E \subseteq I \cup \{f, g\}.$$

Depending on whether f and g are in I or not, we are therefore in one of the four cases (a), (b), (c) or (d). □

9.1.3 The compactness theorem

Theorem 9.5. *Let E be a Hausdorff totally ordered set. Then*

a) $\overline{E}$ is Hausdorff.
b) $\overline{\overline{E}} \cong \overline{E}$.
c) $\overline{E}$ is connected.
d) $\overline{E}$ is compact.

Proof. In order to show that $\overline{E}$ is Hausdorff, let $\overline{x} < \overline{y}$ be in $\overline{E}$. Choose $u \in \overline{y} \setminus \overline{x}$. Since $\overline{y}$ has no maximal element, there exist $v, w \in \overline{y}$ with $u < v < w$. It follows that $\overline{x} \leqslant u < v < w \leqslant \overline{y}$, which proves (a).

From (a) it follows that the natural mapping $\overline{\iota}: \overline{E} \rightarrow \overline{\overline{E}}$ is injective. In order to see that $\overline{\iota}$ is also surjective, consider an open initial segment $\overline{I} \subseteq \overline{E}$ without a maximal element, and consider $\overline{u} = \sup \overline{I}$. We claim that $\overline{\iota}(\overline{u}) = \overline{I}$. Indeed, if $\overline{x} \in \overline{\iota}(\overline{u})$, so that $\overline{x} < \overline{u}$, then there exists a $\overline{y} \in \overline{I}$ with $\overline{x} \leqslant \overline{y}$, by the definition of $\overline{u}$. Hence $\overline{x} \in \overline{I}$, since $\overline{I}$ is an initial segment. Conversely, if $\overline{x} \in \overline{I}$, then there exists a $\overline{y} \in \overline{I}$ with $\overline{x} < \overline{y}$, since $\overline{I}$ has no maximal element. We have $\overline{x} < \overline{y} \leqslant \overline{u}$, so $\overline{x} \in \overline{\iota}(\overline{u})$. This proves our claim and (b).

Let us now show that $\overline{E}$ is connected. Assume the contrary. Then $\overline{E}$ is the disjoint union of two open sets. By (9.1), it follows that

$$\overline{E} = \coprod_{\overline{I} \in \mathscr{I}} \overline{I},$$

where $\mathscr{I}$ is a set of at least two open intervals. Let $\overline{K} \in \mathscr{I}$ be non-maximal. Then we also have a decomposition of $\overline{E}$ as the disjoint union of two non-empty open intervals

$$\overline{E} = \overline{I}_1 \amalg \overline{I}_2 := \left(\coprod_{\overline{J} \in \mathscr{I}, \overline{J} \leqslant \overline{K}} \overline{J} \right) \amalg \left(\coprod_{\overline{J} \in \mathscr{I}, \overline{J} > \overline{K}} \overline{J} \right).$$

Now consider $\overline{u} = \sup \overline{I}_1$. We have either $\overline{u} \in \overline{I}_1$ or $\overline{u} \in \overline{I}_2$. In the first case, $\overline{u} \neq \top_{\overline{E}}$ would be a maximal element of $\overline{I}_1$. In the second case, $u \neq \bot_{\overline{E}}$ would be a minimal element of $\overline{I}_2$. This gives us the desired contradiction which proves (c).

Let us finally show that $\overline{E}$ is compact. In view of (9.1), it suffices to show that from any covering $(\overline{I}_\alpha)_{\alpha \in A}$ of $\overline{E}$ with open intervals we can extract a finite subcovering. Consider the sequence $\overline{x}_0 \leqslant \overline{x}_1 \leqslant \cdots \in \overline{E}$ which is inductively defined by $\overline{x}_0 = \overline{\varnothing}$ and

$$\overline{x}_{k+1} = \sup \bigcup_{\alpha \in A, \overline{x}_k \in \overline{I}_\alpha} \overline{I}_\alpha$$

for all $k \geqslant 0$. If α is such that $\overline{x}_k \in \overline{I}_\alpha$ then we notice that either $\overline{x}_k \in \overline{I}_\alpha < \overline{x}_{k+1}$ or $\overline{x}_{k+1} = \top_{\overline{E}}$, since $\overline{I}_\alpha$ is an open interval.

We claim that $\overline{x}_k = \top_{\overline{E}}$ for all sufficiently large k. Assuming the contrary, consider $\overline{u} = \sup \{\overline{x}_0, \overline{x}_1, \ldots\}$. There exists an α with $\overline{u} \in \overline{I}_\alpha$. Since $\overline{I}_\alpha$ is open, there exists an $\overline{y} < \overline{u}$ in $\overline{I}_\alpha$. Now take $k \in \mathbb{N}$ with $\overline{y} \leqslant \overline{x}_k$. Then $\overline{x}_k$ and $\overline{x}_{k+1} < \top_{\overline{E}}$ are both in $\overline{I}_\alpha$, which contradicts the fact that $\overline{x}_{k+1} = \top_{\overline{E}}$ or $\overline{I}_\alpha < \overline{x}_{k+1}$. This proves the claim.

Denoting by l the minimal number with $x_l = \top_{\overline{E}}$, let us now show how to choose $\alpha_l, \ldots, \alpha_0 \in A$ with $\overline{x}_k \in \overline{I}_{\alpha_k}$ $(0 \leqslant k \leqslant l)$, and $\overline{I}_{\alpha_k} \cap \overline{I}_{\alpha_{k+1}} \neq \varnothing$ $(0 \leqslant k < l)$. This is clear for $k = l$. Having constructed $a_l, \ldots, a_{k+1}$, pick an element $\overline{y} \in (\overline{x}_k, \overline{x}_{k+1}) \cap \overline{I}_{\alpha_{k+1}}$. Then there exists an $\alpha_k \in A$ with $\overline{x}_k \in \overline{I}_{\alpha_k}$ and $\overline{y} \leqslant \overline{z}$ for some $\overline{z} \in \overline{I}_{\alpha_k}$. Since $\overline{I}_{\alpha_k}$ is an interval, it follows that $\overline{y} \in \overline{I}_{\alpha_k}$, whence $\overline{I}_{\alpha_k} \cap \overline{I}_{\alpha_{k+1}} \neq \varnothing$. This completes our construction.

We contend that $\overline{E} = \overline{I}_{\alpha_0} \cup \cdots \cup \overline{I}_{\alpha_k}$. Indeed, given $\overline{y} \in \overline{E}$, we either have $\overline{y} \in \{\overline{x}_0, \ldots, \overline{x}_l\} \subseteq \overline{I}_{\alpha_0} \cup \cdots \cup \overline{I}_{\alpha_k}$, or there exists there exists a unique k with $\overline{y} \in (\overline{x}_k, \overline{x}_{k+1})$. In the second case, let $\overline{z} \in \overline{I}_{\alpha_k} \cap \overline{I}_{\alpha_{k+1}}$. Then we have either $\overline{y} \leqslant \overline{z}$ and $\overline{y} \in \overline{I}_{\alpha_k}$, or $\overline{y} > \overline{z}$ and $\overline{y} \in \overline{I}_{\alpha_{k+1}}$. □

Exercise 9.1. Let E be a totally ordered set. Given $\overline{x} < \overline{y} \in \overline{E}$, show that $\overline{y} \setminus \overline{x}$ contains infinitely many elements.

Exercise 9.2.

a) Determine $\overline{\alpha}$ for all ordinals α.
b) Determine $\overline{\alpha^{\mathrm{op}}}$ for all ordinals α.

9.2 Compactification of totally ordered fields

Let C be a totally ordered field. A natural question is to see whether the algebraic structure on C can be extended to its compactification $\overline{C}$ and which algebraic properties are preserved under this extension. In section 9.2.1, we first show that increasing and decreasing mappings naturally extend when compactifying. After that, we will show how this applies to the field operations on C. We will denote $\overline{\infty} = \sup C$.

9.2.1 Functorial properties of compactification

Proposition 9.6. *Let E and F be Hausdorff total orderings and $\varphi: E \to F$.*

a) Any increasing mapping φ: $E \to F$ extends to an increasing mapping $\overline{\varphi}: \overline{E} \to \overline{F}$, given by

$$\begin{aligned} \overline{\varphi}: \overline{E} &\longrightarrow \overline{F} \\ \overline{x} &\longmapsto \sup\{\varphi(x): x \in E \wedge x \leqslant \overline{x}\}. \end{aligned}$$

b) Any decreasing mapping φ: $E \to F$ extends to a decreasing mapping $\overline{\varphi}: \overline{E} \to \overline{F}$, given by

$$\begin{aligned} \overline{\varphi}: \overline{E} &\longrightarrow \overline{F} \\ \overline{x} &\longmapsto \inf\{\varphi(x): x \in E \wedge x \leqslant \overline{x}\}. \end{aligned}$$

Moreover, in both cases, the mapping $\overline{\varphi}$ is injective resp. surjective if and only if φ is. Also, if φ is surjective, then $\overline{\varphi}$ is its unique extension to a monotonic mapping from $\overline{E}$ into $\overline{F}$.

Proof. Assume that φ is increasing (the decreasing case is proved similarly). The mapping $\overline{\varphi}$ defined in (a) is clearly increasing. Assume that φ is injective and let $\overline{x} < \overline{y} \in \overline{E}$. Choosing $u, v \in E$ with $\overline{x} < u < v < \overline{y}$, we have

$$\overline{\varphi}(\overline{x}) \leqslant \overline{\varphi}(u) = \varphi(u) < \varphi(v) = \overline{\varphi}(v) \leqslant \overline{\varphi}(\overline{y}),$$

so $\overline{\varphi}$ is injective.

Assume from now on that φ is surjective and let $\overline{y} \in \overline{F}$. Then $\overline{x} = \{u \in E: \varphi(u) < \overline{y}\}$ is an open initial segment without a maximal element. Indeed, if $u \in \overline{x}$ were maximal, then we may choose $v \in F$ with $\varphi(u) < v < \overline{y}$ and there would exist a $u' \in E$ with $\varphi(u') = v < \overline{y}$ and necessarily $u < u'$. This shows that $\overline{x} \in \overline{E}$. By construction, we have $\overline{\varphi}(\overline{x}) \leqslant \overline{y}$. Given $v \in \overline{y}$, so that $v < \overline{y}$, there exists an $u \in E$ with $\varphi(u) = v$. Consequently, $u \in \overline{x}$ and $v = \varphi(u) = \overline{\varphi}(u) \leqslant \overline{\varphi}(\overline{x})$. This proves that $\overline{y} \leqslant \overline{\varphi}(\overline{x})$.

Now let $\overline{\psi}: \overline{E} \to \overline{F}$ be another increasing mapping which extends φ on E. Assume for contradiction that $\overline{\varphi}(\overline{x}) < \overline{\psi}(\overline{x})$ for some $\overline{x} \in \overline{E} \setminus E$ (the case $\overline{\varphi}(\overline{x}) > \overline{\psi}(\overline{x})$ is treated similarly) and let $v \in (\overline{\varphi}(\overline{x}), \overline{\psi}(\overline{x}))$. Since φ is surjective, there exists a $u \in E$ with $\varphi(u) = v$. But if $u < \overline{x}$, then $\varphi(u) \leqslant \overline{\varphi}(\overline{x})$ and if $u > \overline{x}$, then $\varphi(u) \geqslant \overline{\varphi}(\overline{x})$. This contradiction shows that $\overline{\varphi}$ is the unique increasing extension of φ to a mapping from $\overline{E}$ into $\overline{F}$. □

Corollary 9.7. *Let E be a Hausdorff ordering and E^* the set E ordered by the opposite ordering of $\leqslant$. Then there exists a natural bijection*

$$\begin{aligned} \cdot^*: \overline{E} &\longrightarrow \overline{E}^* \\ \overline{x} &\longmapsto \inf_{\overline{E}^*} \overline{x}. \end{aligned}$$ □

The following proposition is proved in a similar way as proposition 9.6: see exercise 9.3.

Proposition 9.8. *Let E be a Hausdorff ordering and $I \subseteq E$ an interval. Then there exists a natural inclusion*

$$\begin{aligned} \iota: \overline{I} &\longrightarrow \overline{E} \\ \overline{x} &\longmapsto \sup \{\overline{y} \in \overline{E} : \overline{y} \leqslant \overline{x}\}. \end{aligned}$$

This inclusion is unique with the property that $\iota(\overline{I})$ is an interval. □

9.2.2 Compactification of totally ordered fields

6 Opposites and inverses

By proposition 9.6(b), the mapping

$$\begin{aligned} -: C &\rightarrow C \\ f &\mapsto -f \end{aligned}$$

extends to unique decreasing bijection $\overline{C} \rightarrow \overline{C}$, which we also denote by $-$ and the inversion

$$\begin{aligned} \cdot^{-1}: C^{>} &\rightarrow C^{>} \\ f &\mapsto f^{-1} \end{aligned}$$

extends to a unique decreasing bijection $\overline{C^{>}} \rightarrow \overline{C^{>}}$. Notice that $\overline{C^{>}} = \{0\} \cup \overline{C}^{>}$ and $0^{-1} = \overline{\infty}$. For $\overline{f} < 0$, we may also set $(-\overline{f})^{-1} = -\overline{f}^{-1}$, so that $\cdot^{-1}$ is bijective on $\overline{C} \setminus \{-\overline{\infty}, 0, \overline{\infty}\}$.

7 Addition

The addition on C^2 may be extended to an increasing mapping $+: \overline{C}^2 \rightarrow \overline{C}$ by applying proposition 9.6(a) twice: first to mappings of the form $f + \cdot : C \rightarrow C; g \mapsto f + g$ with $f \in \overline{C}$ and next to mappings of the form $\cdot + \overline{g}: C \rightarrow \overline{C}$; $f \mapsto f + \overline{g}$ with $\overline{g} \in \overline{C}$. This is equivalent to setting

$$\begin{aligned} +: \overline{C} \times \overline{C} &\longrightarrow \overline{C} \\ (\overline{x}, \overline{y}) &\longmapsto \sup \{x + y : x, y \in C \wedge x \leqslant \overline{x} \wedge y \leqslant \overline{y}\}. \end{aligned}$$

Notice that the mapping $f + \cdot : \overline{C} \rightarrow \overline{C}; \overline{g} \mapsto f + \overline{g}$ is an isomorphism for each $f \in C$. Subtraction on $\overline{C}^2$ is defined as usual by $\overline{x} - \overline{y} = \overline{x} + (-\overline{y})$. Since the definition of the addition is symmetric in $\overline{x}$ and $\overline{y}$, the addition is commutative. Clearly, we also have $\overline{x} + 0 = \overline{x}$ for all $\overline{x} \in \overline{C}$, and

$$\overline{x} + (\overline{y} + \overline{z}) = \sup \{x + y + z : x, y, z \in C \wedge x \leqslant \overline{x} \wedge y \leqslant \overline{y} \wedge z \leqslant \overline{z}\} = (\overline{x} + \overline{y}) + \overline{z}$$

for all $\overline{x}, \overline{y}, \overline{z} \in \overline{C}$. However, $\overline{C}$ cannot be an additive group, because $\overline{\infty} - \overline{\infty} = -\overline{\infty} \neq 0$. Nevertheless,

$$-(\overline{x} + y) = (-\overline{x}) + (-y)$$

for all $\overline{x} \in \overline{C}$ and $y \in C$. Indeed, given $z \in C$, we have $z < -(\overline{x} + y) \Leftrightarrow -z > \overline{x} + y \Leftrightarrow -z - y > \overline{x} \Leftrightarrow z + y < -\overline{x} \Leftrightarrow z < (-\overline{x}) + (-y)$.

8 Multiplication

The multiplication extends first to $(\overline{C^>})^2$ by

$$\begin{aligned} \cdot : \overline{C^>} \times \overline{C^>} &\longrightarrow \overline{C^>} \\ (\overline{x}, \overline{y}) &\longmapsto \sup \{x y : x, y \in C^> \wedge x \leqslant \overline{x} \wedge y \leqslant \overline{y}\} \end{aligned}$$

and next to $\overline{C}^2$ by

$$\begin{aligned} (-\overline{x})\, \overline{y} &= -(\overline{x}\, \overline{y}) \\ \overline{x}\, (-\overline{y}) &= -(\overline{x}\, \overline{y}) \\ (-\overline{x})\, (-\overline{y}) &= \overline{x}\, \overline{y} \end{aligned}$$

for all $\overline{x}, \overline{y} \in \overline{C^>}$. This definition is coherent if $\overline{x} = 0$ or $\overline{y} = 0$, since $\overline{x}\, 0 = 0\, \overline{x} = 0$ for all $\overline{x}$. We define division on $\overline{C}^2$ as usual by $\overline{x} / \overline{y} = \overline{x}\, \overline{y}^{-1}$. The multiplication is clearly commutative, associative and with neutral element 1. We also have distributivity $\overline{x}\, (\overline{y} + \overline{z}) = \overline{x}\, \overline{y} + \overline{x}\, \overline{z}$ whenever $\overline{x} \geqslant 0$. However, $(-1)\, (\overline{\infty} - \overline{\infty}) = (-1)\, (-\overline{\infty}) = \overline{\infty} \neq -\overline{\infty} = (-\overline{\infty}) + \overline{\infty}$.

Exercise 9.3. Prove proposition 9.8.

Exercise 9.4. Show that $-(-\overline{x}) = \overline{x}$ for all $\overline{x} \in \overline{\mathbb{T}}$.

9.3 Compactification of grid-based algebras

Let C be a totally ordered field and $\mathfrak{M}$ a totally ordered monomial group and consider the algebra $\mathbb{S} = C[\![\mathfrak{M}]\!]$ of grid-based series. In this section we study the different types of cuts which may occur in $\mathbb{S}$. We will denote $\overline{\varepsilon} = \inf C^>$, $\overline{\infty} = \sup C$, $\overline{\Omega} = \sup \mathbb{S}$. We will also denote $\overline{C}^{\#} = \overline{C} \setminus (C \cup \{-\overline{\infty}, \overline{\infty}\})$.

9.3.1 Monomial cuts

Let C be a totally ordered field and $\mathfrak{M}$ a totally ordered monomial group. An element $\overline{\mathfrak{m}} \in \overline{\mathbb{S}} \setminus \mathbb{S}$ is said to be a *monomial* if $\overline{\mathfrak{m}} \geqslant 0$ and $c\, \overline{\mathfrak{m}} = \overline{\mathfrak{m}}$ for all $c \in C^>$. We denote by $\overline{\mathfrak{M}}$ the union of the set of such monomials and the set $\mathfrak{M}$ of usual monomials. The ordering $\preccurlyeq$ on $\mathfrak{M}$ naturally extends to $\overline{\mathfrak{M}}$, by letting it coincide with the usual ordering $\leqslant$.

Given $\overline{f} \in \overline{\mathbb{S}}$, we define the *dominant monomial* $\overline{\mathfrak{d}}_{\overline{f}}$ of $\overline{f}$ as follows. If $|\overline{f}\,| < c\,|\overline{f}\,|$ for no $c \in C$, so that $|\overline{f}\,| \in \overline{\mathfrak{M}} \setminus \mathfrak{M}$, then we take $\overline{\mathfrak{d}}_{\overline{f}} = |\overline{f}\,|$. If $|\overline{f}\,| < c\,|\overline{f}\,|$ for some $c \in C$, then there exists a $g \in \mathbb{T}$ with $|\overline{f}\,| < g < c\,|\overline{f}\,|$. Moreover, $\mathfrak{d}_g$ does not depend on the choice of g and we set $\overline{\mathfrak{d}}_{\overline{f}} = \mathfrak{d}_g$. Thanks to the notion of dominant monomials, we may extend the asymptotic relations $\preccurlyeq$, $\prec$, $\asymp$ and $\sim$ to $\overline{\mathbb{S}}$ by $\overline{f} \preccurlyeq \overline{g} \Leftrightarrow \overline{\mathfrak{d}}_{\overline{f}} \preccurlyeq \overline{\mathfrak{d}}_{\overline{g}}$, $\overline{f} \prec \overline{g} \Leftrightarrow \overline{\mathfrak{d}}_{\overline{f}} \prec \overline{\mathfrak{d}}_{\overline{g}}$, $\overline{f} \asymp \overline{g} \Leftrightarrow \overline{\mathfrak{d}}_{\overline{f}} = \overline{\mathfrak{d}}_{\overline{g}}$ and $\overline{f} \sim \overline{g} \Leftrightarrow \overline{\mathfrak{d}}_{\overline{f}-\overline{g}} \prec \overline{\mathfrak{d}}_f = \overline{\mathfrak{d}}_{\overline{g}}$.

Proposition 9.9. *For any $\overline{f}, \overline{f}_1, \overline{f}_2 \in \overline{\mathbb{S}}$, we have*

$$\overline{\mathfrak{d}}_{-\overline{f}} = \overline{\mathfrak{d}}_{\overline{f}}\,; \tag{9.2}$$

$$\overline{\mathfrak{d}}_{\overline{f}_1+\overline{f}_2} \preccurlyeq \max\{\overline{\mathfrak{d}}_{\overline{f}_1}, \overline{\mathfrak{d}}_{\overline{f}_2}\}. \tag{9.3}$$

Proof. The first relation is clear from the definition of dominant monomials. As to the second one, we first observe that $|\overline{f}_1| \leqslant c_1\,\overline{\mathfrak{d}}_{\overline{f}_1}$ and $|\overline{f}_2| \leqslant c_2\,\overline{\mathfrak{d}}_{\overline{f}_2}$ for sufficiently large $c_1, c_2 \in C$. Hence,

$$|\overline{f}_1 + \overline{f}_2| \leqslant |\overline{f}_1| + |\overline{f}_2| \leqslant (c_1 + c_2)\max\{\overline{\mathfrak{d}}_{\overline{f}_1}, \overline{\mathfrak{d}}_{\overline{f}_2}\}.$$

Since we also have $|\overline{f}_1 + \overline{f}_2| \geqslant c\,\overline{\mathfrak{d}}_{\overline{f}_1+\overline{f}_2}$ for a sufficiently small $c \in C^>$, it follows that $\overline{\mathfrak{d}}_{\overline{f}_1+\overline{f}_2} \preccurlyeq \max\{\overline{\mathfrak{d}}_{\overline{f}_1}, \overline{\mathfrak{d}}_{\overline{f}_2}\}$. □

9.3.2 Width of a cut

Let $\overline{f} \in \overline{\mathbb{S}}$. We define the *width* of $\overline{f}$ by

$$\overline{\mathfrak{w}}_{\overline{f}} = \inf\{\overline{\mathfrak{d}}_{\overline{f}-g} : g \in \mathbb{S}\} \in \overline{\mathfrak{M}}.$$

Notice that $\overline{f} \in \mathbb{S} \Leftrightarrow \overline{\mathfrak{w}}_{\overline{f}} = 0$.

Proposition 9.10. *For any $\overline{f}, \overline{f}_1, \overline{f}_2 \in \overline{\mathbb{S}}$, we have*

$$\overline{\mathfrak{w}}_{-\overline{f}} = \overline{\mathfrak{w}}_{\overline{f}}\,; \tag{9.4}$$

$$\overline{\mathfrak{w}}_{\overline{f}_1+\overline{f}_2} = \max\{\overline{\mathfrak{w}}_{\overline{f}_1}, \overline{\mathfrak{w}}_{\overline{f}_2}\}. \tag{9.5}$$

Proof. We have

$$\begin{aligned}
\overline{\mathfrak{w}}_{-\overline{f}} &= \inf\{\overline{\mathfrak{d}}_{-\overline{f}-g} : g \in \mathbb{S}\}\\
&= \inf\{\overline{\mathfrak{d}}_{-\overline{f}+g} : g \in \mathbb{S}\}\\
&= \inf\{\overline{\mathfrak{d}}_{-(\overline{f}-g)} : g \in \mathbb{S}\}\\
&= \inf\{\overline{\mathfrak{d}}_{\overline{f}-g} : g \in \mathbb{S}\} = \overline{\mathfrak{w}}_{\overline{f}}
\end{aligned}$$

which proves (9.4). Similarly, we have

$$\begin{aligned}
\overline{\mathfrak{w}}_{\overline{f}_1+\overline{f}_2} &= \inf\{\overline{\mathfrak{d}}_{\overline{f}_1+\overline{f}_2-g} : g \in \mathbb{S}\}\\
&= \inf\{\overline{\mathfrak{d}}_{\overline{f}_1+\overline{f}_2-g_1-g_2} : g_1, g_2 \in \mathbb{S}\}\\
&\preccurlyeq \inf\{\max\{\overline{\mathfrak{d}}_{\overline{f}_1-g_1}, \overline{\mathfrak{d}}_{\overline{f}_2-g_2}\} : g_1, g_2 \in \mathbb{S}\}\\
&= \max\{\overline{\mathfrak{w}}_{\overline{f}_1}, \overline{\mathfrak{w}}_{\overline{f}_2}\}.
\end{aligned}$$

Conversely, given $g \in \mathbb{S}$ with $g \leqslant \overline{f}_1 + \overline{f}_2$, let $g_1, g_2 \in \mathbb{S}$ be such that $g_1 \leqslant \overline{f}_1$, $g_2 \leqslant \overline{f}_2$ and $g = g_1 + g_2$. Then $\overline{f}_1 - g_1 \succcurlyeq \mathfrak{w}_{f_1}$ and $\overline{f}_2 - g_2 \succcurlyeq \mathfrak{w}_{f_2}$, whence

$$\overline{f}_1 + \overline{f}_2 - g = \overline{f}_1 - g_1 + \overline{f}_2 - g_2 \succcurlyeq \max\{\mathfrak{w}_{f_1}, \mathfrak{w}_{f_2}\}.$$

The case $g \geqslant \overline{f}_1 + \overline{f}_2$ is treated in a similar way. □

9.3.3 Initializers

Let $\overline{f} \in \overline{\mathbb{S}}$. Given $\mathfrak{m} \in \mathfrak{M}$ with $\mathfrak{m} \succ \mathfrak{w}_{\overline{f}}$, there exists a $g \in \mathbb{S}$ with $\overline{f} - g \prec \mathfrak{m}$. Moreover, $g_{\mathfrak{m}}$ does not depend on the choice of g, and we set $\overline{f}_{\mathfrak{m}} = g_{\mathfrak{m}}$. We define the *initializer* $\varphi_{\overline{f}}$ of $\overline{f}$ by

$$\varphi_{\overline{f}} = \overline{f}_{\succ \mathfrak{w}_{\overline{f}}} = \sum_{\mathfrak{m} \succ \mathfrak{w}_{\overline{f}}} \overline{f}_{\mathfrak{m}} \, \mathfrak{m}.$$

We claim that $\varphi_{\overline{f}} \in C[[\mathfrak{M}]]$, where we recall that $C[[\mathfrak{M}]]$ stands for the set of well-based series in $\mathfrak{M}$ over C. Indeed, consider $\mathfrak{m} \in \operatorname{supp} \varphi_{\overline{f}}$. Then there exists a $g \in \mathbb{S}$ with $\overline{f} - g \prec \mathfrak{m}$ and we have $(\varphi_{\overline{f}})_{\succ \mathfrak{m}} = g_{\succ \mathfrak{m}} \in \mathbb{S}$. In particular, there exists no infinite sequence $\mathfrak{n}_1 \prec \mathfrak{n}_2 \prec \cdots$ in $\operatorname{supp} \varphi_{\overline{f}}$ with $\mathfrak{m} = \mathfrak{n}_1$.

Proposition 9.11. *For any* $\overline{f}, \overline{f}_1, \overline{f}_2 \in \overline{\mathbb{S}}$*, we have*

$$\varphi_{-\overline{f}} = -\varphi_{\overline{f}} \tag{9.6}$$

$$\varphi_{\overline{f}_1 + \overline{f}_2} = (\varphi_{\overline{f}_1} + \varphi_{\overline{f}_2})_{\succ \mathfrak{w}_{\overline{f}_1 + \overline{f}_2}}. \tag{9.7}$$

Proof. In order to prove (9.6), let $\mathfrak{m} \in \mathfrak{M}$ be such that $\mathfrak{m} \succ \mathfrak{w}_{\overline{f}} = -\mathfrak{w}_{\overline{f}}$, and let $g \in \mathbb{S}$ be such that $\overline{f} - g \prec \mathfrak{m}$. Then $(-\overline{f}) - (-g) \prec \mathfrak{m}$, $\overline{f}_{\mathfrak{m}} = g_{\mathfrak{m}}$ and $(-\overline{f})_{\mathfrak{m}} = -g_{\mathfrak{m}}$.

Similarly, given $\mathfrak{m} \in \mathfrak{M}$ with $\mathfrak{m} \succ \mathfrak{w}_{\overline{f}_1 + \overline{f}_2} = \max\{\mathfrak{w}_{\overline{f}_1}, \mathfrak{w}_{\overline{f}_2}\}$, let $g_1, g_2 \in \mathbb{S}$ be such that $\overline{f}_1 - g_1 \prec \mathfrak{m}$ and $\overline{f}_2 - g_2 \prec \mathfrak{m}$. Then we have

$$(\overline{f}_1 + \overline{f}_2) - (g_1 + g_2) = (\overline{f}_1 - g_1) + (\overline{f}_2 - g_2) \prec \mathfrak{m}$$

and

$$(\overline{f}_1 + \overline{f}_2)_{\mathfrak{m}} = (g_1 + g_2)_{\mathfrak{m}} = g_{1,\mathfrak{m}} + g_{2,\mathfrak{m}} = \overline{f}_{1,\mathfrak{m}} + \overline{f}_{2,\mathfrak{m}}.$$

This proves (9.7). □

9.3.4 Serial cuts

Let $\hat{f} \in \overline{\mathbb{S}} \setminus \mathbb{S}$ be a cut with $\varphi_{\hat{f}} \notin \mathbb{S}$. Then for any $\psi \lhd \varphi_{\hat{f}}$ and $\mathfrak{m} = \mathfrak{d}(\varphi_{\hat{f}} - \psi)$, there exists a $g \in \mathbb{S}$ with $\overline{f} - g \prec \mathfrak{m}$ and we have $(\varphi_{\overline{f}})_{\succ \mathfrak{m}} = \psi = g_{\succ \mathfrak{m}} \in \mathbb{S}$. In other words, we always have $\varphi_{\hat{f}} \in \hat{\mathbb{S}} \setminus \mathbb{S}$, where

$$\hat{\mathbb{S}} = \{\hat{f} \in C[[\mathfrak{M}]] : \forall g \in C[[\mathfrak{M}]], g \lhd \hat{f} \Rightarrow g \in \mathbb{S}\}.$$

A cut $\hat{f} \in \overline{\mathbb{S}} \setminus \mathbb{S}$ is said to be *serial*, if there exists a $\psi \in C[[\mathfrak{M}]]$ with

$$\hat{f} = \overline{\iota}(\psi) = \sup\{g \in \mathbb{S}: g < \psi\}. \tag{9.8}$$

From the proposition below it follows that we may always replace ψ by $\varphi_{\overline{\mathfrak{w}}_{\hat{f}}} \in \hat{\mathbb{S}} \setminus \mathbb{S}$ and obtain the same serial cut. For this reason, we will identify the set of serial cuts with $\hat{\mathbb{S}} \setminus \mathbb{S}$.

Proposition 9.12. *Given a serial cut* $\hat{f} = \overline{\iota}(\psi)$, *we have*

a) $\varphi_{\hat{f}} = \psi_{\succ \overline{\mathfrak{w}}_{\hat{f}}}$.

b) $\overline{\iota}(\varphi_{\hat{f}}) = \overline{\iota}(\psi)$.

Proof. The equation (9.8) implies $g < \hat{f} \Leftrightarrow g < \psi$ for $g \in \mathbb{S}$. Now, given $\mathfrak{m} \succ \overline{\mathfrak{w}}_{\hat{f}}$, let $\mathfrak{n} \prec \mathfrak{m}$ and $g \in \mathbb{S}$ be such that $\hat{f} - g \prec \mathfrak{n}$. Then $g - \mathfrak{n} < \hat{f} < g + \mathfrak{n}$ and $g - \mathfrak{n} < \psi < g + \mathfrak{n}$, so that $\varphi_{\hat{f},\mathfrak{m}} = \hat{f}_{\mathfrak{m}} = g_{\mathfrak{m}} = \psi_{\mathfrak{m}}$. This proves (*a*).

Given $g \in \mathbb{S}$, we have $g - \psi \succcurlyeq \overline{\mathfrak{w}}_{\hat{f}}$, since otherwise $g - \mathfrak{n} < \psi < g + \mathfrak{n}$ for some $\mathfrak{n} \prec \overline{\mathfrak{w}}_{\hat{f}}$, whence $g - \hat{f} \preccurlyeq \mathfrak{n} \prec \overline{\mathfrak{w}}_{\hat{f}}$. We even have $g - \psi \succ \overline{\mathfrak{w}}_{\hat{f}}$, since $g - \psi \asymp \overline{\mathfrak{w}}_{\hat{f}}$ would imply $\psi_{\succcurlyeq \overline{\mathfrak{w}}_{\hat{f}}} - \psi \prec \overline{\mathfrak{w}}_{\hat{f}}$ and $\psi_{\succcurlyeq \overline{\mathfrak{w}}_{\hat{f}}} \in g_{\succcurlyeq \overline{\mathfrak{w}}_{\hat{f}}} + C\,\overline{\mathfrak{w}}_{\hat{f}} \subseteq \mathbb{S}$. Consequently, $g < \psi \Leftrightarrow g < \psi_{\succ \overline{\mathfrak{w}}_{\hat{f}}} = \varphi_{\hat{f}}$ so that $\overline{\iota}(\varphi_{\hat{f}}) = \overline{\iota}(\psi)$. □

9.3.5 Decomposition of non-serial cuts

Proposition 9.13. *For any* $\bar{f} \in \overline{\mathbb{S}} \setminus \hat{\mathbb{S}}$, *we have either*

1. $\overline{\mathfrak{w}}_{\bar{f}} \in \mathfrak{M}$ *and for some* $\bar{c} \in C^{\#}$ *we have*

$$\bar{f} = \varphi_{\bar{f}} + \bar{c}\,\overline{\mathfrak{w}}_{\bar{f}}.$$

2. $\overline{\mathfrak{w}}_{\bar{f}} \in \overline{\mathfrak{M}} \setminus \mathfrak{M}$ *and*

$$\bar{f} = \varphi_{\bar{f}} \pm \overline{\mathfrak{w}}_{\bar{f}}.$$

Proof. Modulo substitution of $\bar{f} - \varphi_{\bar{f}}$ for $\bar{f}$, we may assume without loss of generality that $\varphi_{\bar{f}} = 0$, since $\overline{\mathfrak{w}}_{\bar{f} - \varphi_{\bar{f}}} = \overline{\mathfrak{w}}_{\bar{f}}$.

Suppose that $\overline{\mathfrak{w}}_{\bar{f}} = \mathfrak{m} \in \mathfrak{M}$ and consider

$$\bar{c} = \sup\{c \in C: c\,\mathfrak{m} < \bar{f}\} \in \overline{C}.$$

We must have $\bar{c} \in \overline{C} \setminus C$, since otherwise $\bar{f} - c\,\mathfrak{m} \prec \mathfrak{m} = \overline{\mathfrak{w}}_{\bar{f}}$. We also cannot have $\bar{c} = \pm\bar{\infty}$, since otherwise $\overline{\mathfrak{w}}_{\bar{f}} = \bar{\infty}\mathfrak{m}$. Hence $\bar{c} \in C^{\#}$. If $\bar{c}\,\mathfrak{m} < \bar{f}$, then there exists a $\psi \in \mathbb{S}$ with $\bar{c}\,\mathfrak{m} < \psi < \bar{f}$. If $\psi \preccurlyeq \mathfrak{m}$, then $c'\mathfrak{m} < \bar{f}$ for some $c' \in C$ with $\bar{c} < c' < \psi_{\mathfrak{m}}$. If $\psi \succ \mathfrak{m}$, then $\bar{f}_{\succ \mathfrak{m}} = \psi_{\succ \mathfrak{m}} \neq 0$, which is again impossible. This proves that $\bar{c}\,\mathfrak{m} \geqslant \bar{f}$. Applying the same argument for $-\bar{f}$, we also obtain $\bar{c}\,\mathfrak{m} \leqslant \bar{f}$, whence $\bar{f} = \bar{c}\,\mathfrak{m}$.

Assume now that $\overline{\mathfrak{w}}_{\bar{f}} \in \overline{\mathfrak{M}} \setminus \mathfrak{M}$ and let us show that $\bar{f} = \pm \overline{\mathfrak{w}}_{\bar{f}}$. Replacing $\bar{f}$ by $-\bar{f}$ in the case when $\bar{f} < 0$, we may assume without loss of generality that $\bar{f} > 0$. For $g \in \mathbb{S}$ we now have

$$0 \leqslant g < \bar{f} \Leftrightarrow 0 \leqslant g_{\succ \overline{\mathfrak{w}}_{\bar{f}}} \leqslant \bar{f}_{\succ \overline{\mathfrak{w}}_{\bar{f}}} = 0 \Leftrightarrow 0 \leqslant g < \overline{\mathfrak{w}}_{\bar{f}}. \qquad \square$$

The above proposition allows us to extend the notions of dominant coefficients and terms to $\overline{\mathbb{S}}$. Indeed, given $\bar{f} \in \overline{\mathbb{S}}^{\neq}$, we have either $\varphi_{\bar{f}} \neq 0$, in which case we set $\bar{c}_{\bar{f}} = c_{\varphi_{\bar{f}}}$ and $\bar{\tau}_{\bar{f}} = \bar{c}_{\bar{f}}\, \bar{\mathfrak{d}}_{\bar{f}} = \tau_{\varphi_{\bar{f}}}$, or $\varphi_{\bar{f}} = 0$, in which case $\bar{f} = \bar{c}\, \bar{\mathfrak{w}}_{\bar{f}} = \bar{c}\, \bar{\mathfrak{d}}_{\bar{f}}$ for some $\bar{c} \in \overline{C}^{\#} \cup \{-1, 1\}$, and we set $\bar{c}_{\bar{f}} = \bar{c}$ and $\bar{\tau}_{\bar{f}} = \bar{f}$. By convention, we also set $c_0 = \tau_0 = 0$.

Exercise 9.5. Show that for all $\bar{\mathfrak{m}}, \bar{\mathfrak{n}} \in \overline{\mathfrak{M}}$ we have

$$\begin{aligned} \bar{\mathfrak{m}} \in \mathfrak{M} \wedge \bar{\mathfrak{n}} \in \mathfrak{M} &\Rightarrow \bar{\mathfrak{m}}\bar{\mathfrak{n}} \in \mathfrak{M} \\ \bar{\mathfrak{m}} \in \overline{\mathfrak{M}} \vee \bar{\mathfrak{n}} \in \overline{\mathfrak{M}} &\Rightarrow \bar{\mathfrak{m}}\bar{\mathfrak{n}} \in \overline{\mathfrak{M}} \setminus \mathfrak{M}. \end{aligned}$$

Exercise 9.6. Show that $\overline{\mathfrak{M}}$ is stable under $\cdot^{-1}$ and show that one may extend the flatness relation $\preceq\!\!\!\prec$ to $\overline{\mathfrak{M}}$.

Exercise 9.7. Given $\bar{f}, \bar{g} \in \overline{\mathbb{S}}$, what can be said about $\bar{\mathfrak{d}}_{\bar{f}\bar{g}}$ and $\bar{\mathfrak{w}}_{\bar{f}\bar{g}}$?

Exercise 9.8. If $C = \mathbb{R}$, then show that $\bar{\mathfrak{w}}_{\bar{f}} \in \overline{\mathfrak{M}} \setminus \mathfrak{M}$.

Exercise 9.9. Given $\bar{f} \in \overline{\mathbb{S}}$, compute $(-\bar{f}) + \bar{f}$.

Exercise 9.10. Given $\bar{f}, \bar{g} \in \overline{\mathbb{S}}$, show that

$$\begin{aligned} \bar{f} \preccurlyeq \bar{g} &\Longleftrightarrow \exists c \in C, |\bar{f}| \leqslant c\bar{g} \\ \bar{f} \prec \bar{g} &\Longleftrightarrow \forall c \in C, c\bar{f} < |\bar{g}|. \end{aligned}$$

Exercise 9.11. Generalize the theory of section 9.3.4 to other types of supports, like those from exercise 2.1. Show that there exist no serial cuts in the well-based setting.

Exercise 9.12. Characterize the embeddings of $\overline{C[[\mathfrak{M}]]}$ into $\overline{C[[\mathfrak{M}]]}$.

Exercise 9.13. Given $\bar{f} \in \overline{\mathbb{S}}$ and $\bar{\mathfrak{m}} \in \overline{\mathfrak{M}}$, we may define the coefficient $\bar{f}_{\bar{\mathfrak{m}}}$ of $\bar{\mathfrak{m}}$ in $\bar{f}$ as follows. If $\bar{\mathfrak{m}} \prec \bar{\mathfrak{w}}_{\bar{f}}$, then $\bar{f}_{\bar{\mathfrak{m}}} = 0$. If $\bar{\mathfrak{m}} \succ \bar{\mathfrak{w}}_{\bar{f}}$, then we have already defined $\bar{f}_{\bar{\mathfrak{m}}}$ if $\bar{\mathfrak{m}} \in \mathfrak{M}$ and we set $\bar{f}_{\bar{\mathfrak{m}}} = 0$ if $\bar{\mathfrak{m}} \notin \mathfrak{M}$. If $\bar{\mathfrak{m}} = \bar{\mathfrak{w}}_{\bar{f}}$ and $\bar{f} = \varphi_{\bar{f}} + \bar{c}\,\mathfrak{m}$ with $\bar{c} \in \overline{C}^{\#} \cup \{-1, 0, 1\}$, then $\bar{f}_{\bar{\mathfrak{m}}} = \bar{c}$. Show that we may see $\overline{\mathbb{S}}$ as a subset of $\overline{C}[[\overline{\mathfrak{M}}]]$. Also give a characterization of the elements in $\overline{\mathbb{S}}$.

Exercise 9.14. If $C = \mathbb{R}$, then define a "symmetric addition" on $\overline{C[[\mathfrak{M}]]}$ by $\bar{f} + \bar{g} = \varphi_{\bar{f}} + \varphi_{\bar{g}, \succ \bar{\mathfrak{w}}_{\bar{f}}} \pm \bar{\mathfrak{w}}_{\bar{f}}$ if $\bar{\mathfrak{w}}_{\bar{f}} \succ \bar{\mathfrak{w}}_{\bar{g}}$, likewise if $\bar{\mathfrak{w}}_{\bar{f}} \prec \bar{\mathfrak{w}}_{\bar{g}}$, $\bar{f} + \bar{g} = \varphi_{\bar{f}} + \varphi_{\bar{g}}$ if $\bar{\mathfrak{w}}_{\bar{f}} = \bar{\mathfrak{w}}_{\bar{g}}$ but $(\bar{f} - \varphi_{\bar{f}})(\bar{g} - \varphi_{\bar{g}}) < 0$, and $\bar{f} + \bar{g} = \varphi_{\bar{f}} + \varphi_{\bar{g}} \pm \bar{\mathfrak{w}}_{\bar{f}}$ for equal signs. Show that this addition is commutative and that $\bar{f} + (-\bar{f}) = 0$ for all $\bar{f} \in \overline{\mathbb{S}}$. Show also that the symmetric addition is not necessarily associative.

9.4 Compactification of the transline

Let us now consider the field $\mathbb{T} = C[[\mathfrak{T}]]$ of grid-based transseries. Given a transseries cut $\bar{f}$, the aim of this section is to find an explicit expression for $\bar{f}$ in terms of cuts in C, the field operations, seriation and exponentiation. We will denote $\bar{\varkappa}_k = \sup\{f \in \mathbb{T}: \operatorname{expo}(f) = k\}$ for all $k \in \mathbb{Z}$.

9.4.1 Exponentiation in $\overline{\mathbb{T}}$

By proposition 9.6(a), the functions $\exp: \mathbb{T} \to \mathbb{T}^>$ and $\log: \mathbb{T}^> \to \mathbb{T}$ uniquely extend to increasing bijections $\exp: \overline{\mathbb{T}} \to \overline{\mathbb{T}^>}$ and $\log: \overline{\mathbb{T}^>} \to \overline{\mathbb{T}}$, which are necessarily each others inverses.

Proposition 9.14.

a) For all $\bar{c} \in \overline{C} \setminus \{-\bar{\infty}\}$*, we have*

$$\exp_{\overline{\mathbb{T}}} \bar{c} = \exp_{\overline{C}} \bar{c}.$$

b) For all $\bar{f}, \bar{g} \in \overline{\mathbb{T}}$*, we have*

$$\exp(\bar{f} + \bar{g}) = \exp(\bar{f}) \exp(\bar{g}).$$

c) For any $\bar{\mathfrak{m}} \in \overline{\mathbb{T}^>}$*, we have*

$$\bar{\mathfrak{m}} \in \overline{\mathfrak{T}} \setminus \mathfrak{T} \Leftrightarrow \bar{\mathfrak{w}}_{\log \bar{\mathfrak{m}}} \succ 1.$$

Proof. Let $\bar{c} \in \overline{C} \setminus \{-\bar{\infty}\}$. If $\bar{c} \in C$, then $\exp_{\overline{\mathbb{T}}} \bar{c} = \exp_{\overline{C}} \bar{c} \in C^>$. Assume that $\bar{c} \notin C$. Then it follows from $\log_{\overline{\mathbb{T}}} \exp_{\overline{\mathbb{T}}} \bar{c} = \bar{c}$ that $\exp_{\overline{\mathbb{T}}} \bar{c} \notin C$ and similarly $\exp_{\overline{C}} \bar{c} \notin C$. For any $\lambda \in C$ with $\lambda < \bar{c}$, we have $e^{\lambda} < \exp_{\overline{\mathbb{T}}} \bar{c}$. It follows that $\exp_{\overline{C}} \bar{c} \leqslant \exp_{\overline{\mathbb{T}}} \bar{c}$. Conversely, for any $g \in \mathbb{T}$ with $-\bar{\infty} < g < \bar{c}$, there exists a $c' \in C$ with $g_{\asymp} < c' < \bar{c}$, so that $e^g = e^{g_{\asymp}} < e^{c'} \leqslant \exp_{\overline{C}} \bar{c}$. This shows that we also have $\exp_{\overline{\mathbb{T}}} \bar{c} \leqslant \exp_{\overline{C}} \bar{c}$.

Now consider $\bar{f}, \bar{g} \in \overline{\mathbb{T}}$. We have

$$\begin{aligned} e^{\bar{f} + \bar{g}} &= \sup \{e^{\varphi + \psi} : \varphi \in \mathbb{T} \wedge \varphi < \bar{f} \wedge \psi \in \mathbb{T} \wedge \psi < \bar{g}\} \\ &= \sup \{e^{\varphi} : \varphi \in \mathbb{T} \wedge \varphi < \bar{f}\} \sup \{e^{\psi} : \psi \in \mathbb{T} \wedge \psi < \bar{g}\} \\ &= e^{\bar{f}} e^{\bar{g}}. \end{aligned}$$

This proves (b).

Let $\bar{\mathfrak{m}} \in \overline{\mathfrak{T}}$. If $\bar{\mathfrak{w}}_{\log \bar{\mathfrak{m}}} \succ 1$, then assume for contradiction that there exists a $c \in C^>$ with $c\,\bar{\mathfrak{m}} \neq \bar{\mathfrak{m}}$, and take $c > 1$. Then there exists a $g \in \mathbb{T}$ with $\bar{\mathfrak{m}} < g < c\,\bar{\mathfrak{m}}$. But then $\log \bar{\mathfrak{m}} < \log g < \log \bar{\mathfrak{m}} + \log c$ and $\log \bar{\mathfrak{m}} - \log g \preccurlyeq 1$, which contradicts our assumption. We conclude that $\bar{\mathfrak{m}} \in \overline{\mathfrak{T}} \setminus \mathfrak{T}$. Similarly, if $\bar{\mathfrak{w}}_{\log \bar{\mathfrak{m}}} \preccurlyeq 1$, then let $g \in \mathbb{T}$ and $c_1, c_2 \in C$ be such that $c_1 < \log \bar{\mathfrak{m}} - g < c_2$. Then $\bar{\mathfrak{m}} < e^{g + c_2} < e^{c_2 - c_1} \bar{\mathfrak{m}}$, so that $\bar{\mathfrak{m}} \notin \overline{\mathfrak{T}} \setminus \mathfrak{T}$. This completes the proof of (c). □

9.4.2 Classification of transseries cuts

Let $\bar{f} \in \overline{\mathbb{T}}$. The *nested sequence* for $\bar{f}$ is the possibly finite sequence $\bar{f}_0, \bar{f}_1, \ldots \in \overline{\mathbb{T}}$ defined as follows. We take $\bar{f}_0 = \bar{f}$. Given $\bar{f}_i$, we distinguish two cases for the construction of $\bar{f}_{i+1}$:

NS1. If $\bar{f}_i \in \hat{\mathbb{T}} \cup \pm\{\bar{\mho}\} \cup \pm \bar{\varkappa}_{\mathbb{Z}} \cup \overline{C}$, then the construction has been completed.

NS2. Otherwise, we let $\varphi_i = \varphi_{\bar{f}_i}$, $\epsilon_i = \operatorname{sign}(\bar{f}_i - \varphi_i)$ and $\bar{f}_{i+1} = \log \epsilon_i (\bar{f}_i - \varphi_i)$, so that

$$\bar{f}_i = \varphi_i + \epsilon_i e^{\bar{f}_{i+1}}. \tag{9.9}$$

We will denote by $l \in \mathbb{N}$ the number such that $\overline{f}_l$ is the last term of the nested sequence; if no such term exists, then we let $l = +\infty$.

For any $0 \leqslant i < j \leqslant l$, repeated application of (9.9) entails

$$\overline{f}_i = \varphi_i + \epsilon_i \, \mathrm{e}^{\varphi_{i+1} + \epsilon_{i+1} \mathrm{e}^{\cdot^{\cdot^{\varphi_{j-1} + \epsilon_{j-1} \mathrm{e}^{\overline{f}_j}}}}}. \tag{9.10}$$

In particular, if $l < +\infty$, then we call

$$\overline{f} = \varphi_0 + \epsilon_0 \, \mathrm{e}^{\varphi_1 + \epsilon_1 \mathrm{e}^{\cdot^{\cdot^{\varphi_{l-1} + \epsilon_{l-1} \mathrm{e}^{\overline{f}_l}}}}} \tag{9.11}$$

the *nested expansion* of $\overline{f}$. If $l = +\infty$, then the nested expansion of $\overline{f}$ is defined to be

$$\overline{f} = \varphi_0 + \epsilon_0 \, \mathrm{e}^{\varphi_1 + \epsilon_1 \mathrm{e}^{\varphi_2 + \epsilon_2 \mathrm{e}^{\cdot^{\cdot^{\cdot}}}}}. \tag{9.12}$$

In this latter case, the nested expansion of each $\overline{f}_i$ is given by

$$\overline{f}_i = \varphi_i + \epsilon_i \, \mathrm{e}^{\varphi_{i+1} + \epsilon_{i+1} \mathrm{e}^{\varphi_{i+2} + \epsilon_{i+2} \mathrm{e}^{\cdot^{\cdot^{\cdot}}}}}.$$

The following proposition is a direct consequence of our construction:

Proposition 9.15. *Each $\overline{f} \in \overline{\mathbb{T}}$ admits a unique nested expansion of one and only one of the following forms:*

$$\overline{f} \in \mathbb{T}; \tag{9.13}$$

$$\overline{f} = \pm \overline{\mho}; \tag{9.14}$$

$$\overline{f} = \varphi_0 + \epsilon_0 \, \mathrm{e}^{\varphi_1 + \epsilon_1 \mathrm{e}^{\cdot^{\cdot^{\varphi_{l-1} + \epsilon_{l-1} \mathrm{e}^{\overline{\varkappa}_k}}}}} \quad (k \in \mathbb{Z}); \tag{9.15}$$

$$\overline{f} = \varphi_0 + \epsilon_0 \, \mathrm{e}^{\varphi_1 + \epsilon_1 \mathrm{e}^{\cdot^{\cdot^{\varphi_{l-1} + \epsilon_{l-1} \mathrm{e}^{\overline{c}}}}}} \quad (\overline{c} \in \overline{C} \setminus C); \tag{9.16}$$

$$\overline{f} = \varphi_0 + \epsilon_0 \, \mathrm{e}^{\varphi_1 + \epsilon_1 \mathrm{e}^{\cdot^{\cdot^{\varphi_{l-1} + \epsilon_{l-1} \mathrm{e}^{\hat{g}}}}}} \quad (\hat{g} \in \hat{\mathbb{T}} \setminus \mathbb{T}); \tag{9.17}$$

$$\overline{f} = \varphi_0 + \epsilon_0 \, \mathrm{e}^{\varphi_1 + \epsilon_1 \mathrm{e}^{\varphi_2 + \epsilon_2 \mathrm{e}^{\cdot^{\cdot^{\cdot}}}}}. \tag{9.18}$$

In order to completely classify the elements in $\overline{\mathbb{T}}$, we still need to determine under which conditions on the φ_i, ϵ_i, $\varkappa_k$, $\overline{c}$ and $\hat{g}$, the expressions (9.15), (9.16), (9.17) and (9.18) are the nested expansion of a cut $\overline{f} \in \overline{\mathbb{T}} \setminus \mathbb{T}$. This problem will be addressed in the next sections.

9.4.3 Finite nested expansions

Proposition 9.16. *Assume that $\overline{f} \in \overline{\mathbb{T}}$ admits a finite nested expansion. Then*

a) $l \geqslant 2 \Rightarrow \varphi_1 \in \mathbb{T}_{\succ}$ *and* $1 < i < l \Rightarrow \varphi_i \in \mathbb{T}_{\succ}^{\gtrless}$.

b) $1 < i < l \wedge \varphi_i = 0 \Rightarrow \epsilon_i = 1$ *and*
$l > 0 \wedge \varphi_{l-1} = 0 \Rightarrow \overline{f}_l \in \hat{\mathbb{T}}_{\succ} \setminus \mathbb{T} \vee (l = 1 \wedge \overline{f} \in -\overline{\varkappa}_{\mathbb{Z}})$.

c) $\mathrm{e}^{\varphi_{i+1}+\epsilon_{i+1}\mathrm{e}^{\cdot^{\cdot^{\cdot^{\varphi_{l-1}+\epsilon_{l-1}\mathrm{e}^{\bar{f}_l}}}}}} \prec \operatorname{supp}\varphi_i$ *for all* $0\leqslant i<l$.
d) $l\geqslant 1\Rightarrow \bar{f}_l\notin \mathbb{T}\cup\pm\{\overline{\mho}\}$.
e) $l\geqslant 2\Rightarrow \bar{f}_l>0\vee \bar{f}_l\in\overline{\mathcal{C}}\setminus\{-\overline{\mathfrak{m}}\}$.

Proof. Given $0<i<l$, proposition 9.13 implies that either $\mathrm{e}^{\bar{f}_i}=\mathfrak{w}_{\bar{f}_{i-1}}\in\overline{\mathfrak{T}}\setminus\mathfrak{T}$ or $\mathrm{e}^{\bar{f}_i}=\bar{c}\,\mathfrak{m}$ for some $\bar{c}\in\overline{\mathcal{C}}^{\#}$ and $\mathfrak{m}\in\mathfrak{T}$. In the first case, proposition 9.14(*c*) implies $\mathfrak{w}_{\bar{f}_i}\succ 1$ whence $\varphi_i\in\mathbb{T}_{\succ}$ and $\bar{f}_{i+1}>0$. In the second case, we obtain $\bar{f}_i=\log\mathfrak{m}+\log\bar{c}$ with $\log\bar{c}\in\overline{\mathcal{C}}^{\#}$. We cannot have $\mathfrak{m}=1$, since otherwise $l=i$. Therefore, $\varphi_i=\log\mathfrak{m}\in\mathbb{T}_{\succ}^{\neq}$, $\epsilon_i=\operatorname{sign}(\log\bar{c})$, $\bar{f}_i=\log|\log\bar{c}|$ and $l=i+1$. This proves (*a*). Similarly, if $1<i=l$, then either $\bar{f}_l\in\overline{\mathcal{C}}^{\#}$ or $\mathrm{e}^{\bar{f}_l}\in\overline{\mathfrak{T}}\setminus\mathfrak{T}$. In the second case, $\mathfrak{w}_{\bar{f}_l}\succ 1$ and $\mathfrak{w}_{\bar{f}_{l-1}}\succ 1$ yield either $\bar{f}_l\in(\hat{\mathbb{T}}_{\succ}\setminus\mathbb{T})^{>}$, $\bar{f}_l\in\overline{\varkappa}_{\mathbb{Z}}$ or $\bar{f}_l=\overline{\mathfrak{m}}$. This proves (*e*).

Now let $1<i<l$. By what precedes, we necessarily have $\mathrm{e}^{\bar{f}_{i-1}}=\mathfrak{w}_{\bar{f}_{i-2}}$ and $\bar{f}_i>0$. If $\varphi_i=0$, then it follows that $\epsilon_i=1$, since $\epsilon_i=\mathrm{e}^{\bar{f}_{i+1}}/\bar{f}_i\geqslant 0$. This proves the first part of (*b*). Assume that $l\geqslant 1$. We cannot have $\bar{f}_l\in\mathbb{T}$, since otherwise $\bar{f}_{l-1}=\varphi_{l-1}+\epsilon_{l-1}\mathrm{e}^{\bar{f}_l}\in\mathbb{T}$. Similarly, $\bar{f}_l=\overline{\mho}$ would imply $\bar{f}_{l-1}=\overline{\mho}$ and $\bar{f}_l=-\overline{\mho}$ would imply $\bar{f}_{l-1}=\varphi_{l-1}\in\mathbb{T}$. If $\varphi_{l-1}=0$ and $\bar{f}_l=\overline{\varkappa}_k$, then $\bar{f}_{l-1}=\epsilon_{l-1}\overline{\varkappa}_{k+1}\notin\overline{\varkappa}_{\mathbb{Z}}$, whence $l=1$ and $\epsilon_{l-1}=-1$. We cannot have $\varphi_{l-1}=0$ and $\bar{f}_l\in\overline{\mathcal{C}}$, since this would imply $\bar{f}_{l-1}=\epsilon_{l-1}\mathrm{e}^{\bar{f}_l}\in\overline{\mathcal{C}}$. Finally, if $\bar{f}_l\in\hat{\mathbb{T}}$, then we have shown above that $\mathfrak{w}_{\bar{f}_l}\succ 1$, so that $\bar{f}_l\in\hat{\mathbb{T}}_{\succ}\setminus\mathbb{T}$. This completes the proof of (*b*) and also proves (*d*).

In order to prove (*c*), let $0\leqslant i<l$ and $\overline{\mathfrak{m}}=\mathrm{e}^{\varphi_{i+1}+\epsilon_{i+1}\mathrm{e}^{\cdot^{\cdot^{\cdot^{\varphi_{l-1}+\epsilon_{l-1}\mathrm{e}^{\bar{f}_l}}}}}}$, so that $\bar{f}_i=\varphi_i+\epsilon_i\overline{\mathfrak{m}}$. We conclude that $\overline{\mathfrak{m}}=\mathfrak{w}_{\overline{\mathfrak{m}}}=\mathfrak{w}_{\bar{f}_i}\prec\operatorname{supp}\varphi_i$. □

Proposition 9.17. *Let* $\bar{f}\in\mathbb{T}$ *be as in* (9.11)*, where* $\varphi_0,\ldots,\varphi_{l-1}\in\mathbb{T}$, $\epsilon_0,\ldots,\epsilon_{l-1}\in\{-1,1\}$ *and* $\bar{f}_l\in\hat{\mathbb{T}}\cup\pm\{\overline{\mho}\}\cup\pm\overline{\varkappa}_{\mathbb{Z}}\cup\overline{\mathcal{C}}$ *are such that the conditions (a–e) of proposition* 9.16 *are satisfied. Then,* $\bar{f}$ *admits* (9.11) *as its nested expansion.*

Proof. Let us prove by induction over $i=l,l-1,\ldots,0$ that

$$\bar{f}_i=\varphi_i+\epsilon_i\mathrm{e}^{\varphi_{i+1}+\epsilon_{i+1}\mathrm{e}^{\cdot^{\cdot^{\cdot^{\varphi_{l-1}+\epsilon_{l-1}\mathrm{e}^{\bar{f}_l}}}}}} \tag{9.19}$$

satisfies

A) $l\geqslant 1\Rightarrow\bar{f}_i\notin\mathbb{T}\cup\pm\{\overline{\mho}\}$.
B) $i\geqslant 2\Rightarrow\bar{f}_i>0\vee(l=i\wedge\bar{f}_i\in\overline{\mathcal{C}}\setminus\{-\overline{\mathfrak{m}}\})$.
C) $1\leqslant i<l\Rightarrow\bar{f}_i\succ 1$.
D) $0\leqslant i<l\Rightarrow\mathfrak{w}_{\bar{f}_i}\asymp\mathrm{e}^{\bar{f}_{i+1}}$.
E) $\bar{f}_i$ admits (9.19) as its nested expansion.

These properties are is trivially satisfied for $i=l$. So assume that they hold for $i+1$ and let us show that they again hold for i.

From (A) at order $i+1$, we get $\overline{f}_{i+1} \notin \mathbb{T} \cup \pm\{\overline{\mho}\}$. Since $\overline{f}_{i+1} = \log(\epsilon_i(\overline{f}_i - \varphi_i))$, we have $\overline{f}_i \in \mathbb{T} \cup \pm\{\overline{\mho}\} \Rightarrow \overline{f}_{i+1} \in \mathbb{T} \cup \pm\{\overline{\mho}\}$. This proves (A) at order i. For $i \geqslant 2$, we have either $\varphi_i \neq 0$, in which case $\varphi_i \in \mathbb{T}^{\gtrless}$ implies $\overline{f}_i > 0$, or $\varphi_i = 0$, in which case $\epsilon_i = 1$ and $\overline{f}_{i+1} > 0$ imply $\overline{f}_i = \mathrm{e}^{\overline{f}_{i+1}} > 0$. This proves (B).

As to (C), if $1 \leqslant i < l$ and $\varphi_i \neq 0$, then $\varphi_i \in \mathbb{T}_{\succ}$ implies $\overline{f}_i \sim \varphi_i \succ 1$. If $1 \leqslant i < l-1$ and $\varphi_i = 0$, then $\overline{f}_{i+1} \succ 1$ and $\overline{f}_{i+1} > 0$ imply $\overline{f}_i = \mathrm{e}^{\overline{f}_{i+1}} \succ 1$. If $1 \leqslant i = l-1$ and $\varphi_i = 0$, then $\overline{f}_l \in \hat{\mathbb{T}}_{\succ} \setminus \mathbb{T}$ and $\overline{f}_{l-1} = \mathrm{e}^{\overline{f}_l} \succ 1$.

Now let $0 \leqslant i < l$. In order to prove (D), it suffices to show that $\mathrm{e}^{\overline{f}_{i+1}} \in (\overline{\mathfrak{T}} \setminus \mathfrak{T}) \cup \overline{\mathcal{C}}^{\#}\, \mathfrak{T}$. Assume first that $i < l-1$, so that $\overline{\mathfrak{w}}_{\overline{f}_{i+1}} \asymp \mathrm{e}^{\overline{f}_{i+2}}$. If $\overline{f}_{i+2} \in \overline{\mathcal{C}} \setminus \{\overline{\varpi}\}$, then $\varphi_{i+1} \neq 0$ and $\mathrm{e}^{\overline{f}_{i+1}} \in \overline{\mathcal{C}}^{\#}\mathfrak{T}$. If $\overline{f}_{i+2} \notin \overline{\mathcal{C}}$ or $\overline{f}_{i+2} = \overline{\varpi}$, then $\overline{f}_{i+2} \succ 1$ and $\overline{\mathfrak{w}}_{\overline{f}_{i+1}} \asymp \mathrm{e}^{\overline{f}_{i+2}} \succ 1$. Hence $\mathrm{e}^{\overline{f}_{i+1}} \in \overline{\mathfrak{T}} \setminus \mathfrak{T}$, by proposition 9.14($c$). Assume now that $i = l-1$. Then either $\overline{f}_l \in \overline{\mathcal{C}}^{\#}$ and $\mathrm{e}^{\overline{f}_l} \in \overline{\mathcal{C}}^{\#}$, or $\overline{f}_l = \overline{\varkappa}_k$ for some $k \in \mathbb{Z}$ and $\mathrm{e}^{\overline{f}_l} = \overline{\varkappa}_{k+1} \in \overline{\mathfrak{T}} \setminus \mathfrak{T}$, or $\overline{f}_l \in \hat{\mathbb{T}}_{\succ} \setminus \mathbb{T}$ and $\mathrm{e}^{\overline{f}_{i+1}} \in \overline{\mathfrak{T}} \setminus \mathfrak{T}$, since $\overline{\mathfrak{w}}_{\overline{f}_{i+1}} \succ 1$. This proves (D). The last property (E) follows from (D) and (E) at stage $i+1$. □

9.4.4 Infinite nested expansions

To any $f \in \mathbb{T}$, we may associate a natural interval

$$\mathbb{T}_f = \{\overline{g} \in \overline{\mathbb{T}} : f \trianglelefteq \overline{g}\} = [f - \mathfrak{t}_f, f + \mathfrak{t}_f],$$

where $f \trianglelefteq \overline{g} \Leftrightarrow f \trianglelefteq \varphi_{\overline{g}}$ and $\mathfrak{t}_f = \inf\{\mathfrak{m} \in \mathfrak{T} : \operatorname{supp} f \succ \mathfrak{m}\}$. Given a sequence $(\varphi_0, \epsilon_0), (\varphi_1, \epsilon_1), \ldots$ with $\varphi_0, \varphi_1, \ldots \in \mathbb{T}$ and $\epsilon_1, \epsilon_2, \ldots \in \{-1, 1\}$, we denote

$$\overline{\Delta}_{i,j} = \varphi_i + \epsilon_i \mathrm{e}^{\varphi_{i+1} + \epsilon_{i+1} \mathrm{e}^{\cdot^{\cdot^{\mathbb{T}_{\varphi_j}}}}}$$

for all $i \leqslant j$ and $\overline{\Delta}_i = \overline{\Delta}_{0,i}$ for all i. We also denote

$$\mathrm{I}_i = \overline{\Delta}_{i,i} \cap \overline{\Delta}_{i,i+1} \cap \overline{\Delta}_{i,i+2} \cap \cdots$$

for all $i \geqslant 0$ and $\mathrm{I} = \mathrm{I}_0$. Given $f \in \mathbb{T}$, we finally define $\eta(f) \in \mathbb{N}$ by

$$\eta(f) = \begin{cases} 0 & \text{if } f = 0 \\ 1 & \text{if } f \in \exp_{\mathbb{Z}} x \\ \max_{\mathfrak{m} \in \operatorname{supp} f} \eta(\log \mathfrak{m}) + 1 & \text{otherwise.} \end{cases}$$

Proposition 9.18. *Assume that $\overline{f} \in \overline{\mathbb{T}}$ admits an infinite nested expansion. Then*

a) $\varphi_1 \in \mathbb{T}_{\succ}$ and $\varphi_2, \varphi_3, \ldots \in \mathbb{T}^{\gtrless}_{\succ}$.
b) We have $\varphi_i \neq 0$ for infinitely many i, and $\varphi_i = 0 \Rightarrow \epsilon_i = 1$ for all $i > 1$.
c) For every $i \geqslant 0$, we have $\overline{\Delta}_0 \cap \cdots \cap \overline{\Delta}_i \neq \varnothing$.

Proof. Property (a) is proved in a similar way as in proposition 9.16, as well as the fact that $\varphi_i = 0 \Rightarrow \epsilon_i = 1$ for all $i > 1$. Property (c) is obvious, since $\overline{f} \in \overline{\Delta}_0 \cap \cdots \cap \overline{\Delta}_i$ for all $i \geqslant 0$.

Let us prove that $\varphi_i \neq 0$ for infinitely many i. It suffices to prove that $\varphi_i \neq 0$ for one i, modulo repetition of the same argument for $\overline{f}_{i+1}$ instead of $\overline{f}$. Considering $\overline{f}_1$ instead of $\overline{f}$, we may also assume without loss of generality that $\overline{f} > 0$ and $\overline{f} \succ 1$. Since $\overline{f} \notin \pm\{\overline{\mho}\} \cup \pm \overline{\varkappa}_{\mathbb{Z}} \cup \pm \{\overline{\varpi}\}$, there exist $g, h \in \mathbb{T}$ with $g < \overline{f} < h$ and $\operatorname{expo}(g) = \operatorname{expo}(h) = k$. For a sufficiently large r, we now have $\log_r g = \exp_{k-r} x + o(1)$ and $\log_r h = \exp_{k-r} x + o(1)$. But then $\exp_{k-r} x \trianglelefteqslant \log_r \overline{f}$ so that $\varphi_i \neq 0$ for some $i \leqslant r$. This completes the proof of (b). □

Proposition 9.19. *Consider* $\varphi_0, \varphi_1, \ldots \in \mathbb{T}$ *and* $\epsilon_0, \epsilon_1, \ldots \in \{-1, 1\}$, *which satisfy conditions (a–c) of proposition* 9.18. *Then* $\overline{\mathrm{I}}_1 \cap \mathbb{T} = \overline{\mathrm{I}}_1 \cap \mathbb{T} + \mathbb{T}^{\preccurlyeq}$.

Proof. Let $f_1 \in \overline{\mathrm{I}}_1 \cap \mathbb{T}$ and define $f_2 = \log(\epsilon_1 (f_1 - \varphi_1))$, $f_3 = \log(\epsilon_2 (f_2 - \varphi_2))$ and so on. We claim that $f_i - \varphi_i \succ 1$ for all $i \geqslant 1$. Indeed, let k be such that $\varphi_{i+1} = \cdots = \varphi_{i+k-1} = 0$ but $\varphi_{i+k} \neq 0$. Then $\log_k (\epsilon_i (f_i - \varphi_i)) \trianglerighteqslant \varphi_{i+k} \in \mathbb{T}^{>}_{\succ}$, whence $f_i - \varphi_i = \epsilon_i \exp_k (\varphi_{i+k} + \cdots) \succ 1$.

Given $\delta_1 \in \mathbb{T}^{\preccurlyeq}$, we have to prove that $f_1 + \delta_1 \in \overline{\mathrm{I}}_1$. Let us construct a sequence $\delta_2, \delta_3, \ldots$ of elements in $\mathbb{T}^{\preccurlyeq}$ as follows. Assuming that we have constructed δ_i, we deduce from $f_i - \varphi_i \succ 1$ that $f_i + \delta_i - \varphi_i \succ 1$, so, taking

$$\delta_{i+1} = \log\left(1 + \frac{\delta_i}{f_i - \varphi_i}\right),$$

we indeed have $\delta_{i+1} \prec 1 \preccurlyeq 1$ as well as

$$f_{i+1} + \delta_{i+1} = \log(\epsilon_i (f_i + \delta_i - \varphi_i)). \tag{9.20}$$

Now $f_i \in \overline{\Delta}_{i,i}$ and $\delta \preccurlyeq 1 \prec \operatorname{supp} \varphi_i$ imply $f_i + \delta_i \in \overline{\Delta}_{i,i}$. By induction over $j - i$, the formula (9.20) therefore yields $f_i + \delta_i \in \overline{\Delta}_{i,j}$ for all $1 \leqslant i \leqslant j$. In other words, $f_i + \delta_i \in \overline{\mathrm{I}}_i$ for all $i \geqslant 1$ and in particular for $i = 1$. □

Proposition 9.20. *Consider* $\varphi_0, \varphi_1, \ldots \in \mathbb{T}$ *and* $\epsilon_0, \epsilon_1, \ldots \in \{-1, 1\}$, *which satisfy conditions (a–c) of proposition* 9.18. *Then* $\overline{\mathrm{I}} = \{\overline{f}\}$ *for some* $\overline{f} \in \overline{\mathbb{T}} \setminus \mathbb{T}$ *with nested expansion* (9.18).

Proof. Since $\overline{\mathrm{I}} = \overline{\Delta}_0 \cap (\overline{\Delta}_0 \cap \overline{\Delta}_1) \cap (\overline{\Delta}_0 \cap \overline{\Delta}_1 \cap \overline{\Delta}_2) \cap \cdots$ is a decreasing intersection of compact non-empty intervals, $\overline{\mathrm{I}}$ contains at least one element. If $\overline{\mathrm{I}}$ contains more than one element, then it contains in particular an element $f \in \mathbb{T}$. Assume for contradiction that $\overline{\mathrm{I}} \cap \mathbb{T} \neq \varnothing$. Then we may choose (φ_0, ϵ_0), $(\varphi_1, \epsilon_1), \ldots$ and $f \in \overline{\mathrm{I}}$ such that $\eta(f)$ is minimal.

Let $\mathfrak{m} = \mathfrak{d}(f - \varphi_0)$ and $g = \log \mathfrak{m}$. From $\varphi_0 \trianglelefteqslant f$, it follows that $\mathfrak{m} \in \operatorname{supp} f$ and $\eta(g) \leqslant \max\{1, \eta(f) - 1\}$. Since $\log(\epsilon_0 (f - \varphi_0)) - \log \mathfrak{m} \preccurlyeq 1$, we also have $g \in \overline{\mathrm{I}}_1$, by proposition 9.19. Hence $\eta(g) \geqslant \eta(f)$ and $\eta(f) \leqslant 1$, by the minimality of the counterexample f. Now $f = \varphi_0$ is impossible, since otherwise $\varphi_0 - f = 0 \in \epsilon_0 \exp \overline{\mathrm{I}}_1$. It follows that $f \neq 0$, since $f \trianglerighteqslant \varphi_0$, whence $\eta(f) = \eta(g) = 1$. We cannot have $f \in C$, since otherwise $\mathfrak{m} = 1$, $g = 0$ and $\eta(g) = 0$. Therefore, there exists an $l \in \mathbb{Z}$ with $f = \exp_l x$, $\varphi_0 = 0$ and $g = \exp_{l-1} x$. Repeating the same argument, we conclude that $\varphi_0 = \varphi_1 = \cdots = 0$, which is impossible.

Now that we have proved that $\bar{\mathrm{I}} = \{\bar{f}\}$ for some $\bar{f} \in \bar{\mathbb{T}} \setminus \mathbb{T}$, let us show that $\bar{f}$ admits (9.18) as its nested expansion. Indeed, we also have $\bar{\mathrm{I}}_1 = \{\bar{g}\}$ for $\bar{g} = \log(\epsilon_0 (\bar{f} - \varphi_0))$ and proposition 9.19 implies $\mathrm{e}^{\bar{g}} \in \bar{\mathfrak{T}} \setminus \mathfrak{T}$. Consequently, $\overline{\mathfrak{w}}_{\bar{f}} = \overline{\mathfrak{w}}_{\epsilon_0 \mathrm{e}^{\bar{g}}} = \overline{\mathfrak{w}}_{\mathrm{e}^{\bar{g}}} = \mathrm{e}^{\bar{g}}$, since $\mathrm{e}^{\bar{g}} \prec \operatorname{supp} \varphi_0$. This shows that $\bar{g} = \bar{f}_1$. Using the same argument, it follows by induction that $\bar{\mathrm{I}}_k = \{\bar{f}_k\}$ for all k. □

Proposition 9.21. *Assume that $\bar{f} \in \bar{\mathbb{T}}$ admits an infinite nested expansion. Then for every $i \geqslant 0$ and $\mathfrak{m} \in \operatorname{supp} \varphi_i$, there exists a $j > i$ with $\overline{\Delta}_{i,j} - \varphi_i \prec \mathfrak{m}$.*

Proof. Let $\mathfrak{S}_{\bar{f}}$ be the set of monomials $\mathfrak{m} \in \operatorname{supp} \varphi_0$, such that for all $i > 0$ there exists a $\bar{g} \in \overline{\Delta}_i$ with $\bar{g} - \varphi_0 \succcurlyeq \mathfrak{m}$. Let $\mathfrak{S}$ be the union of all $\mathfrak{S}_{\bar{f}}$, for nested expansions $\bar{f}$ of the form (9.12). If $\mathfrak{S} = \varnothing$, then we are clearly done, since we would in particular have $\mathfrak{S}_{\bar{f}_i} = \varnothing$ for each $\bar{f}_i = \varphi_i + \epsilon_i \mathrm{e}^{\varphi_{i+1} + \epsilon_{i+1} \mathrm{e}^{\cdot^{\cdot^{\cdot}}}}$. So let us assume for contradiction that $\mathfrak{S}$ is non-empty and choose $\bar{f}$ and $\mathfrak{m} \in \mathfrak{S}_{\bar{f}} \subseteq \mathfrak{S}$ such that $\eta(\mathfrak{m})$ is minimal. Let $i > 0$ be minimal such that $\varphi_i \neq 0$. If $\delta = 1$ or $\mathfrak{m} \succ 1$, then let $\delta = 1$. Otherwise, let $\delta = -1$. Setting $\psi = \log_i \mathfrak{m}_i^{\delta}$ and $\mathfrak{n} = \mathfrak{d}_{\varphi_i - \psi}$ (whenever $\varphi_i \neq \psi$), we distinguish the following four cases:

Case $\varphi_i = \psi$. We first observe that $\epsilon_{i+1} = -\delta$. Now let $j > i$ be minimal such that $\varphi_j \neq 0$. Then $\exp_{j-i} \bar{h} \succ 1$ and $\mathfrak{m} \succ \exp_i^{\delta}(\varphi_i - \delta \exp_{j-i} \bar{h})$ for all $\bar{h} \trianglerighteq \varphi_j$. This contradicts the fact that $\mathfrak{m} \in \mathfrak{S}_{\bar{f}}$.

Case $\mathfrak{n} \notin \operatorname{supp} \psi$. For all $\bar{g} \trianglerighteq \varphi_i$, we have $\bar{g} - \psi \sim \varphi_i - \psi \sim \varphi_{i,\mathfrak{n}} \mathfrak{n}$, so the sign of $\bar{g} - \psi$ does not depend on the choice of $\bar{g}$. Since $\mathfrak{m} \in \mathfrak{S}_{\bar{f}}$, we may choose $\bar{g}$ such that $\mathfrak{m} \preccurlyeq \exp_i \bar{g}$. But then $\operatorname{sign}(\bar{g} - \psi) \neq \operatorname{sign}(\bar{f}_i - \psi)$.

Case $\mathfrak{n} \in \operatorname{supp} \psi \setminus \mathfrak{S}_{\bar{f}_i}$. Let $j > i$ be such that $\mathfrak{n} \succ \bar{g} - \varphi_i$ for all $\bar{g} \in \overline{\Delta}_{i,j}$. Given $\bar{g} \in \overline{\Delta}_{i,j}$, it follows that $\bar{g} - \psi \sim \varphi_i - \psi$, so the sign on $\bar{g} - \psi$ does not depend on the choice of $\bar{g}$. We obtain a contraction as in the previous case.

Case $\mathfrak{n} \in \operatorname{supp} \psi \cap \mathfrak{S}_{\bar{f}_i}$. The minimality hypothesis entails $\eta(\mathfrak{n}) \geqslant \eta(\mathfrak{m})$. By the construction of $\mathfrak{n}$, we thus must have $\eta(\mathfrak{m}) \leqslant 1$. Since $\mathfrak{m} = 1$ implies $\psi = 0$ and $\mathfrak{n} \notin \operatorname{supp} \psi$, it follows that $\mathfrak{m} = \exp_k x$ for some $k \in \mathbb{Z}$ and $\psi = \exp_{k-i} x$. Since $\operatorname{supp} \psi$ is a singleton, we also must have $\mathfrak{n} = \psi = \exp_{k-i} x$. Now if $\tau_{\varphi_i} > \mathfrak{n}$, then we would have $\exp_i \bar{f}_i \succ \mathfrak{m}$, which is impossible. If $\tau_{\varphi_i} < \mathfrak{n}$, then $\exp_i \bar{g} \prec \mathfrak{m}$ for all $\bar{g} \trianglerighteq \varphi_i$, which contradicts the fact that $\mathfrak{m} \in \mathfrak{S}_{\bar{f}}$.

In all cases, we thus obtain a contradiction, so we conclude that $\mathfrak{S} = \varnothing$. □

Exercise 9.15. Prove that $\mathrm{e}^{-\overline{\varpi}} = \overline{\jmath}$ and $\mathrm{e}^{\overline{\varpi}} = \overline{\varpi}$. In the case when $C = \mathbb{R}$, show that (modulo suitable adjustments of the theory) the "halting condition" **NS1** may be replaced by the alternative condition that

$$\bar{f}_i \in \hat{\mathbb{T}} \cup \pm\{\overline{\mho}\} \cup \pm\overline{\varkappa}_{\mathbb{Z}} \cup \pm\{\overline{\jmath}, \overline{\varpi}\}\mathfrak{T}.$$

Exercise 9.16. Show that the condition (d) is needed in proposition 9.20.

Exercise 9.17. Show that the conclusion of proposition 9.21 may be replaced by the stronger statement that for all $i \geqslant 0$, there exists a $j > i$ with $\overline{\Delta}_j \subseteq \overline{\Delta}_i$. Does this still hold in the case of well-based transseries?

9.5 Integral neighbourhoods of cuts

9.5.1 Differentiation and integration of cuts

Let $\mathfrak{J}$ be an interval of $\mathfrak{T}$. Any cut $\bar{f} = \sup_{\overline{C[[\mathfrak{J}]]}} I \in \overline{C[[\mathfrak{J}]]} \setminus C[[\mathfrak{J}]]$ (where I is an open initial segment without maximal element) naturally induces an element $\iota(\bar{f}) = \sup_{\mathbb{T}} I$ in $\overline{\mathbb{T}}$. Identifying $\bar{f}$ with $\iota(\bar{f})$, this yields a natural inclusion of $\overline{C[[\mathfrak{J}]]}$ into $\overline{\mathbb{T}}$, which extends the inclusion of $C[[\mathfrak{J}]]$ into $\mathbb{T}$. For any $g \in I$ with $g < \bar{f}$, there exists a $h \in I$ with $g < h < \bar{f}$ so that $\bar{f} - g \geqslant h - g \in C[[\mathfrak{J}]]^{>}$. In other words, $\bar{f}$ is a cut in $\overline{\mathbb{T}} \setminus \mathbb{T}$ whose width lies in $\bar{\mathfrak{J}}$. From proposition 9.13 it now follows that either $\bar{f} = \varphi_{\bar{f}} \in \widehat{C[[\mathfrak{J}]]}$ or $\bar{f} = \varphi_{\bar{f}} + \bar{c}\, \varpi_{\bar{f}}$ for some $\varphi_{\bar{f}} \in C[[\mathfrak{J}]]$ and $\bar{c} \in \bar{C}_{\asymp} = \bar{C} \setminus \{-\bar{\varpi}, \bar{\varpi}\}$. In other words,

$$\overline{C[[\mathfrak{J}]]} = \bar{C}_{\asymp}[[\bar{\mathfrak{J}}]] \cap \overline{\mathbb{T}}.$$

In particular, each element $\bar{f} \in \overline{\mathbb{T}}$ admits a canonical decomposition

$$\bar{f} = \bar{f}_{\succ} + \bar{f}_{\asymp} + \bar{f}_{\prec}, \tag{9.21}$$

with $\bar{f}_{\succ} \in \overline{\mathbb{T}_{\succ}} = \overline{C[[\mathfrak{T}_{\succ}]]}$, $\bar{f}_{\asymp} \in \bar{C}_{\asymp}$ and $\bar{f}_{\prec} \in \overline{\mathbb{T}_{\prec}} = \overline{C[[\mathfrak{T}_{\prec}]]}$.

Denote $\bar{\gamma} = (x \log x \log_2 x \cdots)^{-1}$ and consider the differential operator ∂ on $\mathbb{T}$. The restrictions of ∂ to $\mathbb{T}_{\succ}$ and $\mathbb{T}_{\prec}$ respectively yield increasing and decreasing bijections

$$\begin{aligned} \partial_{\succ}: \mathbb{T}_{\succ} &\longrightarrow C[[\mathfrak{T}_{\succ \bar{\gamma}}]] \\ \partial_{\prec}: \mathbb{T}_{\prec} &\longrightarrow C[[\mathfrak{T}_{\prec \bar{\gamma}}]] \end{aligned}$$

By proposition 9.6, we may extend $\partial_{\succ}$ and $\partial_{\prec}$ to the compactifications of $\mathbb{T}_{\succ}$ and $\mathbb{T}_{\prec}$. This allows us to extend ∂ to $\overline{\mathbb{T}}$ by setting $\partial \bar{f} = \partial_{\succ} \bar{f}_{\succ} + \partial_{\prec} \bar{f}_{\prec}$ for all $\bar{f} \in \overline{\mathbb{T}}$. Notice that $\bar{\varpi}' = (-\bar{\jmath})' = \bar{\gamma}$ and $(-\bar{\varpi})' = \bar{\jmath}' = -\bar{\gamma}$. The logarithmic derivative of $\bar{f} \in \overline{\mathbb{T}}^{\neq}$ is defined by $\bar{f}^{\dagger} = (\log |\bar{f}|)'$.

Similarly, the inverses of $\partial_{\succ}$ and $\partial_{\prec}$, which coincide with restrictions of the distinguished integration, extend to the compactifications of $C[[\mathfrak{T}_{\succ \bar{\gamma}}]]$ and $C[[\mathfrak{T}_{\prec \bar{\gamma}}]]$. By additivity, the distinguished integration therefore extends to $\overline{\mathbb{T}} \setminus (\mathbb{T}_{\succ \bar{\gamma}} \pm \bar{\gamma})$. The distinguished integrals of $\bar{\gamma}$ and $-\bar{\gamma}$ are undetermined, since $\int \pm \bar{\gamma}$ can be chosen among $\pm \bar{\varpi}$ and $\mp \bar{\jmath}$.

9.5.2 Integral nested expansions

Let $\bar{f} \in \overline{\mathbb{T}} \setminus \mathbb{T}$ be a cut. We say that $\bar{f}$ has *integral height* l, if either

- $l = 0$ and $\bar{f} \in \hat{\mathbb{T}}$.
- $l = 0$ and $\bar{f} = \varphi_{\bar{f}} + \bar{c}\, \mathfrak{m}$ for some $\bar{c} \in (\bar{C} \cup \{-\bar{\jmath}, \bar{\jmath}\})$ and $\mathfrak{m} \in \mathfrak{T}$.
- $\bar{f} \notin \hat{\mathbb{T}}$ and $\varpi_{\bar{f}} \notin \{\bar{\jmath}, 1, \bar{\varpi}\}\, \mathfrak{T}$, so that $\bar{f} = \varphi + \epsilon\, e^{\int \bar{\bar{f}}}$ for $\varphi = \varphi_{\bar{f}} \in \mathbb{T}$, $\epsilon = \operatorname{sign}(\bar{f} - \varphi)$ and $\bar{\bar{f}} = (\bar{f} - \varphi)^{\dagger}$, and $\bar{\bar{f}}$ has integral height $l - 1$.

The integral height of $\overline{f}$ is defined to be $l=\infty$, if none of the above conditions holds for a finite $l\in\mathbb{N}$.

We say that $\overline{f}$ is *right-oriented* (resp. *left-oriented*) if

- $l=0$ and $\overline{f}=\varphi_{\overline{f}}+\overline{\varpi}\,\mathfrak{m}$ (resp. $\overline{f}=\varphi_{\overline{f}}-\overline{\varpi}\,\mathfrak{m}$) for some $\mathfrak{m}\in\mathfrak{T}$.
- $l=0$ and $\overline{f}=\varphi_{\overline{f}}-\daleth\,\mathfrak{m}$ (resp. $\overline{f}=\varphi_{\overline{f}}+\daleth\,\mathfrak{m}$) for some $\mathfrak{m}\in\mathfrak{T}$.
- $l>0$ and $\overline{f}=\varphi_{\overline{f}}+\mathrm{e}^{\int\overline{\overline{f}}}$ (resp. $\overline{f}=\varphi_{\overline{f}}-\mathrm{e}^{\int\overline{\overline{f}}}$), where $\overline{\overline{f}}$ is a right-oriented cut of height $l-1$.
- $l>0$ and $\overline{f}=\varphi_{\overline{f}}-\mathrm{e}^{\int\overline{\overline{f}}}$ (resp. $\overline{f}=\varphi_{\overline{f}}+\mathrm{e}^{\int\overline{\overline{f}}}$), where $\overline{\overline{f}}$ is a left-oriented cut of height $l-1$.
- $l=\infty$ and $\overline{f}=-\overline{\mho}$ (resp. $\overline{f}=\overline{\mho}$).

An *oriented* cut is a cut which is either left- or right-oriented. A cut $\overline{f}$ is said to be *pathological* if $\overline{f}=\varphi_{\overline{f}}+\overline{c}\,\mathfrak{m}$ for some $\overline{c}\in\overline{C}^{\#}$ and $\mathfrak{m}\in\mathfrak{T}$, or $\overline{f}=\varphi_{\overline{f}}\pm\mathrm{e}^{\int\overline{\overline{f}}}$, where $\overline{\overline{f}}$ is a pathological cut. If $C=\mathbb{R}$, then there are no pathological cuts. If $\overline{f}$ is neither an oriented nor a pathological cut, then $\overline{f}$ is said to be *regular*.

For each $k<l$, we recursively define $\varphi_k\in\mathbb{T}$, $\epsilon_k\in\{-1,1\}$ and $\overline{f}_{k+1}\in\overline{\mathbb{T}}\setminus\mathbb{T}$ by taking $\varphi_k=\varphi_{\overline{f}_k}$ (starting with $\overline{f}_0=\overline{f}$), $\epsilon_k=\operatorname{sign}(\overline{f}_k-\varphi_k)$ and $\overline{f}_{k+1}=(\overline{f}_k-\varphi_k)^{\dagger}$. The sequence $\overline{f}_0,\overline{f}_1,\ldots$ is called the *integral nested sequence* of $\overline{f}$ and the sequence $\varphi_0,\varphi_1,\ldots$ its *integral guiding sequence*. For each $k\in\mathbb{N}$ with $k\leqslant l$, we call

$$\overline{f}=\varphi_0+\epsilon_0\,\mathrm{e}^{\int\varphi_1+\epsilon_1\mathrm{e}^{\int\cdot^{\cdot^{\cdot\varphi_{k-1}+\epsilon_{k-1}\mathrm{e}^{\int\overline{f}_k}}}}}$$

the *integral nested expansion* of $\overline{f}$ at height k. If $\overline{f}$ is an irregular cut of height $l<\infty$, so that $\overline{f}_l=\varphi_{\overline{f}_l}+\overline{c}\,\mathfrak{m}$ for certain $\overline{c}\in\overline{C}\cup\{-\daleth,\daleth\}\setminus C$ and $\mathfrak{m}\in\mathfrak{T}$, then we also define $\varphi_l=\varphi_{\overline{f}_l}$ and $\varphi_{l+1}=\log\mathfrak{m}$. In that case, we call $l+2$ the *extended integral height* of $\overline{f}$ and φ_0, ..., φ_{l+1} the *extended integral guiding sequence*. If $\overline{f}$ is a regular cut, then the extended integral height and guiding sequence are defined to be same as the usual ones.

9.5.3 Integral neighbourhoods

Let $\overline{f}\in\overline{\mathbb{T}}\setminus\mathbb{T}$ be a cut of integral height l and with extended integral guiding sequence $\varphi_0,\varphi_1,\ldots$. Let $g<h$ be transseries in $\mathbb{T}\cup\{\leftarrow,\rightarrow\}$, where $\leftarrow$ and $\rightarrow$ are formal symbols with $\leftarrow<\mathbb{T}<\rightarrow$. Then the set

$$\mathcal{L}_{\varphi_0,\ldots,\varphi_{k-1},g,h}=\left\{\varphi_0+c_0\mathrm{e}^{\int\varphi_1+c_1\mathrm{e}^{\int\cdot^{\cdot^{\cdot\varphi_{k-1}+c_{k-1}\mathrm{e}^{\int f_k}}}}}:c_0,\ldots,c_{k-1}\in C^{\neq},g<f_k<h\right\}$$

is called a *basic integral neighbourhood* of extended height k, if either one of the following conditions holds:

- $k=0$ and $g<\overline{f}<h$. This must be the case if $\overline{f}\in\hat{\mathbb{T}}$.
- $k=1$, $l=0$, $\overline{f}$ is irregular and $g<\varphi_1-\overline{\gamma}<\varphi_1+\overline{\gamma}<h$.
- $k=2$, $l=0$, $\overline{f}$ is irregular and $g<\overline{\gamma}^{\dagger}<h$.
- $k>0$, $l>0$ and $\mathcal{L}_{\varphi_1,\ldots,\varphi_{k-1},g,h}$ is a basic integral neighbourhood of $\overline{f}_1$.

The height of $\mathcal{L}_{\varphi_0,\ldots,\varphi_{k-1},g,h}$ is the minimum of k and l. An *integral neighbourhood* of $\overline{f}$ is a superset $\mathcal{V}$ of a finite intersection of basic integral neighbourhoods. The (extended) height of such a neighbourhood is the maximal (extended) height of the components in the intersection.

Let $\mathcal{V}$ be an integral neighbourhood of f of height k and consider a transseries $f\in\mathcal{V}$ close to $\overline{f}$. We define the *integral coordinates* of f by

$$\begin{aligned} f_0 &= f \\ f_1 &= (f_0-\varphi_0)^\dagger \\ &\vdots \\ f_k &= (f_{k-1}-\varphi_{k-1})^\dagger \end{aligned}$$

If $\mathcal{W}$ is an integral neighbourhood of $\overline{f}_1$, then we notice that $\mathcal{V}=\varphi_0+C^{\neq}\,\mathrm{e}^{\int\mathcal{W}}$ is an integral neighbourhood of $\overline{f}$, and it is convenient to denote the integral coordinates of $f_1\in\mathcal{W}$ by $f_1,\ldots,f_k$.

Example 9.22. Let $\bar{c}\in\overline{C}\cup\{-\daleth,\daleth\}\setminus C$ and consider a basic integral neighbourhood $\mathcal{V}$ of $\bar{c}$ of height $k>0$.

If $k=1$, then $\mathcal{V}=\mathcal{L}_{0,g,h}$, with $g<-\overline{\gamma}<\overline{\gamma}<h$. In particular, there exists an $l\in\mathbb{N}$ with $g<-(\log_l x)'$ and $h>(\log_l x)'$. For any $f\in\mathbb{T}$ with $f\not\asymp 1$ and $f\prec\!\!\prec\log_{l-1}x$, it follows that $f^\dagger=(\log|f|)'\prec(\log_l x)'$, whence $f\in\mathcal{V}$. For any $f\in\mathbb{T}$ with $f\asymp 1$, we also have $|f^\dagger|\asymp|f'|<\overline{\gamma}$, whence $f\in\mathcal{V}$. By distinguishing the cases $\bar{c}=\pm\daleth$, $\bar{c}=\pm\overline{\varpi}$ and $\bar{c}\in\overline{C}^{\#}$, it follows that $\mathcal{V}\supseteq(\tilde{g},\tilde{h})$ for certain $\tilde{g},\tilde{h}\in\mathbb{T}$ with $\tilde{g}<\bar{c}<\tilde{h}$.

If $k=2$, then $\mathcal{V}=\mathcal{L}_{0,0,g,h}$, where $\mathcal{V}^\dagger=\mathcal{L}_{0,g,h}$ is an integral neighbourhood of both $\overline{\gamma}$ and $-\overline{\gamma}$. Hence,

$$g<(\log\overline{\gamma})'=-\frac{1}{x}-\frac{1}{x\log x}-\frac{1}{x\log x\log_2 x}-\cdots<h,$$

so there exists an $l\in\mathbb{N}$ with

$$g<(\log_l x)^{\dagger\dagger}-(\log_l x)^\dagger=-\frac{1}{x}-\cdots-\frac{1}{x\cdots\log_{l-1}x}-\frac{2}{x\cdots\log_l x}$$

and

$$h>(\log_l x)^{\dagger\dagger}=-\frac{1}{x}-\cdots-\frac{1}{x\cdots\log_l x}.$$

It follows that for any $f\not\asymp 1$ with $f\prec\!\!\prec\log_l x$, we have

$$f^\dagger=(\log f)'\prec(\log_{l+1}x)'=(\log_l x)^\dagger$$

and

$$f^{\dagger\dagger}=(\log f^\dagger)'<(\log(\log_l x)^\dagger)'=(\log_l x)^{\dagger\dagger},$$

so that $f\in\mathcal{V}$. Similarly, if $f=c+\varepsilon$ with $c\in C^{\neq}$ and $(\log_l x)^{-1}\prec\varepsilon\prec 1$, then

$$f^\dagger\asymp\varepsilon'\succ((\log_l x)^{-1})'\asymp(\log_l x)^\dagger/\log_l x,$$

whence

$$f^{\dagger\dagger} = (\log f^{\dagger})' > (\log ((\log_l x)^{\dagger}/\log_l x))' = (\log_l x)^{\dagger\dagger} - (\log_l x)^{\dagger}$$

and $f \in \mathcal{V}$.

9.5.4 On the orientation of integral neighbourhoods

Let $\bar{f} \in \bar{\mathbb{T}} \setminus \mathbb{T}$ be a cut. A *one-sided neighbourhood* $\mathcal{U}$ of $\bar{f}$ is either a superset of an interval $(\bar{f}, g)$ with $g \in \mathbb{T}$ and $g > \bar{f}$ (and we say that $\mathcal{U}$ is a *right neighbourhood* of $\bar{f}$) or a superset of an interval $(g, \bar{f})$ with $g \in \mathbb{T}$ and $g < \bar{f}$ (and we say that $\mathcal{U}$ is a *left neighbourhood* of $\bar{f}$). A *neighbourhood* of $\bar{f}$ is a set $\mathcal{U}$ which is both a left neighbourhood of $\bar{f}$ (unless $\bar{f} = -\bar{\mho}$) and a right neighbourhood of $\bar{f}$ (unless $\bar{f} = \bar{\mho}$).

Proposition 9.23. *Let $\bar{f} \in \bar{\mathbb{T}} \setminus \mathbb{T}$ be a non-pathological cut and let $\mathcal{V}$ be an integral neighbourhood of $\bar{f}$.*

a) If $\bar{f}$ is regular, then there exists a neighbourhood $\mathcal{U}$ of $\bar{f}$ with $\mathcal{U} \subseteq \mathcal{V}$.
b) If $\bar{f}$ is right-oriented, then $\bar{f}$ admits a right neighbourhood $\mathcal{U}$ with $\mathcal{U} \subseteq \mathcal{V}$.
c) If $\bar{f}$ is left-oriented, then $\bar{f}$ admits a left neighbourhood $\mathcal{U}$ with $\mathcal{U} \subseteq \mathcal{V}$.

Proof. We prove the proposition by induction over the height k of $\mathcal{V}$. If $\bar{f} = \pm \bar{\mho}$, or $k = 0$ and $\bar{f}$ is regular, then we may take $\mathcal{U} = \mathcal{V}$. If $k = 0$ and $\bar{f} \neq \pm \bar{\mho}$ is oriented, then the result follows from what has been said in example 9.22. Assume therefore that $k > 0$ and let $\bar{f} = \varphi_0 + \epsilon_0 \, \mathrm{e}^{\bar{f}_1}$ be the integral expansion of $\bar{f}$ at height 1.

We have $\mathcal{V} \supseteq \mathcal{V}_0 \cap \cdots \cap \mathcal{V}_k$, where each $\mathcal{V}_i$ is a basic integral neighbourhood of $\bar{f}$ of height k. Modulo a final adjustment of $\mathcal{U}$, we may assume without loss of generality that $\mathcal{V}_0 = \mathbb{T}$. We have $\mathcal{V}_i = \varphi_0 + C^{\neq} \, \mathrm{e}^{\int \mathcal{W}_i}$ for all $i > 0$, where each $\mathcal{W}_i$ is a basic integral neighbourhood of $\bar{f}_1$. Let $\mathcal{W} = \mathcal{W}_1 \cap \cdots \cap \mathcal{W}_k$.

a) If $\bar{f}$ is regular, then so is $\bar{f}_1$, hence the induction hypothesis implies that there exist $g, h \in \mathbb{T}$ with $g < \bar{f}_1 < h$ and $(g, h) \subseteq \mathcal{W}$. We conclude that either $\epsilon_0 = 1$ and $(\varphi + \mathrm{e}^{\int g}, \varphi + \mathrm{e}^{\int h}) \subseteq \mathcal{V}$ or $\epsilon_0 = -1$ and $(\varphi - \mathrm{e}^{\int h}, \varphi - \mathrm{e}^{\int g}) \subseteq \mathcal{V}$.
b) If $\bar{f}$ is right-oriented, then either $\epsilon_0 = 1$ and $\bar{f}_1$ is right-oriented, or $\epsilon_0 = -1$ and $\bar{f}_1$ is left-oriented. In the first case, the induction hypothesis implies that there exists a $g \in \mathbb{T}$ with $\bar{f}_1 < g$ and $(\bar{f}_1, g) \subseteq \mathcal{W}$, so that $(\bar{f}, \varphi + \mathrm{e}^{\int g}) \subseteq \mathcal{V}$. In the second case, there exists a $g \in \mathbb{T}$ with $g < \bar{f}_1$ and $(g, \bar{f}_1) \subseteq \mathcal{W}$, so that $(\varphi - \mathrm{e}^{\int g}, \bar{f}) \subseteq \mathcal{V}$.
c) The case when $\bar{f}$ is left-oriented is treated in a similar way as (b). □

Proposition 9.24. *Let $\bar{f} \in \bar{\mathbb{T}} \setminus \mathbb{T}$ be a cut and $\mathcal{V}$ an integral neighbourhood of $\bar{f}$, of height k. Then there exists an integral neighbourhood $\mathcal{W}$ of $\bar{f}$ of height k, such that $\mathcal{W} \subseteq \mathcal{V}$ and $f_0 - \varphi_0, \ldots, f_{k-1} - \varphi_{k-1}$ have constant sign for $f \in \mathcal{W}$.*

Proof. We prove the proposition by induction over k. If $k=0$, then we may take $\mathcal{W}=\mathcal{V}$. So assume that $k>0$ and write $\bar{f}=\varphi_0+\epsilon_0\,\mathrm{e}^{\int \bar{f}_1}$. We have $\mathcal{V}\supseteq\mathcal{V}_0\cap\mathcal{V}_*$, where $\mathcal{V}_0$ is a basic integral neighbourhood of height 0 of $\bar{f}$ and $\mathcal{V}_*$ an intersection of basic integral neighbourhoods of heights >0. By the induction hypothesis, there exists an integral neighbourhood $\mathcal{X}$ of $\bar{f}_1$, such that $\mathcal{X}\subseteq(\mathcal{V}_*-\varphi_0)^\dagger$ and $f_1-\varphi_1,\ldots,f_{k-1}-\varphi_{k-1}$ have constant sign for all $f_1\in\mathcal{X}$. Now take

$$\mathcal{W}=\begin{cases}\mathcal{V}_0\cap(\varphi_0,\rightarrow)\cap(\varphi_0+C^{\neq}\,\mathrm{e}^{\int\mathcal{X}}) & \text{if } \epsilon_0=1\\ \mathcal{V}_0\cap(\leftarrow,\varphi_0)\cap(\varphi_0+C^{\neq}\,\mathrm{e}^{\int\mathcal{X}}) & \text{if } \epsilon_0=-1\end{cases}$$ □

Exercise 9.18. Show that $\bar{\gamma}=\inf\{f'\colon f\in\mathbb{T}^{>,\succ}\}$.

Exercise 9.19. Show that ∂ maps $\hat{\mathbb{T}}$ into $\hat{\mathbb{T}}$.

Exercise 9.20. If $\bar{f}\in\overline{\mathbb{T}}$, then show that either $\bar{f}_\succ\notin\mathbb{T}$ and $\bar{f}_\preccurlyeq=0$, or $\bar{f}_\succ\in\mathbb{T}$, $\bar{f}_\asymp\notin C$ and $\bar{f}_\prec=0$, or $\bar{f}_\succcurlyeq\in\mathbb{T}$.

Exercise 9.21. Show that the extension of ∂ to $\overline{\mathbb{T}}$ is not additive.

Exercise 9.22.

a) Show that the operators $\circ\colon\mathbb{T}\times\mathbb{T}^{\succ,>}\rightarrow\mathbb{T}$ and $\cdot^{\circ-1}\colon\mathbb{T}^{\succ,>}\rightarrow\mathbb{T}^{\succ,>}$ naturally extend to $\overline{\mathbb{T}}\times\overline{\mathbb{T}^{\succ,>}}$ resp. $\overline{\mathbb{T}^{\succ,>}}$.
b) Give an explicit formula for $\bar{f}\circ\bar{\omega}$, where $\bar{f}\in\overline{\mathbb{T}}$.
c) Does the post-composition operator $\circ_g\colon\overline{\mathbb{T}}\rightarrow\overline{\mathbb{T}}$ with $g\in\mathbb{T}$ preserve addition and/or multiplication?

Exercise 9.23.

a) Compute the nested integral sequences for $\bar{\mho}$, $\bar{\omega}$ and $\bar{\varkappa}_0$.
b) Prove analogues of the results from section 9.4 for nested integral sequences.

9.6 Differential polynomials near cuts

Let $P\in\mathbb{T}\{F\}^{\neq}$ and $\bar{f}\in\overline{\mathbb{T}}\setminus\mathbb{T}$. In this section, we study the asymptotic behaviour of $P(f)$ for f close to $\bar{f}$. In particular, we study the sign of $P(f)$ for f close to $\bar{f}$.

9.6.1 Differential polynomials near serial cuts

Lemma 9.25. *Let $\hat{f}\in\hat{\mathbb{T}}\setminus\mathbb{T}$. Then there exist $g,h\in\mathbb{T}$ with $g<\hat{f}<h$ and $\tau\in C^{\neq}\mathfrak{T}$, such that $P(f)\sim\tau$ for all $f\in(g,h)$. Moreover, if $\bar{\mathfrak{w}}_{\hat{f}}\succ\bar{\gamma}$, then g and h may be chosen such that $\deg_{\prec\bar{\gamma}}P_{+f}=0$ for all $f\in(g,h)$.*

Proof. If there exists a $\varphi \triangleleft \hat{f}$ with $\deg_{\prec \operatorname{supp} \varphi} P_{+\varphi} = 0$, then the lemma follows for $\tau = P_{+\varphi,0}$ and any $g, h \in \mathbb{T}$ with $g < \hat{f} < h$ and $h - g \prec \operatorname{supp} \varphi$. Assume for contradiction that $d = \min_{\varphi \triangleleft \hat{f}} \deg_{\prec \operatorname{supp} \varphi} P_{+\varphi} > 0$.

If $d = 1$, then each $\varphi \triangleleft \hat{f}$ with $\deg_{\prec \operatorname{supp} \varphi} P_{+\varphi} = 1$ induces a solution $f_\varphi = \varphi + h$ to $P(f) = 0$, by letting h be the distinguished solution to the equation $P_{+\varphi}(h) = 0$ $(h \prec \operatorname{supp} \varphi)$. Now pick $\varphi_1 \triangleleft \varphi_2 \triangleleft \cdots \triangleleft \hat{f}$ such that

$$(f_{\varphi_j} - f_{\varphi_i})_{|\{\mathfrak{m}: \exists \mathfrak{n} \in \operatorname{supp} \varphi_j, \mathfrak{m} \succcurlyeq \mathfrak{n}\}} \neq 0$$

for all $j > i$. This is possible, since $\operatorname{supp} \hat{f}$ would be a subset of the grid-based set $\operatorname{supp} f_{\varphi_i}$, if $(f_\psi - f_{\varphi_i})_{|\{\mathfrak{m}: \exists \mathfrak{n} \in \operatorname{supp} \psi, \mathfrak{m} \succcurlyeq \mathfrak{n}\}} = 0$ for some i and all $\varphi_i \triangleleft \psi \triangleleft \hat{f}$. Now $\mathfrak{d}(f_{\varphi_{r+2}} - f_{\varphi_1}), \ldots, \mathfrak{d}(f_{\varphi_{r+2}} - f_{\varphi_{r+1}})$ are pairwise distinct starting monomials for the linear differential equation $P_{+\varphi_{r+2},1}(h) = 0$, which is impossible.

Assume now that $d > 1$ and choose $\chi \triangleleft \hat{\psi}$ with $d = \deg_{\prec \operatorname{supp} \chi} P_{+\chi}$. Consider the set $\mathscr{S}$ of all partial unravellings

$$f = \xi + \tilde{f} \qquad (\tilde{f} \in C[\![\tilde{\mathfrak{V}}]\!]) \tag{9.22}$$

relative to the equation $P_{+\chi}(\tilde{f}) = 0$ $(\tilde{f} \prec \operatorname{supp} \chi)$, such that $\varphi = \xi_{\mathfrak{T} \setminus \tilde{\mathfrak{V}}} \triangleleft \hat{f}$ and $\deg_{\tilde{\mathfrak{V}}} P_{+\varphi} = d$. Since $\mathscr{S}$ contains the identity refinement, we may choose (9.22) to be finest in $\mathscr{S}$, by corollary 8.32. We claim that $\varphi \trianglelefteq \hat{f}$ is maximal for $\trianglelefteq$, such that $\deg_{\prec \operatorname{supp} \varphi} P_{+\varphi} = d$.

Indeed, assume for contradiction that some $\psi \triangleright \varphi$ also satisfies

$$\deg_{\prec \operatorname{supp} \psi} P_{+\psi} = d,$$

and let $\tau = \tau(\psi - \varphi)$. By proposition 8.27, there exists a partial unravelling

$$f = \xi + \tilde{\xi} + \tilde{\tilde{f}} \qquad (\tilde{\tilde{f}} \prec \tilde{\xi}),$$

which is finer than (9.22), and such that $\tilde{\xi} \sim \tau$. But then $\varphi + \tau = (\xi + \tilde{\xi})_{\succcurlyeq \tilde{\xi}} \triangleleft \hat{f}$ and $\deg_{\prec \tilde{\xi}} P_{+\varphi+\tau} = d$, which contradicts the maximality of (9.22).

Our claim implies that $\deg_{\prec \operatorname{supp} \psi} P_{+\psi} < d$ for any $\psi \triangleleft \hat{f}$ with $\varphi \triangleleft \psi$. This contradicts the definition of d. □

9.6.2 Differential polynomials near constants

Lemma 9.26. *Let $\bar{f} \in \bar{C} \cup \{-\mathfrak{d}, \mathfrak{d}\} \setminus C$ and $\mathfrak{m} \in \mathfrak{T}_{\prec}$. Then there exist an integral neighbourhood $\mathcal{V}$ of $\bar{f}$ and $\mathfrak{n} \in \mathfrak{T}$, such that*

$$P(f) \sim N_P(f)\, \mathfrak{n}$$

and $\deg_{\prec \mathfrak{m}} P_{+f} = 0$ for all $f \in \mathcal{V}$.

Proof. Let $l \geqslant 0$ be such that $P \uparrow_l$ is exponential, $N_P = D_{P \uparrow_l}$ and $\log_l x \lll \mathfrak{m}$. Let $Q \in C[F]$ and $\nu \in \mathbb{N}$ be such that $N_P = Q\,(F')^\nu$.

Take $\mathcal{V} = \mathcal{L}_{0,0,(\log_l x)^{\dagger\dagger}-(\log_l x)^{\dagger},(\log_l x)^{\dagger\dagger}}$ and let $f \in \mathcal{V}$. If $f^{\dagger\dagger} \succ \overline{\gamma}$, then $f^{\dagger\dagger} < (\log_l x)^{\dagger\dagger}$, so $f^{\dagger} \preccurlyeq (\log_l x)^{\dagger}$ and $f \preceq\!\!\!\preccurlyeq \log_l x$. If $f^{\dagger\dagger} \prec \overline{\gamma}$, then $f^{\dagger\dagger} > (\log_l x)^{\dagger\dagger} - (\log_l x)^{\dagger}$, whence $f^{\dagger} \succcurlyeq (\log_l x)^{\dagger}/\log_l x$, $\log(f/f_{\asymp}) \succcurlyeq 1/\log_l x$ and $f - f_{\asymp} \succcurlyeq 1/\log_l x$. This proves that either $f{\uparrow_l} \succ 1$ and $f{\uparrow_l} \preceq\!\!\!\preccurlyeq x$, or $f{\uparrow_l} \preccurlyeq 1$ and $(f{\uparrow_l})_{\not\asymp} \preceq\!\!\!\preccurlyeq x$.

If $f{\uparrow_l} \succ 1$, then $N_P(f{\uparrow_l}) \sim c\, f{\uparrow_l^d}\, (f{\uparrow_l'})^{\nu}$, where $c \neq 0$ is the leading coefficient of Q and $d = \deg Q$. Since $(f{\uparrow_l'})^{\nu} \preceq\!\!\!\preccurlyeq x$, it follows that $N_P(f{\uparrow_l}) \succcurlyeq_{e^x} 1$, whence $P(f) \sim N_P(f)\, \mathfrak{n}$ for $\mathfrak{n} = \mathfrak{d}_{N_P \downarrow_l}\, \mathfrak{d}_{P\uparrow_l} \downarrow_l$. Moreover, $\mathfrak{d}(f{\uparrow_l})$ is not a starting monomial for $P{\uparrow_l}(\tilde{f}) = N_P(\tilde{f})\, \mathfrak{d}_{P\uparrow_l} + \cdots = 0$, since $\mathfrak{d}(f{\uparrow_l}) \neq 1$. Consequently $\deg_{\prec \mathfrak{m}} P_{+f} \leqslant \deg_{\prec f} P_{+f} = 0$.

Similarly, if $f{\uparrow_l} \preccurlyeq 1$, then $N_P(f{\uparrow_l}) \sim c\, f{\uparrow_l^{\mu}}\, (f{\uparrow_l'})^{\nu}$, where $c \neq 0$ and μ are such that $Q(f_{\asymp} + \varepsilon) = c\,\varepsilon^{\mu} + \cdots$. Again, we have $(f{\uparrow_l'})^{\nu} \preceq\!\!\!\preccurlyeq x$, $N_P(f{\uparrow_l}) \succcurlyeq_{e^x} 1$ and $P(f) \sim N_P(f)\, \mathfrak{n}$. Furthermore, $\mathfrak{d}((f{\uparrow_l})_{\not\asymp}) \neq 1$, so $\mathfrak{d}((f{\uparrow_l})_{\not\asymp})$ is not a starting monomial for $P_{+f_{\asymp}}{\uparrow_l}(\tilde{f}) = N_{P_{+f_{\asymp}}}(\tilde{f})\, \mathfrak{d}_{P\uparrow_l} + \cdots = 0$. Therefore, $\deg_{\prec \mathfrak{m}} P_{+f} \leqslant \deg_{\prec f_{\not\asymp}} P_{+f} = 0$. □

Corollary 9.27. *Let $\overline{f} = \varphi_0 + \overline{c}\, e^{\int \varphi_1}$ be an irregular cut of height 0. Then there exist an integral neighbourhood $\mathcal{V}$ of $\overline{f}$, $Q \in C[F]^{\neq} (F')^{\mathbb{N}}$, and $\mathfrak{n} \in \mathfrak{T}$, such that for all $f \in \mathcal{V}$, we have*

$$P(f) \sim Q\Big(\frac{f_0 - \varphi_0}{e^{\int \varphi_1}}\Big)\, \mathfrak{n}.$$

Moreover, if $e^{\int \varphi_1} \succ \overline{\gamma}$, then we may take $\mathcal{V}$ such that $\deg_{\prec \overline{\gamma}} P_{+f} = 0$ for all $f \in \mathcal{V}$.

9.6.3 Differential polynomials near nested cuts

Lemma 9.28. *Let $\overline{f} = \varphi_0 + \epsilon_0\, e^{\int \overline{f}_1} \in \overline{\mathbb{T}} \setminus \mathbb{T}$ be a cut of integral height $\geqslant 1$. Then there exist $g, h \in \mathbb{T} \cup \{\leftarrow, \rightarrow\}$ with $g < \overline{f} < h$ and $i \in \mathbb{N}$, such that for all $f \in (g, h)$, so that $\mathfrak{d}_{f - \varphi_0}$ is not a starting monomial for $P_{+\varphi_0}(\tilde{f}) = 0$, we have*

$$P(f) \sim R_{P_{+\varphi_0}, i}((f - \varphi_0)^{\dagger})\, (f - \varphi_0)^i.$$

Moreover, if $\overline{\mathfrak{w}}_{\overline{f}} \succ \overline{\gamma}$, then g and h may be chosen such that $\deg_{\prec \overline{\gamma}} P_{+f} = 0$ for all f as above.

Proof. Let $\tilde{P} = P_{+\varphi_0}$. By proposition 8.17, there exists a unique integer i such that for each equalizer $\mathfrak{e}_{j,k}$ for $\tilde{P}(\tilde{f}) = 0$, we have either $\mathfrak{e}_{j,k} \prec \overline{\mathfrak{w}}_{\overline{f}}$ and $k \leqslant i$ or $\mathfrak{e}_{j,k} \succ \overline{\mathfrak{w}}_{\overline{f}}$ and $j \leqslant i$. Now let $f = \varphi_0 + \tilde{f} \in \mathbb{T}$ be such that $\tilde{\mathfrak{m}} = \mathfrak{d}_{\tilde{f}}$ is not a starting monomial for $\tilde{P}(\tilde{f}) = 0$, and $\mathfrak{e}_{j,k} \prec \tilde{f}$ if $k \leqslant i$ and $\tilde{f} \prec \mathfrak{e}_{j,k}$ if $i \leqslant j$ for all equalizers $\mathfrak{e}_{j,k}$ for $\tilde{P}(\tilde{f}) = 0$. Then $N_{\tilde{P}_{\times \tilde{\mathfrak{m}}}} = c\, F^i$ for some $c \in C^{\neq}$ and $\tilde{P}_{\times \tilde{\mathfrak{m}}}{\uparrow_l} = c\, \mathfrak{n}\, F^i + o_{e^x}(\mathfrak{n})$ for some sufficiently large l and $\mathfrak{n} \in \mathfrak{T}$. Consequently,

$$\begin{aligned} P(f) = \tilde{P}(\tilde{f}) &= (\tilde{P}_{\times \tilde{\mathfrak{m}}}{\uparrow_l})((\tilde{f}/\tilde{\mathfrak{m}}){\uparrow_l}){\downarrow_l} \sim (c\,\mathfrak{n}){\downarrow_l} \\ R_{\tilde{P}_i}(\tilde{f}^{\dagger})\, \tilde{f}^i = \tilde{P}_i(\tilde{f}) &= (\tilde{P}_{\times \tilde{\mathfrak{m}}, i}{\uparrow_l})((\tilde{f}/\tilde{\mathfrak{m}}){\uparrow_l}){\downarrow_l} \sim (c\,\mathfrak{n}){\downarrow_l}, \end{aligned}$$

which proves the first statement of the lemma. Moreover, since $\tilde{\mathfrak{m}}$ is not a starting monomial for $P_{+\varphi_0}(\tilde{f})=0$, we have $\deg_{\prec\tilde{\mathfrak{m}}} P_{+f}=0$. If $\overline{\mathfrak{w}}_{\bar{f}} \succ \bar{\gamma}$, it follows that $\deg_{\prec\bar{\gamma}} P_{+f}=0$ whenever f is chosen such that $\tilde{f} \succ \bar{\gamma}$. □

9.6.4 Differential polynomials near arbitrary cuts

Theorem 9.29. *Let $P \in \mathbb{T}\{F\}^{\neq}$ and let $\bar{f} \in \overline{\mathbb{T}} \setminus \mathbb{T}$ be a cut of height l with integral guiding sequence $\varphi_0, \varphi_1, \ldots$. Then there exists an integral neighbourhood $\mathcal{V}$ of $\bar{f}$ of height $k \leqslant \min\{l, r\}$, such that one of the following holds:*

- *There exist $i_0, \ldots, i_{k-1} \in \mathbb{N}$ and $\tau \in C^{\neq}\mathfrak{T}$, such that for all $f \in \mathcal{V}$, we have*

$$P(f) \sim (f_0 - \varphi_0)^{i_0} \cdots (f_{k-1} - \varphi_{k-1})^{i_{k-1}} \tau. \tag{9.23}$$

- *The cut $\bar{f}$ is irregular, $k = l$, and there exist $i_0, \ldots, i_{k-1} \in \mathbb{N}$, $Q \in C[F](F')^{\mathbb{N}} \setminus C$ and $\mathfrak{n} \in \mathfrak{T}$, such that for all $f \in \mathcal{V}$, we have*

$$P(f) \sim (f_0 - \varphi_0)^{i_0} \cdots (f_{k-1} - \varphi_{k-1})^{i_{k-1}} Q\left(\frac{f_k - \varphi_k}{e^{\int \varphi_{k+1}}}\right) \mathfrak{n}. \tag{9.24}$$

Moreover, if $\overline{\mathfrak{w}}_{\bar{f}} \succ \bar{\gamma}$, then $\mathcal{V}$ may be chosen such that $\deg_{\prec\bar{\gamma}} P_{+f} = 0$ for all $f \in \mathcal{V}$.

Proof. We prove the theorem by induction over r. So assume that we proved the theorem for all smaller r (for $r<0$, there is nothing to prove). If $\bar{f} \in \hat{\mathbb{T}}$, then the result follows from lemma 9.25. If $\bar{f} = \varphi + \bar{c}\,\mathfrak{m}$ with $\mathfrak{m} \prec \operatorname{supp} \varphi$ and $\bar{c} \in \overline{C} \cup \{-\mathfrak{z}, \mathfrak{z}\} \setminus C$, then we are done by corollary 9.27.

In the last case, we have $\bar{f} = \varphi_0 + \epsilon_0 e^{\int \bar{f}_1}$ for some $\epsilon_0 = \pm 1$. By lemma 9.28, there exists an i_0 and an integral neighbourhood $\mathcal{V}_0$ of $\bar{f}$ of height 0, such that for all $f \in \mathcal{V}_0$ so that $\mathfrak{d}_{f-\varphi_0}$ is not a starting monomial for $P_{+\varphi_0}(\tilde{f})=0$, we have

$$P(f) \sim R_{P_{+\varphi_0}, i_0}(f_0 - \varphi_0)\,(f_0 - \varphi_0)^{i_0}. \tag{9.25}$$

By the induction hypothesis, there exists an integral neighbourhood $\mathcal{W}$ of $\bar{f}_1$ of height k', such that $k := k' + 1 \leqslant \min\{l, r\}$ and one of the following holds:

- There exist $i_1, \ldots, i_{k-1} \in \mathbb{N}$, and $\tau \in C^{\neq}\mathfrak{T}$, such that for all $f_1 \in \mathcal{W}$, we have

$$R_{P_{+\varphi_0}, i_0}(f_1) \sim (f_1 - \varphi_1)^{i_1} \cdots (f_{k-1} - \varphi_{k-1})^{i_{k-1}} \tau. \tag{9.26}$$

- The cut $\bar{f}_1$ is irregular, $k = l$, and there exist $i_1, \ldots, i_{k-1} \in \mathbb{N}$, $Q \in C[F](F')^{\mathbb{N}} \setminus C$ and $\mathfrak{n} \in \mathfrak{T}$, such that for all $f_1 \in \mathcal{W}$, such that

$$R_{P_{+\varphi_0}, i_0}(f_1) \sim (f_1 - \varphi_1)^{i_1} \cdots (f_{k-1} - \varphi_{k-1})^{i_{k-1}} Q\left(\frac{f_k - \varphi_k}{e^{\int \varphi_{k+1}}}\right) \mathfrak{n}. \tag{9.27}$$

Moreover, for $f_1 \in \mathcal{W}$, the induction hypothesis and proposition 8.16 also imply that $e^{\int f_1}$ is not a starting monomial for $P_{+\varphi_0}(\tilde{f})=0$, since $\overline{\mathfrak{w}}_{\bar{f}_1} \succ \bar{\gamma}$.

Now take $\mathcal{V} = \mathcal{V}_0 \cap (\varphi_0 + C^{\neq} e^{\int \mathcal{W}})$. Then the relations (9.25) and (9.26) resp. (9.27) entail (9.23) resp. (9.24) for all $f \in \mathcal{V}$. Moreover, if $\mathfrak{w}_{\bar{f}} \succ \bar{\gamma}$, then $\mathcal{V}_0$ may be chosen such that $\deg_{\prec \bar{\gamma}} P_{+f} = 0$ for all $f \in \mathcal{V} \subseteq \mathcal{V}_0$, by lemma 9.28. □

9.6.5 On the sign of a differential polynomial

Let $P \in \mathbb{T}\{F\}$ be a differential polynomial. We denote by $\sigma_P: \mathbb{T} \to \{-1, 0, 1\}$ the sign function associated to P:

$$\sigma_P(f) = \operatorname{sign} P(f) = \begin{cases} -1, & \text{if } P(f) < 0 \\ 0, & \text{if } P(f) = 0 \\ 1, & \text{if } P(f) > 0 \end{cases}.$$

We say that σ_P is *constant at the right* of $\bar{f} \in \bar{\mathbb{T}}$, if there exist $\epsilon \in \{-1, 0, 1\}$ and $g > \bar{f}$ such that $\sigma_P(f) = \epsilon$ for all $f \in (\bar{f}, g)$. In that case, we denote $\sigma_P^+(\bar{f}) = \epsilon$. We say that σ_P is *constant at the left* of $\bar{f} \in \bar{\mathbb{T}}$, if there exist $\epsilon \in \{-1, 0, 1\}$ and $g < \bar{f}$ such that $\sigma_P(f) = \epsilon$ for all $f \in (\bar{f}, g)$, and we denote $\sigma_P^-(\bar{f}) = \epsilon$. If σ_P is constant at the left and at the right of $\bar{f}$, then we say that σ_P is *constant at both sides* of $\bar{f}$.

Proposition 9.30. *Let* $Q \in \mathcal{Q} = (Q_d F^d + \cdots + Q_v F^v)(F')^\nu \in C[F](F')^{\mathbb{N}}$ *with* $Q_d \neq 0$ *and* $Q_v \neq 0$. *Then*

$$\begin{aligned} \sigma_Q^+(\varpi) &= \operatorname{sign} Q_d && (9.28)\\ \sigma_Q^-(-\varpi) &= (-1)^{d+\nu} \operatorname{sign} Q_d && (9.29)\\ \sigma_Q^-(\eth) &= (-1)^\nu \operatorname{sign} Q_v && (9.30)\\ \sigma_Q^+(-\eth) &= (-1)^v \operatorname{sign} Q_v && (9.31)\end{aligned}$$

Proof. For $f \in \mathbb{T}^{>,\succ}$, we have

$$Q(f) \sim Q_d f^d (f')^\nu$$

and $f' > 0$. That proves (9.28). The other properties follow by considering $Q(-f)$ and $Q(\pm 1/f) f^{\deg Q}$ instead of $Q(f)$. □

Theorem 9.31. *Let* $P \in \mathbb{T}\{F\}$ *and* $\bar{f} \in \bar{\mathbb{T}}$. *Then*

a) If $\bar{f}$ is regular, then σ_P is constant on both sides of $\bar{f}$, and $\sigma_P^+(\bar{f}) = \sigma_P^-(\bar{f})$.
b) If $\bar{f}$ is left-oriented, then σ_P is constant at the left of $\bar{f}$.
c) If $\bar{f}$ is right-oriented, then σ_P is constant at the right of $\bar{f}$.
d) If $\bar{f} \in \mathbb{T}$, then P is constant at both sides of $\bar{f}$.

Proof. Propositions 9.24, 9.30 and theorem 9.29 imply (*a*), (*b*) and (*c*). Property (*d*) follows by considering $P(1/f) f^{\deg P}$ instead of $P(f)$. □

Proposition 9.32. *Let $P \in \mathbb{T}\{F\}^{\neq}$, $\mathfrak{m} \in \mathfrak{T}$ and denote $i = \operatorname{val} N_{P_{\times \mathfrak{m}}} \leqslant j = \deg N_{P_{\times \mathfrak{m}}}$. Then*

$$\begin{aligned}\sigma_P^+(\overline{\varpi}\,\mathfrak{m}) &= \sigma_{P_j}^+(\overline{\varpi}\,\mathfrak{m}) = \sigma_{R_{P_j}}^+(\mathfrak{m}^\dagger + \overline{\gamma})\\ \sigma_P^-(\underline{\varpi}\,\mathfrak{m}) &= \sigma_{P_i}^-(\underline{\varpi}\,\mathfrak{m}) = \sigma_{R_{P_i}}^-(\mathfrak{m}^\dagger - \overline{\gamma})\end{aligned}$$

Proof. From (9.28), it follows that $\sigma_{N_{P_{\times \mathfrak{m}}}}^+(\overline{\varpi}) = \sigma_{N_{P_{j,\times \mathfrak{m}}}}^+(\overline{\varpi})$. Consequently, $P_{\times \mathfrak{m}}(f) \sim P_{j,\times \mathfrak{m}}(f)$ for all sufficiently small $f \in \mathbb{T}^{\succ,>}$, so that $\sigma_P^+(\overline{\varpi}\,\mathfrak{m}) = \sigma_{P_j}^+(\overline{\varpi}\,\mathfrak{m})$. Similarly, we obtain $\sigma_P^-(\underline{\varpi}\,\mathfrak{m}) = \sigma_{P_i}^-(\underline{\varpi}\,\mathfrak{m})$. Since

$$\begin{aligned}\sigma_{P_{j,\times \mathfrak{m}}}(f) &= \sigma_{R_{P_j}}(\mathfrak{m}^\dagger + f^\dagger)\\ \sigma_{P_{i,\times \mathfrak{m}}}(f) &= \sigma_{R_{P_i}}(\mathfrak{m}^\dagger + f^\dagger)\end{aligned}$$

for all $f \in \mathbb{T}^>$, we also have

$$\begin{aligned}\sigma_{P_j}^+(\overline{\varpi}\,\mathfrak{m}) &= \sigma_{R_{P_j}}^+(\mathfrak{m}^\dagger + \overline{\gamma})\\ \sigma_{P_i}^-(\underline{\varpi}\,\mathfrak{m}) &= \sigma_{R_{P_i}}^-(\mathfrak{m}^\dagger - \overline{\gamma}).\end{aligned}$$

□

Let $\mathfrak{W}$ be an initial segment of $\mathfrak{T}$. The sign $\sigma_{P,\mathfrak{W}}$ of P modulo $\mathfrak{W}$ at a point $f \in \mathbb{T}$ is defined as follows. If $\deg_{\mathfrak{W}} P_{+f} > 0$, then we set $\sigma_{P,\mathfrak{W}}(f) = 0$. Recall that $\deg_{\mathfrak{W}} P_{+f}$ is the multiplicity of f as a zero of P modulo $\mathfrak{W}$ in this case. If $\deg_{\mathfrak{W}} P_{+f} = 0$, then for all $\delta \in C[[\mathfrak{W}]]$, we have $\sigma_{P_{+f}}(\delta) = \operatorname{sign} P_0$, and we set $\sigma_{P,\mathfrak{W}}(f) = \operatorname{sign} P_0 \in \{-1, 1\}$. Given $\bar{f} \in \overline{\mathbb{T}}$ and $f \in \mathbb{T}$, we write $\bar{f} <_{\mathfrak{W}} f$ if $\bar{f} < f + \delta$ for all $\delta \in C[[\mathfrak{W}]]$. Given $\bar{f}, \bar{g} \in \overline{\mathbb{T}}$, we denote

$$(\bar{f}, \bar{g})_{\mathfrak{W}} = \{h \in \mathbb{T} : f <_{\mathfrak{W}} h <_{\mathfrak{W}} g\}.$$

We say that $\sigma_{P,\mathfrak{W}}$ is *constant at the right* of $\bar{f} \in \overline{\mathbb{T}}$, if there exist $\epsilon \in \{-1, 0, 1\}$ and $g >_{\mathfrak{W}} \bar{f}$ such that $\sigma_{P,\mathfrak{W}}(f) = \epsilon$ for all $f \in (\bar{f}, g)_{\mathfrak{W}}$. In that case, we denote $\sigma_{P,\mathfrak{W}}^+(\bar{f}) = \epsilon$. Constance at the left is defined similarly. If $\mathfrak{W}$ is of the form $\mathfrak{W} = \{\mathfrak{m} \in \mathfrak{T} : \mathfrak{m} \prec \mathfrak{w}\}$, then we also write $\sigma_{P,\prec \mathfrak{w}} = \sigma_{P,\mathfrak{W}}$, $\sigma_{P,\prec \mathfrak{w}}^+ = \sigma_{P,\mathfrak{W}}^+$ and $\sigma_{P,\prec \mathfrak{w}}^- = \sigma_{P,\mathfrak{W}}^-$.

Exercise 9.24. Let $\mathcal{H} \supseteq \mathbb{T}^{\mathrm{cv}}\{F\}$ be a Hardy field. Consider a cut $\bar{f} \in \overline{\mathbb{T}}$ and an element $h \in \mathcal{H}$, such that $g < \bar{f} \Leftrightarrow g < h$ for $g \in \mathbb{T}^{\mathrm{cv}}$. If $\sigma_P^+(\bar{f})$ is defined, then show that there exists a $g \in \mathcal{H}$ with $g > h$ and $\sigma_P(\varphi) = \sigma_P^+(\bar{f})$ for all $\varphi \in (h, g)$.

Exercise 9.25. Show that $\zeta(x)$,

$$\varphi(x) = \frac{1}{x} + \frac{1}{x^\pi} + \frac{1}{x^{\pi^2}} + \cdots$$

and

$$\psi(x) = \frac{1}{x} + \frac{1}{\mathrm{e}^{\log^2 x}} + \frac{1}{\mathrm{e}^{\log^4 x}} + \cdots$$

do not satisfy an algebraic differential equation with coefficients in $\mathbb{T}$. Compare with the technique from exercise 8.25.

Exercise 9.26. Let L be a real analytic solution to $L(\log x) = L(x) - 1$ (for a construction of such a solution, see [Kne50]). Show that $\mathbb{T}^{\mathrm{cv}}\{L\}$ is a Hardy field.

9.7 The intermediate value theorem

In this section, we assume that C is a real closed field. Our main aim is to prove the following intermediate value theorem:

Theorem 9.33. *Let $P \in \mathbb{T}\{F\}$ and $f, g \in \mathbb{T}$ be such that $f < g$ and $P(f)\,P(g) < 0$. Then there exists a $h \in (f, g)$ with $P(h) = 0$.*

In fact, we will prove the following stronger version of the theorem:

Theorem 9.34. *Let $P \in \mathbb{T}\{F\}$ and let $\mathfrak{W}$ be an initial segment of $\mathfrak{T}$. Assume that $f, g \in \mathbb{T}$ are such that $f <_{\mathfrak{W}} g$ and $\sigma_{P,\mathfrak{W}}(f)\,\sigma_{P,\mathfrak{W}}(g) < 0$. Then there exists a $h \in (f, g)_{\mathfrak{W}}$ such that $\deg_{\mathfrak{W}} P_{+h}$ is odd.*

In both theorems, the interval (f, g) may actually be replaced by a more general interval $(\bar{f}, \bar{g})$ with $\bar{f}, \bar{g} \in \bar{\mathbb{T}}$. More precisely, we say that P *changes sign* on $(\bar{f}, \bar{g})$ modulo $\mathfrak{W}$, if $\sigma^{+}_{P,\mathfrak{W}}(\bar{f})$ and $\sigma^{-}_{P,\mathfrak{W}}(\bar{g})$ exist and $\sigma^{+}_{P,\mathfrak{W}}(\bar{f})\;\sigma^{-}_{P,\mathfrak{W}}(\bar{g}) < 0$. Notice that P changes sign on $(\bar{f}, \bar{g})$ modulo $\mathfrak{W}$ if and only if P changes sign on $(\bar{f}, \bar{g})_{\mathfrak{W}}$. We say that P *changes sign* at $h \in \mathbb{T}$ modulo $\mathfrak{W}$ if $\deg_{\mathfrak{W}} P_{+h}$ is odd. Now if P changes sign on $(\bar{f}, \bar{g})$, then it also changes sign on (f, g) for some $f, g \in \mathbb{T}$ with $\bar{f} < f < g < \bar{g}$, $\sigma_{P,\mathfrak{W}}(f) = \sigma^{+}_{P,\mathfrak{W}}(\bar{f})$ and $\sigma_{P,\mathfrak{W}}(g) = \sigma^{-}_{P,\mathfrak{W}}(\bar{g})$. Consequently, if theorem 9.34 holds for all intervals (f, g) with $f, g \in \mathbb{T}$, then it also holds for all intervals $(\bar{f}, \bar{g})$ with $\bar{f}, \bar{g} \in \bar{\mathbb{T}}$.

Remark 9.35. The fact that P changes sign at $h \in \mathbb{T}$ modulo $\mathfrak{W}$ does not necessarily imply $\sigma^{+}_{P,\mathfrak{W}}(h)\,\sigma^{-}_{P,\mathfrak{W}}(h) < 0$. Indeed, $P = F'$ changes sign modulo $o(1)$ at $h = 0$, but $\sigma^{+}_{F',\prec 1}(0)$ and $\sigma^{-}_{F',\prec 1}(0)$ are not defined.

9.7.1 The quasi-linear case

Lemma 9.36. *Let $P \in C\{F\}$ be of order r and let $\mathfrak{W}$ be an initial segment of $\mathfrak{T}$. Assume that the theorem 9.34 holds for all differential polynomials of order $< r$. Let $\mathfrak{v} \in \mathfrak{T}$ be such that the equation*

$$P(f) = 0 \qquad (f \prec \mathfrak{v}) \tag{9.32}$$

is quasi-linear and assume that P changes sign on $(0, \gimel\,\mathfrak{v})_{\mathfrak{W}}$. Then there exists a $h \in (0, \gimel\,\mathfrak{v})_{\mathfrak{W}}$ with $\deg_{\mathfrak{W}} P_{+h} = 1$.

Proof. Modulo an additive conjugation by a sufficiently small $\delta \in (0, \text{ȝ}\,\mathfrak{v})_{\mathfrak{W}}$, we may assume without loss of generality that $\deg_{\mathfrak{W}} P = 0$. Since (9.32) is quasi-linear, it admits only a finite number of starting monomials. Let $\mathfrak{m}$ be the largest such monomial. Modulo a multiplicative conjugation with $\mathfrak{m}$, we may assume without loss of generality that $\mathfrak{m} = 1$. We must have $\mathfrak{W} \prec 1$, since otherwise $1 = \deg_{\preccurlyeq 1} P \leqslant \deg_{\mathfrak{W}} P = 0$. Furthermore, since $N_P \in C[F]\,(F')^{\mathbb{N}}$, we either have $N_P = \alpha\, F + \beta$ with $\alpha, \beta \in C^{\neq}$, or $N_P = \alpha\, F'$ with $\alpha \in C^{\neq}$.

If $N_P = \alpha\ F + \beta$, then the distinguished solution h to (9.32) satisfies $h \sim -\beta / \alpha \neq 0$. Moreover, from proposition 9.32, it follows that

$$\begin{aligned} \sigma_{P,\mathfrak{W}}(0) &= \sigma_P(\text{ȝ}) &= \sigma^-_{R_{P_0}}(-\overline{\gamma}) &= \operatorname{sign} \beta\,; \\ \sigma^+_{P,\mathfrak{W}}(\overline{\varpi}) &= \sigma^+_P(\overline{\varpi}) &= \sigma^+_{R_{P_1}}(\overline{\gamma}) &= \operatorname{sign} \alpha\,; \\ \sigma^-_{P,\mathfrak{W}}(\text{ȝ}\,\mathfrak{v}) &= \sigma^-_P(\text{ȝ}\,\mathfrak{v}) &= \sigma^-_{R_{P_1}}(\mathfrak{v}^\dagger - \overline{\gamma}) &= -\operatorname{sign} \beta. \end{aligned}$$

We claim that $\sigma_{R_{P_1}}(\overline{\gamma}) = \sigma_{R_{P_1}}(\mathfrak{v}^\dagger - \overline{\gamma})$. Otherwise, theorem 9.34 applied to R_{P_1} implies the existence of a $\psi \in (0, \mathfrak{v}^\dagger)_{\prec \overline{\gamma}}$ with

$$\deg_{\prec \overline{\gamma}} R_{P_1,+\psi} \in 2\,\mathbb{N} + 1.$$

Taking ψ such that $\psi_{\prec \overline{\gamma}} = 0$ (whence $\int \psi \in \mathbb{T}_{\succ}$) , it follows that $e^{\int \psi} \succ 1$ would be a starting monomial for (9.32). Our claim implies that $\operatorname{sign} \beta = -\operatorname{sign} \alpha$, so that $h \in (0, \text{ȝ}\,\mathfrak{v})_{\mathfrak{W}}$. Furthermore, $P_{+h,0} = 0$, so

$$1 \leqslant \deg_{\mathfrak{W}} P_{+h} \leqslant \deg_{\prec \mathfrak{v}} P_{+h} = 1.$$

If $N_P = \alpha\, F'$, then $\deg_{\prec 1} P_{+\lambda} = 1$ for any $\lambda \in C$. Let $h = 1 + \varepsilon$, where ε is the distinguished solution to $P_{+1}(\varepsilon) = 0$ ($\varepsilon \prec 1$). Then $h \in (0, \text{ȝ}\,\mathfrak{v})_{\mathfrak{W}}$ and $P_{+h,0} = 0$ again implies $\deg_{\mathfrak{W}} P_{+h} = 1$. □

9.7.2 Preserving sign changes during refinements

Lemma 9.37. *Let $P \in C\{F\}$ and let I be of one of the following forms:*

a) $I = (c_1, c_2)_{\prec 1} = (c_1 + \text{ȝ}, c_2 - \text{ȝ})$ *with* $c_1, c_2 \in C$.
b) $I = (c_1, \overline{\varpi})_{\prec 1} = (c_1 + \text{ȝ}, \overline{\varpi})$ *with* $c_1 \in C$.
c) $I = (-\overline{\varpi}, \overline{\varpi})_{\prec 1} = (-\overline{\varpi}, \overline{\varpi})$.

If P changes sign on I, then there exists a $c \in I \cap C$ with

$$\sigma^+_P(c - \text{ȝ})\, \sigma^-_P(c + \text{ȝ}) < 0.$$

Proof. In cases (b) and (c), we may replace $\overline{\varpi}$ (and $-\overline{\varpi}$) by a sufficiently large $c_2 \in C$ (resp. small $c_1 \in C$). Therefore, it suffices to deal with intervals I of the form (a). From lemma 9.26, it follows that $\sigma^+_P(c - \text{ȝ}) = \sigma^+_{N_P}(c - \text{ȝ})$, $\sigma^-_P(c + \text{ȝ}) = \sigma^-_{N_P}(c + \text{ȝ})$ for all $c \in C$. Without loss of generality, we may therefore assume that $P = N_P = A\,(F')^\nu$ with $A \in C[F]$ and $\nu \in \mathbb{N}$.

If ν is odd, then we choose $c \in I \cap C$ with $A(c) \neq 0$, and obtain

$$\sigma_P^+(c - \overline{\backepsilon})\, \sigma_P^-(c + \overline{\backepsilon}) = \sigma_{(F')^\nu}^+(c - \overline{\backepsilon})\, \sigma_{(F')^\nu}^-(c + \overline{\backepsilon}) = (-1)^\nu < 0.$$

If ν is even, then A changes sign on I. Since C is real closed, it follows that there exists a $c \in I \cap C$ where A admits a root of odd multiplicity μ, and

$$\sigma_P^+(c - \overline{\backepsilon})\, \sigma_P^-(c + \overline{\backepsilon}) = \sigma_A^+(c - \overline{\backepsilon})\, \sigma_A^-(c + \overline{\backepsilon}) = (-1)^\mu < 0. \qquad \square$$

Lemma 9.38. *Let $P \in C\{F\}$ be of order r and let $\mathfrak{W}$ be an initial segment of $\mathfrak{T}$. Assume that the theorem 9.34 holds for all differential polynomials of order $< r$. Let $\mathfrak{m} \in \mathfrak{T}$ be such that $\sigma_{P,\mathfrak{W}}^+(0)\, \sigma_{P,\mathfrak{W}}^-(\overline{\backepsilon}\, \mathfrak{m}) < 0$. Then there exists $c \in C^>$ and $\mathfrak{v} \in \mathfrak{T}$ with $\mathfrak{W} \prec \mathfrak{v} \prec \mathfrak{m}$ and $\sigma_{P_{+c\mathfrak{v}}}^+(-\overline{\backepsilon}\, \mathfrak{v})\, \sigma_{P_{+c\mathfrak{v}}}^-(\overline{\backepsilon}\, \mathfrak{v}) < 0$.*

Proof. Modulo an additive conjugation with a sufficiently small $\delta \in (0, \overline{\backepsilon}\, \mathfrak{m})_{\mathfrak{W}}$, we may assume without loss of generality that

$$\sigma_{P,\mathfrak{W}}^+(0) = \sigma_{P,\mathfrak{W}}(0) = \operatorname{sign} P_0 \neq 0.$$

We prove the lemma by induction over $d = \deg_{\prec \mathfrak{m}} P$. If $d = 0$, then the assumptions cannot be met, so we have nothing to prove. So assume that $d > 0$. Since $P_0 \neq 0$, there exists an equalizer of the form $\mathfrak{e} = \mathfrak{e}_{v,d}$ for the equation $P(f) = 0$ $(f \prec \mathfrak{m})$. We distinguish the following cases:

$\boldsymbol{\sigma_{P,\mathfrak{W}}^+(0)\, \sigma_{P,\mathfrak{W}}^-(\overline{\backepsilon}\, \mathfrak{e}) < 0.}$ Since $\deg_{\prec \mathfrak{e}} P = v < d$, we are done by the induction hypothesis.

$\boldsymbol{\mathfrak{e} \succ \mathfrak{W}}$ **and** $\boldsymbol{\sigma_P^-(\overline{\backepsilon}\, \mathfrak{e})\, \sigma_P^+(\overline{\mho}\, \mathfrak{e}) < 0.}$ The result follows immediately when applying lemma 9.37 to $P_{\times \mathfrak{e}}$ and the interval $(\overline{\backepsilon}, \overline{\mho})$.

$\boldsymbol{\mathfrak{e} \in \mathfrak{W}}$ **or** $\boldsymbol{\sigma_P^+(\overline{\mho}\, \mathfrak{e})\, \sigma_P^-(\overline{\backepsilon}\, \mathfrak{m}) < 0.}$ If $\mathfrak{e} \in \mathfrak{W}$, then let $g >_{\mathfrak{W}} 0$ be such that $\sigma_P(f) = \sigma_{P,\mathfrak{W}}^+(0)$ for all $f \in (0, g)_{\mathfrak{W}}$. Then for any $\mathfrak{n} \in \mathfrak{T}$ with $\mathfrak{W} \prec \mathfrak{n} \prec g$, we have $\sigma_P^+(\overline{\mho}\, \mathfrak{n})\, \sigma_P^-(\overline{\backepsilon}\, \mathfrak{m}) < 0$. So both if $\mathfrak{e} \in \mathfrak{W}$ and if $\sigma_P^+(\overline{\mho}\, \mathfrak{e})\, \sigma_P^-(\overline{\backepsilon}\, \mathfrak{m}) < 0$, there exists an $\mathfrak{n} \in \mathfrak{T}$ with $\mathfrak{W} \prec \mathfrak{n} \prec \mathfrak{m}$, $\mathfrak{n} \succcurlyeq \mathfrak{e}$ and $\sigma_P^+(\overline{\mho}\, \mathfrak{n})\, \sigma_P^-(\overline{\backepsilon}\, \mathfrak{m}) < 0$.

Since $\mathfrak{m} \succ \mathfrak{n} \succcurlyeq \mathfrak{e}$, we must have $\deg N_{P_{\times \mathfrak{n}}} = d$. From proposition 9.32, it follows that

$$\sigma_{R_{P_d}}^+(\mathfrak{n}^\dagger + \overline{\gamma})\, \sigma_{R_{P_d}}^-(\mathfrak{m}^\dagger - \overline{\gamma}) = \sigma_{P_d}^+(\overline{\mho}\, \mathfrak{n})\, \sigma_{P_d}^-(\overline{\backepsilon}\, \mathfrak{m}) = \sigma_P^+(\overline{\mho}\, \mathfrak{n})\, \sigma_P^-(\overline{\backepsilon}\, \mathfrak{m}) < 0.$$

Applying theorem 9.34 to R_{P_d}, we infer that there exists a $g \in (\mathfrak{n}^\dagger, \mathfrak{m}^\dagger)_{\overline{\gamma}}$ with $\deg_{\prec \overline{\gamma}} R_{P_d} \in 2\,\mathbb{N} + 1$. Taking g such that $g_{\prec \overline{\gamma}} = 0$ (whence $\int g \in \mathbb{T}_\succ$), it follows that $\mathfrak{v} = e^{\int g}$ is a starting monomial for $P(f) = 0$. Moreover, $N = N_{P_{\times \mathfrak{v}}}$ is of the form $N = \alpha\, F^{d-\nu} (F')^\nu$ with $\alpha \in C^{\neq}$, since $\deg N = \operatorname{val} N = d$. Furthermore, since $\sigma_{R_{P_d}}^-(g - \overline{\gamma})\, \sigma_{R_{P_d}}^+(g + \overline{\gamma}) < 0$, we have

$$(-1)^\nu = \sigma_N^-(\overline{\backepsilon})\, \sigma_N^+(\overline{\mho}) = \sigma_{P_d}^-(\overline{\backepsilon}\, \mathfrak{v})\, \sigma_{P_d}^+(\overline{\mho}\, \mathfrak{v}) = \sigma_{R_{P_d}}^-(g - \overline{\gamma})\, \sigma_{R_{P_d}}^+(g + \overline{\gamma}) < 0,$$

whence ν is odd. For any $c > 0$, we conclude that

$$\sigma_{P_{+c\mathfrak{v}}}^+(-\overline{\backepsilon}\, \mathfrak{v})\, \sigma_{P_{+c\mathfrak{v}}}^-(\overline{\backepsilon}\, \mathfrak{v}) = \sigma_N^+(c - \overline{\backepsilon})\, \sigma_N^-(c + \overline{\backepsilon}) = (-1)^\nu < 0. \qquad \square$$

9.7.3 Proof of the intermediate value theorem

We will prove the following variant of theorem 9.34:

Theorem 9.39. *Let $P \in \mathbb{T}\{F\}$ and let $\mathfrak{W}$ be an initial segment of $\mathfrak{T}$. Given $\mathfrak{v} \in \mathfrak{T}$, consider an interval I of one of the following forms:*

a) $I = (\xi, \chi)_{\mathfrak{W}}$ with $\xi, \chi \in \mathbb{T}$ and $\chi - \xi \sim \lambda \mathfrak{v}$ with $\lambda \in C^{>}$.
b) $I = (\xi, \xi + \mathfrak{z}\, \mathfrak{v})_{\mathfrak{W}}$ with $\xi \in \mathbb{T}$.
c) $I = (\xi - \mathfrak{z}\, \mathfrak{v}, \xi)_{\mathfrak{W}}$ with $\xi \in \mathbb{T}$.
d) $I = (\xi - \mathfrak{z}\, \mathfrak{v}, \xi + \mathfrak{z}\, \mathfrak{v})_{\mathfrak{W}}$ with $\xi \in \mathbb{T}$.

If P changes sign on I, then there exists a point $h \in I$ such that $\deg_{\mathfrak{W}} P_{+h}$ is odd.

Proof. We prove the theorem by a double induction over the order of P and the Newton degree d of

$$P(f) = 0 \qquad (f \prec \mathfrak{v}).$$

The case when $d = 0$ is contrary to our assumptions. So assume that $d > 0$ and that the hypothesis holds for all smaller orders, as well as for the same order and smaller d. Notice that we must have $\mathfrak{W} \prec \mathfrak{v}$, since P changes sign modulo $\mathfrak{W}$ on I.

Let us first show that cases (a), (c) and (d) can all be reduced to case (b). This is clear for (c) by considering $P(-f)$ instead of $P(f)$. In case (d), there exists a $\chi \in (\xi - \mathfrak{z}\, \mathfrak{v}, \xi + \mathfrak{z}\, \mathfrak{v})_{\mathfrak{W}}$ such that $\sigma_P^{+}(\xi - \mathfrak{z}\, \mathfrak{v}) = \sigma_P(\eta)$ for all $\eta \in \mathbb{T}$ with $\eta \in (\xi - \mathfrak{z}\, \mathfrak{v}, \chi)_{\mathfrak{W}}$. For any such η, it follows that P changes sign on $(\eta, \eta + \mathfrak{z}\, \mathfrak{v})_{\mathfrak{W}} = (\eta, \eta + \mathfrak{z}\, \mathfrak{v})_{\mathfrak{W}}$. As to (a), we observe that P changes sign either on $(\xi, \xi + \mathfrak{z}\, \mathfrak{v})_{\mathfrak{W}}$, on $(\chi - \mathfrak{z}\, \mathfrak{v}, \chi)_{\mathfrak{W}}$, or on $(\xi + \mathfrak{z}\, \mathfrak{v}, \chi - \mathfrak{z}\, \mathfrak{v})_{\mathfrak{W}} = (\xi + \mathfrak{z}\, \mathfrak{v}, \chi - \mathfrak{z}\, \mathfrak{v})$. The first to cases have already been dealt with. The last case reduces to (d) when applying lemma 9.37 to the polynomial $P_{+\xi, \times \mathfrak{v}}$ and the interval $(\mathfrak{z}, (\chi - \xi)_{\mathfrak{v}} - \mathfrak{z})$.

Let us now show how to prove (b). Modulo an additive conjugation, we may assume without loss of generality that $\xi = 0$. If $d = 1$, then we are done by lemma 9.36. So assume that $d > 1$. Consider the set $\mathscr{S}$ of all partial unravellings

$$f = \varphi + \tilde{f} \qquad (\tilde{f} \prec \tilde{\mathfrak{v}}) \tag{9.33}$$

with either $\varphi = 0$ and $\tilde{\mathfrak{v}} = \mathfrak{v}$, or $\varphi \in (0, \mathfrak{z}\, \mathfrak{v})_{\mathfrak{W}}$ and

$$\sigma^{+}_{P_{+\varphi}, \mathfrak{W}}(-\mathfrak{z}\, \tilde{\mathfrak{v}})\, \sigma^{-}_{P_{+\varphi}, \mathfrak{W}}(\mathfrak{z}\, \tilde{\mathfrak{v}}) < 0.$$

By corollary 8.32, we may choose a finest partial unravelling (9.33) in $\mathscr{S}$.

Take $\eta = 0$ if $\varphi = 0$ and $\eta \prec \tilde{\mathfrak{v}}$ such that $\sigma^{+}_{P_{+\varphi}}(-\mathfrak{z}\, \tilde{\mathfrak{v}}) = \sigma^{+}_{P_{+\varphi}, \mathfrak{W}}(\eta)$ otherwise. By lemma 9.38, applied to $P_{+\varphi+\eta}$, there exists a term $c\, \mathfrak{m} \in (\eta, \mathfrak{z}\, \mathfrak{v})_{\mathfrak{W}}$ with $\mathfrak{W} \prec \mathfrak{m}$, and such that

$$\sigma^{+}_{P_{+\varphi+c\mathfrak{m}}}(-\mathfrak{z}\, \mathfrak{m})\, \sigma^{-}_{P_{+\varphi+c\mathfrak{m}}}(\mathfrak{z}\, \mathfrak{m}) < 0.$$

We claim that we cannot have $\deg_{\prec \mathfrak{m}} P_{+\varphi+c\mathfrak{m}} = d$. Indeed, by proposition 8.27, this would imply the existence of a partial unravelling

$$f = \varphi + \tilde{\varphi} + \tilde{\tilde{f}} \qquad (\tilde{\tilde{f}} \prec \mathfrak{m})$$

with $\tilde{\varphi} \sim c\,\mathfrak{m}$, which is finer than (9.33). But then

$$\sigma^{+}_{P_{+\varphi+\tilde{\varphi}}}(-\daleth\,\mathfrak{m})\,\sigma^{-}_{P_{+\varphi+\tilde{\varphi}}}(\daleth\,\mathfrak{m}) = \sigma^{+}_{P_{+\varphi+c\mathfrak{m}}}(-\daleth\,\mathfrak{m})\,\sigma^{-}_{P_{+\varphi+c\mathfrak{m}}}(\daleth\,\mathfrak{m}) < 0$$

contradicts the maximality of (9.33). Consequently, we have

$$\deg_{\prec \mathfrak{m}} P_{+\varphi+c\mathfrak{m}} < d$$

and the theorem follows by applying the induction hypothesis for $P_{+\varphi+c\mathfrak{m}}$ on the interval $(-\daleth\,\mathfrak{m}, \daleth\,\mathfrak{m})$. □

Exercise 9.27.

a) Prove that $|\sigma_{P,\mathfrak{V}}(f)| \leqslant |\sigma_{P,\mathfrak{W}}(f)|$ if $\mathfrak{V} \supseteq \mathfrak{W}$.
b) Prove that $\sigma^{-}_{P,\mathfrak{W}}(\daleth\,\mathfrak{m}) = \sigma^{-}_{P}(\daleth\,\mathfrak{m})$ if $\mathfrak{W} \prec \mathfrak{m}$.
c) Other similar properties.

References

AS26 E. Artin and O. Schreier. Algebraische Konstruction reeller Körper. *Hamb. Abh.*, pages 85–99, 1926.

AvdD01 M. Aschenbrenner and L. van den Dries. Liouville closed H-fields. *J. Pure Appl. Algebra*, 2001. to appear.

AvdD02 M. Aschenbrenner and L. van den Dries. H-fields and their Liouville extensions. *Math. Z.*, 242:543–588, 2002.

AvdD04 M. Aschenbrenner and L. van den Dries. Asymptotic differential algebra. In *Contemporary Mathematics*. Amer. Math. Soc., Providence, RI, 2004. To appear.

AvdDvdH M. Aschenbrenner, L. van den Dries, and J. van der Hoeven. Linear differential equations over H-fields. In preparation.

AvdDvdH05 Matthias Aschenbrenner, Lou van den Dries, and J. van der Hoeven. Differentially algebraic gaps. *Selecta Mathematica*, 11(2):247–280, 2005.

BB56 C.A. Briot and J.C. Bouquet. Propriétés des fonctions définies par des équations différentielles. *Journal de l'École Polytechnique*, 36:133–198, 1856.

BCR87 J. Bochnak, M. Coste, and M.-F. Roy. *Géométrie algébrique réelle*. Ergebnisse der Mathematik und ihrer Grenzgebiete. Springer, Berlin-Heidelberg, 1987. 3. Folge, Band 12.

Bir09 G.D. Birkhoff. Singular points of ordinary differential equations. *Trans. Am. Math. Soc.*, 10:436–470, 1909.

Bor28 É. Borel. *Leçons sur les séries divergentes*. Gauthiers Villards, 2-nd edition, 1928. Reprinted by Jacques Gabay.

Bos81 M. Boshernitzan. An extension of Hardy's class l of 'orders of infinity'. *J. Analyse Math.*, 39:235–255, 1981.

Bos82 M. Boshernitzan. New 'orders of infinity'. *J. Analyse Math.*, 41:130–167, 1982.

Bos87 M. Boshernitzan. Second-order differential equations over hardy fields. *J. London Math. Soc.*, 35:109–120, 1987.

Bou61 N. Bourbaki. *Fonctions d'une variable réelle*. Éléments de Mathématiques (Chap. 5. Hermann, 2-nd edition, 1961.

Bou70 N. Bourbaki. *Théorie des ensembles*. Hermann, 1970.

Bra91 B.L.J. Braaksma. Multisummability and Stokes multipliers of linear meromorphic differential equations. *J. Diff. Eq*, 92:45–75, 1991.

Bra92 B.L.J. Braaksma. Multisummability of formal power series solutions to nonlinear meromorphic differential equations. *Ann. Inst. Fourier de Grenoble*, 42:517–540, 1992.

Can99 G. Cantor. *Sur les fondements de la théorie des ensembles transfinis*. Jacques Gabay, 1899. Reprint from les Mémoires de la Société des Sciences physiques et naturelles de Bordeaux.

Can93 J. Cano. An extension of the Newton-Puiseux polygon construction to give solutions of Pfaffian forms. *Ann. Institut Fourier de Grenoble*, 43(1):125–142, 1993.

Chi86 A.L. Chistov. Polynomial complexity of the newton-puiseux algorithm. In *Proc. Symp. Math. Found. Comp. Sc.*, pages 247–255, Bratislava, 1986. Springer Lect. Notes in Comp. Sc. 233.

CNP93 B. Candelberger, J.C. Nosmas, and F. Pham. *Approche de la résurgence*. Actualités mathématiques. Hermann, 1993.

Con76 J.H. Conway. *On numbers and games*. Academic Press, 1976.

Dah84 B.I. Dahn. The limiting behaviour of exponential terms. *Fund. Math.*, 124:169–186, 1984.

dBR75 P. du Bois-Reymond. Über asymptotische Werte, infinitäre Approximationen und infinitäre Auflösung von Gleichungen. *Math. Ann.*, 8:363–414, 1875.

dBR77 P. du Bois-Reymond. Über die Paradoxen des Infinitärscalcüls. *Math. Ann.*, 11:149–167, 1877.

DG86 B.I. Dahn and P. Göring. Notes on exponential-logarithmic terms. *Fundamenta Mathematicae*, 127:45–50, 1986.

DW84 Bernd I. Dahn and Helmut Wolter. Ordered fields with several exponential functions. *Z. Math. Logik Grundlag. Math.*, 30(4):341–348, 1984.

É03 J. Écalle. Recent advances in the analysis of divergence and singularities. In Yu.S. Ilyashenko C. Rousseau, editor, *Proc. of the July 2003 Montreal Summer School on Singularities and Normal Forms*. Kluwer, 1903. To appear.

É85 J. Écalle. *Les fonctions résurgentes I–III*. Publ. Math. d'Orsay 1981 and 1985, 1985.

É92 J. Écalle. *Introduction aux fonctions analysables et preuve constructive de la conjecture de Dulac*. Hermann, collection: Actualités mathématiques, 1992.

É93 J. Écalle. Six lectures on transseries, analysable functions and the constructive proof of Dulac's conjecture. In D. Schlomiuk, editor, *Bifurcations and periodic orbits of vector fields*, pages 75–184. Kluwer, 1993.

Fab85 E. Fabry. *Sur les intégrales des équations différentielles linéaires à coefficients rationnels*. PhD thesis, Paris, 1885.

Fin89 H.B. Fine. On the functions defined by differential equations, with an extension of the Newton-Puiseux polygon construction to these equations. *Amer. Jour. of Math.*, 12:295–322, 1889.

GG88 K.O. Geddes and G.H. Gonnet. A new algorithm for computing symbolic limits using hierarchical series. In *Proc. ISSAC '88*, volume 358 of *Lect. Notes in Comp. Science*, pages 490–495. Springer, 1988.

GG92 G.H. Gonnet and D. Gruntz. Limit computation in computer algebra. Technical Report 187, ETH, Zürich, 1992.

Gri91 D.Y. Grigoriev. Complexity of factoring and calculating the gcd of linear ordinary differential operators. *J. Symb. Comp*, 10:7–37, 1991.

Gru96 D. Gruntz. *On computing limits in a symbolic manipulation system*. PhD thesis, E.T.H. Zürich, Switzerland, 1996.

GS91 D. Grigoriev and M. Singer. Solving ordinary differential equations in terms of series with real exponents. *Trans. of the AMS*, 327(1):329–351, 1991.

Hah07 H. Hahn. Über die nichtarchimedischen Größensysteme. *Sitz. Akad. Wiss. Wien*, 116:601–655, 1907.

Har10 G.H. Hardy. *Orders of infinity*. Cambridge Univ. Press, 1910.

Har11 G.H. Hardy. Properties of logarithmico-exponential functions. *Proceedings of the London Mathematical Society*, 10(2):54–90, 1911.

Har63 G.H. Hardy. *Divergent series*. Clarendon Press, Oxford, 3rd edition, 1963.

Hau08 F. Hausdorff. Grundzüge einer Theorie der geordneten Mengen. *Math. Ann.*, 65:435–505, 1908.

Hig52 G. Higman. Ordering by divisibility in abstract algebras. *Proc. London Math. Soc.*, 2:326–336, 1952.

Kho91 A.G. Khovanskii. *Fewnomials*, volume 88 of *Translations of Mathematical Monographs*. A.M.S., Providence RI, 1991.

Kne50 H. Kneser. Reelle analytische Lösungen der Gleichung $\phi(\phi(x))=e^x$ und verwandter Funktionalgleichungen. *Jour. f. d. reine und angewandte Math.*, 187(1/2):56–67, 1950.

Kol73 E.R. Kolchin. *Differential algebra and algebraic groups*. Academic Press, New York, 1973.

Kru60 J.B. Kruskal. Well-quasi-ordering, the tree theorem and Vázsoni's conjecture. *Trans. Am. Math. Soc.*, 95:210–225, 1960.

Kuh00 S. Kuhlmann. *Ordered exponential fields*. Fields Institute Monographs. Am. Math. Soc., 2000.

LC93 T. Levi-Civita. Sugli infiniti ed infinitesimi attuali quali elimenti analitici. *Atti ist. Ven.*, 7:1765–1815, 1893.

Lio37 J. Liouville. Mémoire sur la classification des transcendantes, et sur les racines de certaines équations en fonction finie explicite des coefficients. *J. Math. Pures et Appl.*, 2:56–104, 1837.

Lio38 J. Liouville. Suite du mémoire sur la classification des transcendantes, et sur les racines de certaines équations en fonction finie explicite des coefficients. *J. Math. Pures et Appl.*, 3:523–546, 1838.

Mal79 J. Malitz. *Introduction to mathematical logic*. Undergraduate texts in Mathematics. Springer-Verlag, New York, 1979.

New71 I. Newton. *De methodis serierum et Fluxionum*. Manuscript, 1671.

NW63 C. St. J. A. Nash-Williams. On well-quasi-ordering finite trees. *Proc. Cambridge Philos. Soc.*, 59:833–835, 1963.

Poi86 H. Poincaré. Sur les intégrales irrégulières des équations linéaires. *Acta Math.*, 8:295–344, 1886.

Poi90 H. Poincaré. Sur le problème des trois corps et les équations de la dynamique. *Acta Math.*, 13:1–270, 1890.

Poi93 H. Poincaré. *Les méthodes nouvelles de la mécanique céleste*, volume Tôme II. Gauthier-Villars, Paris, 1893.

Pui50 M.V. Puiseux. Recherches sur les fonctions algébriques. *J. Math. Pures et Appliquées*, 15:365–480, 1850.

Rib64 P. Ribenboim. *Théorie des valuations*. Les Presses de l'Université de Montréal, 1964. 2-nd Ed. 1968.

Ric97 D. Richardson. How to recognise zero. *JSC*, 24:627–645, 1997.

Rit50 J.F. Ritt. *Differential algebra*. Amer. Math. Soc., New York, 1950.

Ros80 M. Rosenlicht. Differential valuations. *Pacific J. Math.*, 86:301–319, 1980.

Ros83a M. Rosenlicht. Hardy fields. *Journal of Math. Analysis and Appl.*, 93:297–311, 1983.

Ros83b M. Rosenlicht. The rank of a Hardy field. *Trans. Amer. Math. Soc.*, 280(2):659–671, 1983.

Ros87 M. Rosenlicht. Growth properties of functions in Hardy fields. *Trans. Amer. Math. Soc.*, 299(1):261–272, 1987.

RSSvdH96 D. Richardson, B. Salvy, J. Shackell, and J. van der Hoeven. Expansions of exp-log functions. In Y.N. Lakhsman, editor, *Proc. ISSAC '96*, pages 309–313, Zürich, Switzerland, July 1996.

Sal91 B. Salvy. *Asymptotique automatique et fonctions génératrices*. PhD thesis, École Polytechnique, France, 1991.

Sch01 M.C. Schmeling. *Corps de transséries*. PhD thesis, Université Paris-VII, 2001.

Sei54 A. Seidenberg. A new decision method for elementary algebra. *An. of Math.*, 60:365–374, 1954.

Sei56 A. Seidenberg. An elimination theorem for differential algebra. *Univ. California Publ. Math. (N.S.)*, pages 31–38, 1956.

Sei68 A. Seidenberg. Reduction of singularities of the differentiable equation $Ady = Bdx$. *Amer. J. of Math.*, 90:248–269, 1968.

Sha90 J. Shackell. Growth estimates for exp-log functions. *Journal of Symbolic Computation*, 10:611–632, 1990.

Sha04 J. Shackell. *Symbolic asymptotics*, volume 12 of *Algorithms and computation in Mathematics*. Springer-Verlag, 2004.

Smi75 S. Smith. On the higher singularities of plane curves. *Prod. London Math. Soc.*, 6:153–182, 1875.

Spe99 P. Speissegger. The Pfaffian closure on an o-minimal structure. *Jour. f. d. reine und angewandte Math.*, 508:189–211, 1999.

Sti86 T.J. Stieltjes. Recherches sur quelques séries semi-convergentes. *Ann. Sc. Éc. Norm. Sup. de Paris*, 3:201–258, 1886.

Sti94 T.J. Stieltjes. Recherches sur les fractions continues. *Ann. Fac. Sci. Toulouse*, 8:J1–J122, 1894.

Sti95 T.J. Stieltjes. Recherches sur les fractions continues. *Ann. Fac. Sci. Toulouse*, 9:A1–A47, 1895.

Str77 W. Strodt. A differential algebraic study of the intrusion of logarithms into asymptotic expansions. In H. Bass, P.J. Cassidy, and J. Kovacic, editors, *Contributions to algebra*, pages 355–375. Academic Press, 1977.

SW71 W. Strodt and R.K. Wright. *Asymptotic behaviour of solutions and adjunction fields for nonlinear first order differential equations*, volume 109. Mem. Amer. Math. Soc., 1971.

Tar31 A. Tarski. Sur les ensembles définissables de nombres réels. *Fund. Math.*, 17:210–239, 1931.

Tar51 A. Tarski. *A decision method for elementary algebra and geometry.* Rand coorporation, Berkeley and Los Angelos, 2 edition, 1951.

vdD98 L. van den Dries. *Tame topology and o-minimal structures*, volume 248 of *London Math. Soc. Lect. Note*. Cambridge university press, 1998.

vdD99 L. van den Dries. O-minimal structures and real analytic geometry. In B. Mazur et al., editor, *Current Developments in Mathematics*, pages 105–152. Int. Press, Somerville, MA, 1999.

vdH95 J. van der Hoeven. Automatic numerical expansions. In J.-C. Bajard, D. Michelucci, J.-M. Moreau, and J.-M. Müller, editors, *Proc. of the conference "Real numbers and computers", Saint-Étienne, France*, pages 261–274, 1995.

vdH96 J. van der Hoeven. On the computation of limsups. *Journal of Pure and Applied Algebra*, 117/118:381–394, 1996.

vdH97 J. van der Hoeven. *Automatic asymptotics.* PhD thesis, École polytechnique, France, 1997.

vdH98 J. van der Hoeven. Generic asymptotic expansions. *AAECC*, 9(1):25–44, 1998.

vdH00 J. van der Hoeven. A differential intermediate value theorem. Technical Report 2000-50, Univ. d'Orsay, 2000.

vdH01a J. van der Hoeven. Complex transseries solutions to algebraic differential equations. Technical Report 2001-34, Univ. d'Orsay, 2001.

vdH01b J. van der Hoeven. Fast evaluation of holonomic functions near and in singularities. *JSC*, 31:717–743, 2001.

vdH01c J. van der Hoeven. Formal asymptotics of solutions to certain linear differential equations involving oscillation. Technical Report 2001-60, Prépublications d'Orsay, 2001.

vdH01d J. van der Hoeven. Operators on generalized power series. *Journal of the Univ. of Illinois*, 45(4):1161–1190, 2001.

vdH02a J. van der Hoeven. A differential intermediate value theorem. In B. L. J. Braaksma, G. K. Immink, M. van der Put, and J. Top, editors, *Differential equations and the Stokes phenomenon*, pages 147–170. World Scientific, 2002.

vdH02b J. van der Hoeven. Relax, but don't be too lazy. *JSC*, 34:479–542, 2002.

vK75 S. von Kowalevsky. Zur Theorie der partiellen Differentialgleichungen. *J. Reine und Angew. Math.*, 80:1–32, 1875.

Wil96 A. Wilkie. Model completeness results for expansions of the real field by restricted Pfaffian functions and the exponential function. *J. A.M.S.*, 9:1051–1094, 1996.

Glossary

Chapter 1. Orderings

$\leqslant$	quasi-ordering	12
$\equiv$	equivalent elements w.r.t. quasi-ordering	12
$E \amalg F$	disjoint union of E and F	12
$E \dot{\amalg} F$	ordered union of E and F	13
$E \times F$	Cartesian product of E and F	13
$E \dot{\times} F$	anti-lexicographic product of E and F	13
E^*	set of words over E	13
E^+	set of non-empty words over E	13
$\leqslant^!$	ordering defined by $x \leqslant^! y \Leftrightarrow x = y \vee x < y$	14
$E^\diamond$	set of commutative words over E	15
ω	the ordinal number $\omega = \{0, 1, 2, ...\}$	16
$\mathrm{fin}(A)$	final segment generated by A	17
$\mathrm{in}(A)$	initial segment generated by A	17
$\mathrm{root}(T)$	root of T	19
$E^\top$	set of finite E-labeled trees	20
$S^{\geqslant}$	subset of positive or zero elements in S	22
$S^{>}$	subset of strictly positive elements in S	22
$S^{\neq}$	subset of non-zero elements in S	22
$S^{\leqslant}$	subset of negative or zero elements in S	22
$S^{<}$	subset of strictly negative elements in S	22
$\lvert x \rvert$	absolute value of x	22
R^*	subset of non-zero-divisors in R	25
$x \preccurlyeq y$	x is dominated by y	25
$x = O(y)$	Landau's notation for $x \preccurlyeq y$	25
$x \asymp y$	x is asymptotic to y	25
$x \prec y$	x is negligible w.r.t. y	25
$x = o(y)$	Landau's notation for $x \prec y$	25
$x \sim y$	x is equivalent to y	25

$A^{\times}$	elements of A which can be raised to a power	30
$\\|x\\|$	$\\|x\\| = x$ if $x \succcurlyeq 1$ and $\\|x\\| = x^{-1}$ otherwise	30
$x \mathrel{\underline{\prec\!\!\prec}} y$	x is flatter than or as flat as y	30
$x \prec\!\!\prec y$	x is flatter than y	30
$x \asymp\!\!\asymp y$	x is as flat as y	30
$x \approxeq y$	x and y are similar modulo flatness	30
$A^{\flat}$	flat subring of A	31
$x \preccurlyeq^{\sharp} y$	$x \preccurlyeq y$ modulo flatness	31
$x \prec^{\sharp} y$	$x \prec y$ modulo flatness	31
$x \preccurlyeq_{\varphi} y$	$x \preccurlyeq y$ modulo elements flatter than φ	31
$x \prec_{\varphi} y$	$x \prec y$ modulo elements flatter than φ	31
$x \preccurlyeq_{\varphi}^{*} y$	$x \preccurlyeq y$ modulo elements flatter than or as flat as φ	31
$x \prec_{\varphi}^{*} y$	$x \prec y$ modulo elements flatter than or as flat as φ	31

Chapter 2. Grid-based series

$\mathfrak{M}$	quasi-ordered monomial monoid	34
$\mathfrak{m}, \mathfrak{n}, \mathfrak{v}, \ldots$	monomials	34
C	ring of coefficients	36
$\operatorname{supp} f$	support of a series f	36
$C[[\mathfrak{M}]]$	ring of grid-based series in $\mathfrak{M}$ with coefficients in C	36
$f_{\mathfrak{m}}$	coefficient of $\mathfrak{m}$ in f	36
$C[[z_1, \ldots, z_n]]$	natural multivariate series in $z_1, \ldots, z_n$	38
$C[[z_1; \ldots; z_n]]$	recursive multivariate series in $z_1, \ldots, z_n$	38
$\mathfrak{d}_f$	dominant monomial of f	40
c_f	dominant coefficient of f	40
τ_f	dominant term of f	40
δ_f	series with $f = \tau_f(1 + \delta_f)$	40
$f_{\succ}$	purely infinite part of f	41
$f_{\asymp}$	constant part of f	41
$f_{\prec}$	infinitesimal part of f	41
$f_{\preccurlyeq}$	bounded part of f	41
$C[[\mathfrak{M}]]_{\succ}$	set of purely infinite elements in $C[[\mathfrak{M}]]$	41
$C[[\mathfrak{M}]]_{\prec}$	set of infinitesimal elements in $C[[\mathfrak{M}]]$	41
$C[[\mathfrak{M}]]_{\preccurlyeq}$	set of bounded elements in $C[[\mathfrak{M}]]$	41
π_{λ}	the series $(1 + z)^{\lambda} \in C[[z]]$	42
$f_{\mathfrak{S}}$	restriction of f to $\mathfrak{S}$	42
$f \trianglelefteq g$	f is a truncation of g	43
$\mathscr{F}(S)$	collection of families with values in S	44
$\mathcal{F}, \mathcal{G}, \ldots$	families	44
$(f_i : i \in I)$	alternative notation for family $(f_i)_{i \in I}$	44
$\mathcal{F} \amalg \mathcal{G}$	disjoint union of two families $\mathcal{F}$ and $\mathcal{G}$	44
$\mathcal{F} \times \mathcal{G}$	Cartesian product of families $\mathcal{F}$ and $\mathcal{G}$	44

Chapter 3. The Newton polygon method

Chapter 4. Transseries

Chapter 5. Operations on transseries

Chapter 6. Grid-based operators

$\Psi \circ (\Phi_1, \ldots, \Phi_n)$	composition of multilinear grid-based operators	117
$\operatorname{supp} \Phi$	operator support of Φ	117
$\mathscr{D}_{\mathbb{T}_{\prec\!\!\prec e^x}}, \mathscr{D}_{\mathbb{T}_{\preceq e^x}}$	sets of operators $\sum_{n \in \mathbb{N}} L_n \partial^n$ on $\mathbb{T}_{\prec\!\!\prec e^x}$ resp. $\mathbb{T}_{\preceq e^x}$	118
Φ_i	homogeneous part of degree i of Φ	122
$\Psi \circ \Phi$	composition of grid-based operators Ψ and Φ	124
$\mathscr{L}(M_1, \ldots, M_m, N)$	set of strongly multilinear $\Phi: M_1 \times \cdots \times M_m \to N$	124
$\mathscr{G}(C[[\mathfrak{M}]], C[[\mathfrak{N}]])$	set of grid-based operators from $\Phi: C[[\mathfrak{M}]] \to C[[\mathfrak{N}]]$	125
$\Omega_{\mathfrak{m}_1, \ldots, \mathfrak{m}_m, \mathfrak{n}}$	atomic m-linear grid-based operator	125
$\lvert\alpha\rvert$	arity of an atomic operator α	126
$\mathfrak{i}_{\alpha,1}, \ldots, \mathfrak{i}_{\alpha,\lvert\alpha\rvert}$	inputs of an atomic operator α	126
$\mathfrak{o}_\alpha$	output of an atomic operator α	126
$\mathcal{D}_{\mathfrak{M}}$	atomic decomposition of $\mathrm{Id}_{C[[\mathfrak{M}]]}$	126
$\mathcal{A} \circ \mathcal{B}$	composition of atomic families $\mathcal{A}$ and $\mathcal{B}$	126

Chapter 7. Linear differential equations

L_*, f_*	alternative notation for dominant coefficient c_L, c_f	136
$\operatorname{supp}_{\mathrm{ser}} L$	serial support of L	136
$L_{\times h}$	multiplicative conjugate of L by h	137
$L\uparrow, L\downarrow$	upward and downward shiftings of L	137
$s_{i,j}$	Stirling numbers of the first kind	137
$S_{i,j}$	Stirling numbers of the second kind	137
$\uparrow_l, \downarrow_l$	l-th iterates of $\uparrow$ and $\downarrow$	138
$\mathscr{D}_{\mathbb{T}^\flat}$	space of operators of the form $\sum_{n \in \mathbb{N}} L_n \partial^n$ on $\mathbb{T}^\flat$	138
U_i	differential polynomial with $f^{(i)} = U_i(f^\dagger) f$	139
R_L	differential Riccati polynomial associated to L	139
$R_{L,\mathrm{alg}}$	algebraic part of R_L	140
$P_{+\varphi}$	additive conjugate of P by φ	140
L'	derivative of differential operator L	140
$\mathfrak{R}_L$	set of regular monomials for L	141
$\mathfrak{H}_L$	set of irregular monomials for L	141
T_L	trace of L	141
$T_{L;\mathfrak{B}}$	trace of L w.r.t. $\mathfrak{B}$	141
L^{-1}	distinguished right inverse of L	146
H_L	set of solutions to $Lh = 0$	146
$\tilde{\mathbb{T}}$	complexification $\mathbb{T} \oplus \mathrm{i}\,\mathbb{T}$ of $\mathbb{T}$	158
$\mathbb{O}$	field of oscillating transseries	159
$f_{;\psi}$	spectral part of f associated to $e^{\mathrm{i}\psi}$	159
$L_{;\psi}$	action of L on the spectral component $\tilde{\mathbb{T}}\, e^{\mathrm{i}\psi}$	159

Chapter 8. Algebraic differential equations

Chapter 9. The intermediate value theorem

$\overline{\mathfrak{M}}$	set of monomials and monomial cuts	208
$\bar{\mathfrak{d}}_{\bar{f}}$	dominant monomial of $\bar{f}$	209
$\overline{\mathfrak{w}}_{\bar{f}}$	width of $\bar{f}$	209
$\varphi_{\bar{f}}$	initializer of $\bar{f}$	210
$\hat{\mathbb{S}}$	serial completion of $\mathbb{S}$	210
$\hat{f}$	typical serial cut	210
$\overline{\varkappa}_k$	shorthand for sup $\{f \in \mathbb{T}: \exp(f) = k\}$	212
$\bar{\gamma}$	the cut $(x \log x \log_2 x \cdots)^{-1}$	219
$\mathcal{L}_{\varphi_0, \ldots, \varphi_{k-1}, g, h}$	basic integral neighbourhood	220
$f_0, \ldots, f_k$	integral coordinates of f	221
σ_P	sign of P	227
σ_P^+	sign of P at the right of a point	227
σ_P^-	sign of P at the left of a point	227
$\sigma_{P,\mathfrak{W}}$	sign of P modulo $\mathfrak{W}$	228
$\sigma_{P,\mathfrak{W}}^+$	sign of P at the right of a point modulo $\mathfrak{W}$	228
$\sigma_{P,\mathfrak{W}}^-$	sign of P at the left of a point modulo $\mathfrak{W}$	228

Index

Lecture Notes in Mathematics

For information about earlier volumes
please contact your bookseller or Springer
LNM Online archive: springerlink.com

Vol. 1682: H.-D. Alber, Materials with Memory (1998)
Vol. 1683: A. Pomp, The Boundary-Domain Integral Method for Elliptic Systems (1998)
Vol. 1684: C. A. Berenstein, P. F. Ebenfelt, S. G. Gindikin, S. Helgason, A. E. Tumanov, Integral Geometry, Radon Transforms and Complex Analysis. Firenze, 1996. Editors: E. Casadio Tarabusi, M. A. Picardello, G. Zampieri (1998)
Vol. 1685: S. König, A. Zimmermann, Derived Equivalences for Group Rings (1998)
Vol. 1686: J. Azéma, M. Émery, M. Ledoux, M. Yor (Eds.), Séminaire de Probabilités XXXII (1998)
Vol. 1687: F. Bornemann, Homogenization in Time of Singularly Perturbed Mechanical Systems (1998)
Vol. 1688: S. Assing, W. Schmidt, Continuous Strong Markov Processes in Dimension One (1998)
Vol. 1689: W. Fulton, P. Pragacz, Schubert Varieties and Degeneracy Loci (1998)
Vol. 1690: M. T. Barlow, D. Nualart, Lectures on Probability Theory and Statistics. Editor: P. Bernard (1998)
Vol. 1691: R. Bezrukavnikov, M. Finkelberg, V. Schechtman, Factorizable Sheaves and Quantum Groups (1998)
Vol. 1692: T. M. W. Eyre, Quantum Stochastic Calculus and Representations of Lie Superalgebras (1998)
Vol. 1694: A. Braides, Approximation of Free-Discontinuity Problems (1998)
Vol. 1695: D. J. Hartfiel, Markov Set-Chains (1998)
Vol. 1696: E. Bouscaren (Ed.): Model Theory and Algebraic Geometry (1998)
Vol. 1697: B. Cockburn, C. Johnson, C.-W. Shu, E. Tadmor, Advanced Numerical Approximation of Nonlinear Hyperbolic Equations. Cetraro, Italy, 1997. Editor: A. Quarteroni (1998)
Vol. 1698: M. Bhattacharjee, D. Macpherson, R. G. Möller, P. Neumann, Notes on Infinite Permutation Groups (1998)
Vol. 1699: A. Inoue,Tomita-Takesaki Theory in Algebras of Unbounded Operators (1998)
Vol. 1700: W. A. Woyczyński, Burgers-KPZ Turbulence (1998)
Vol. 1701: Ti-Jun Xiao, J. Liang, The Cauchy Problem of Higher Order Abstract Differential Equations (1998)
Vol. 1702: J. Ma, J. Yong, Forward-Backward Stochastic Differential Equations and Their Applications (1999)
Vol. 1703: R. M. Dudley, R. Norvaiša, Differentiability of Six Operators on Nonsmooth Functions and p-Variation (1999)
Vol. 1704: H. Tamanoi, Elliptic Genera and Vertex Operator Super-Algebras (1999)
Vol. 1705: I. Nikolaev, E. Zhuzhoma, Flows in 2-dimensional Manifolds (1999)
Vol. 1706: S. Yu. Pilyugin, Shadowing in Dynamical Systems (1999)
Vol. 1707: R. Pytlak, Numerical Methods for Optimal Control Problems with State Constraints (1999)
Vol. 1708: K. Zuo, Representations of Fundamental Groups of Algebraic Varieties (1999)
Vol. 1709: J. Azéma, M. Émery, M. Ledoux, M. Yor (Eds.), Séminaire de Probabilités XXXIII (1999)
Vol. 1710: M. Koecher, The Minnesota Notes on Jordan Algebras and Their Applications (1999)
Vol. 1711: W. Ricker, Operator Algebras Generated by Commuting Projećtions: A Vector Measure Approach (1999)
Vol. 1712: N. Schwartz, J. J. Madden, Semi-algebraic Function Rings and Reflectors of Partially Ordered Rings (1999)
Vol. 1713: F. Bethuel, G. Huisken, S. Müller, K. Steffen, Calculus of Variations and Geometric Evolution Problems. Cetraro, 1996. Editors: S. Hildebrandt, M. Struwe (1999)
Vol. 1714: O. Diekmann, R. Durrett, K. P. Hadeler, P. K. Maini, H. L. Smith, Mathematics Inspired by Biology. Martina Franca, 1997. Editors: V. Capasso, O. Diekmann (1999)
Vol. 1715: N. V. Krylov, M. Röckner, J. Zabczyk, Stochastic PDE's and Kolmogorov Equations in Infinite Dimensions. Cetraro, 1998. Editor: G. Da Prato (1999)
Vol. 1716: J. Coates, R. Greenberg, K. A. Ribet, K. Rubin, Arithmetic Theory of Elliptic Curves. Cetraro, 1997. Editor: C. Viola (1999)
Vol. 1717: J. Bertoin, F. Martinelli, Y. Peres, Lectures on Probability Theory and Statistics. Saint-Flour, 1997. Editor: P. Bernard (1999)
Vol. 1718: A. Eberle, Uniqueness and Non-Uniqueness of Semigroups Generated by Singular Diffusion Operators (1999)
Vol. 1719: K. R. Meyer, Periodic Solutions of the N-Body Problem (1999)
Vol. 1720: D. Elworthy, Y. Le Jan, X-M. Li, On the Geometry of Diffusion Operators and Stochastic Flows (1999)
Vol. 1721: A. Iarrobino, V. Kanev, Power Sums, Gorenstein Algebras, and Determinantal Loci (1999)
Vol. 1722: R. McCutcheon, Elemental Methods in Ergodic Ramsey Theory (1999)
Vol. 1723: J. P. Croisille, C. Lebeau, Diffraction by an Immersed Elastic Wedge (1999)
Vol. 1724: V. N. Kolokoltsov, Semiclassical Analysis for Diffusions and Stochastic Processes (2000)
Vol. 1725: D. A. Wolf-Gladrow, Lattice-Gas Cellular Automata and Lattice Boltzmann Models (2000)
Vol. 1726: V. Marić, Regular Variation and Differential Equations (2000)
Vol. 1727: P. Kravanja M. Van Barel, Computing the Zeros of Analytic Functions (2000)
Vol. 1728: K. Gatermann Computer Algebra Methods for Equivariant Dynamical Systems (2000)
Vol. 1729: J. Azéma, M. Émery, M. Ledoux, M. Yor (Eds.) Séminaire de Probabilités XXXIV (2000)
Vol. 1730: S. Graf, H. Luschgy, Foundations of Quantization for Probability Distributions (2000)
Vol. 1731: T. Hsu, Quilts: Central Extensions, Braid Actions, and Finite Groups (2000)

Vol. 1732: K. Keller, Invariant Factors, Julia Equivalences and the (Abstract) Mandelbrot Set (2000)
Vol. 1733: K. Ritter, Average-Case Analysis of Numerical Problems (2000)
Vol. 1734: M. Espedal, A. Fasano, A. Mikelić, Filtration in Porous Media and Industrial Applications. Cetraro 1998. Editor: A. Fasano. 2000.
Vol. 1735: D. Yafaev, Scattering Theory: Some Old and New Problems (2000)
Vol. 1736: B. O. Turesson, Nonlinear Potential Theory and Weighted Sobolev Spaces (2000)
Vol. 1737: S. Wakabayashi, Classical Microlocal Analysis in the Space of Hyperfunctions (2000)
Vol. 1738: M. Émery, A. Nemirovski, D. Voiculescu, Lectures on Probability Theory and Statistics (2000)
Vol. 1739: R. Burkard, P. Deuflhard, A. Jameson, J.-L. Lions, G. Strang, Computational Mathematics Driven by Industrial Problems. Martina Franca, 1999. Editors: V. Capasso, H. Engl, J. Periaux (2000)
Vol. 1740: B. Kawohl, O. Pironneau, L. Tartar, J.-P. Zolesio, Optimal Shape Design. Tróia, Portugal 1999. Editors: A. Cellina, A. Ornelas (2000)
Vol. 1741: E. Lombardi, Oscillatory Integrals and Phenomena Beyond all Algebraic Orders (2000)
Vol. 1742: A. Unterberger, Quantization and Nonholomorphic Modular Forms (2000)
Vol. 1743: L. Habermann, Riemannian Metrics of Constant Mass and Moduli Spaces of Conformal Structures (2000)
Vol. 1744: M. Kunze, Non-Smooth Dynamical Systems (2000)
Vol. 1745: V. D. Milman, G. Schechtman (Eds.), Geometric Aspects of Functional Analysis. Israel Seminar 1999-2000 (2000)
Vol. 1746: A. Degtyarev, I. Itenberg, V. Kharlamov, Real Enriques Surfaces (2000)
Vol. 1747: L. W. Christensen, Gorenstein Dimensions (2000)
Vol. 1748: M. Ruzicka, Electrorheological Fluids: Modeling and Mathematical Theory (2001)
Vol. 1749: M. Fuchs, G. Seregin, Variational Methods for Problems from Plasticity Theory and for Generalized Newtonian Fluids (2001)
Vol. 1750: B. Conrad, Grothendieck Duality and Base Change (2001)
Vol. 1751: N. J. Cutland, Loeb Measures in Practice: Recent Advances (2001)
Vol. 1752: Y. V. Nesterenko, P. Philippon, Introduction to Algebraic Independence Theory (2001)
Vol. 1753: A. I. Bobenko, U. Eitner, Painlevé Equations in the Differential Geometry of Surfaces (2001)
Vol. 1754: W. Bertram, The Geometry of Jordan and Lie Structures (2001)
Vol. 1755: J. Azéma, M. Émery, M. Ledoux, M. Yor (Eds.), Séminaire de Probabilités XXXV (2001)
Vol. 1756: P. E. Zhidkov, Korteweg de Vries and Nonlinear Schrödinger Equations: Qualitative Theory (2001)
Vol. 1757: R. R. Phelps, Lectures on Choquet's Theorem (2001)
Vol. 1758: N. Monod, Continuous Bounded Cohomology of Locally Compact Groups (2001)
Vol. 1759: Y. Abe, K. Kopfermann, Toroidal Groups (2001)
Vol. 1760: D. Filipović, Consistency Problems for Heath-Jarrow-Morton Interest Rate Models (2001)
Vol. 1761: C. Adelmann, The Decomposition of Primes in Torsion Point Fields (2001)
Vol. 1762: S. Cerrai, Second Order PDE's in Finite and Infinite Dimension (2001)
Vol. 1763: J.-L. Loday, A. Frabetti, F. Chapoton, F. Goichot, Dialgebras and Related Operads (2001)
Vol. 1764: A. Cannas da Silva, Lectures on Symplectic Geometry (2001)
Vol. 1765: T. Kerler, V. V. Lyubashenko, Non-Semisimple Topological Quantum Field Theories for 3-Manifolds with Corners (2001)
Vol. 1766: H. Hennion, L. Hervé, Limit Theorems for Markov Chains and Stochastic Properties of Dynamical Systems by Quasi-Compactness (2001)
Vol. 1767: J. Xiao, Holomorphic Q Classes (2001)
Vol. 1768: M.J. Pflaum, Analytic and Geometric Study of Stratified Spaces (2001)
Vol. 1769: M. Alberich-Carramiñana, Geometry of the Plane Cremona Maps (2002)
Vol. 1770: H. Gluesing-Luerssen, Linear Delay-Differential Systems with Commensurate Delays: An Algebraic Approach (2002)
Vol. 1771: M. Émery, M. Yor (Eds.), Séminaire de Probabilités 1967-1980. A Selection in Martingale Theory (2002)
Vol. 1772: F. Burstall, D. Ferus, K. Leschke, F. Pedit, U. Pinkall, Conformal Geometry of Surfaces in S^4 (2002)
Vol. 1773: Z. Arad, M. Muzychuk, Standard Integral Table Algebras Generated by a Non-real Element of Small Degree (2002)
Vol. 1774: V. Runde, Lectures on Amenability (2002)
Vol. 1775: W. H. Meeks, A. Ros, H. Rosenberg, The Global Theory of Minimal Surfaces in Flat Spaces. Martina Franca 1999. Editor: G. P. Pirola (2002)
Vol. 1776: K. Behrend, C. Gomez, V. Tarasov, G. Tian, Quantum Comohology. Cetraro 1997. Editors: P. de Bartolomeis, B. Dubrovin, C. Reina (2002)
Vol. 1777: E. García-Río, D. N. Kupeli, R. Vázquez-Lorenzo, Osserman Manifolds in Semi-Riemannian Geometry (2002)
Vol. 1778: H. Kiechle, Theory of K-Loops (2002)
Vol. 1779: I. Chueshov, Monotone Random Systems (2002)
Vol. 1780: J. H. Bruinier, Borcherds Products on O(2,1) and Chern Classes of Heegner Divisors (2002)
Vol. 1781: E. Bolthausen, E. Perkins, A. van der Vaart, Lectures on Probability Theory and Statistics. Ecole d' Eté de Probabilités de Saint-Flour XXIX-1999. Editor: P. Bernard (2002)
Vol. 1782: C.-H. Chu, A. T.-M. Lau, Harmonic Functions on Groups and Fourier Algebras (2002)
Vol. 1783: L. Grüne, Asymptotic Behavior of Dynamical and Control Systems under Perturbation and Discretization (2002)
Vol. 1784: L.H. Eliasson, S. B. Kuksin, S. Marmi, J.-C. Yoccoz, Dynamical Systems and Small Divisors. Cetraro, Italy 1998. Editors: S. Marmi, J.-C. Yoccoz (2002)
Vol. 1785: J. Arias de Reyna, Pointwise Convergence of Fourier Series (2002)
Vol. 1786: S. D. Cutkosky, Monomialization of Morphisms from 3-Folds to Surfaces (2002)
Vol. 1787: S. Caenepeel, G. Militaru, S. Zhu, Frobenius and Separable Functors for Generalized Module Categories and Nonlinear Equations (2002)
Vol. 1788: A. Vasil'ev, Moduli of Families of Curves for Conformal and Quasiconformal Mappings (2002)
Vol. 1789: Y. Sommerhäuser, Yetter-Drinfel'd Hopf algebras over groups of prime order (2002)
Vol. 1790: X. Zhan, Matrix Inequalities (2002)

Vol. 1791: M. Knebusch, D. Zhang, Manis Valuations and Prüfer Extensions I: A new Chapter in Commutative Algebra (2002)

Vol. 1792: D. D. Ang, R. Gorenflo, V. K. Le, D. D. Trong, Moment Theory and Some Inverse Problems in Potential Theory and Heat Conduction (2002)

Vol. 1793: J. Cortés Monforte, Geometric, Control and Numerical Aspects of Nonholonomic Systems (2002)

Vol. 1794: N. Pytheas Fogg, Substitution in Dynamics, Arithmetics and Combinatorics. Editors: V. Berthé, S. Ferenczi, C. Mauduit, A. Siegel (2002)

Vol. 1795: H. Li, Filtered-Graded Transfer in Using Noncommutative Gröbner Bases (2002)

Vol. 1796: J.M. Melenk, hp-Finite Element Methods for Singular Perturbations (2002)

Vol. 1797: B. Schmidt, Characters and Cyclotomic Fields in Finite Geometry (2002)

Vol. 1798: W.M. Oliva, Geometric Mechanics (2002)

Vol. 1799: H. Pajot, Analytic Capacity, Rectifiability, Menger Curvature and the Cauchy Integral (2002)

Vol. 1800: O. Gabber, L. Ramero, Almost Ring Theory (2003)

Vol. 1801: J. Azéma, M. Émery, M. Ledoux, M. Yor (Eds.), Séminaire de Probabilités XXXVI (2003)

Vol. 1802: V. Capasso, E. Merzbach, B.G. Ivanoff, M. Dozzi, R. Dalang, T. Mountford, Topics in Spatial Stochastic Processes. Martina Franca, Italy 2001. Editor: E. Merzbach (2003)

Vol. 1803: G. Dolzmann, Variational Methods for Crystalline Microstructure – Analysis and Computation (2003)

Vol. 1804: I. Cherednik, Ya. Markov, R. Howe, G. Lusztig, Iwahori-Hecke Algebras and their Representation Theory. Martina Franca, Italy 1999. Editors: V. Baldoni, D. Barbasch (2003)

Vol. 1805: F. Cao, Geometric Curve Evolution and Image Processing (2003)

Vol. 1806: H. Broer, I. Hoveijn. G. Lunther, G. Vegter, Bifurcations in Hamiltonian Systems. Computing Singularities by Gröbner Bases (2003)

Vol. 1807: V. D. Milman, G. Schechtman (Eds.), Geometric Aspects of Functional Analysis. Israel Seminar 2000-2002 (2003)

Vol. 1808: W. Schindler, Measures with Symmetry Properties (2003)

Vol. 1809: O. Steinbach, Stability Estimates for Hybrid Coupled Domain Decomposition Methods (2003)

Vol. 1810: J. Wengenroth, Derived Functors in Functional Analysis (2003)

Vol. 1811: J. Stevens, Deformations of Singularities (2003)

Vol. 1812: L. Ambrosio, K. Deckelnick, G. Dziuk, M. Mimura, V. A. Solonnikov, H. M. Soner, Mathematical Aspects of Evolving Interfaces. Madeira, Funchal, Portugal 2000. Editors: P. Colli, J. F. Rodrigues (2003)

Vol. 1813: L. Ambrosio, L. A. Caffarelli, Y. Brenier, G. Buttazzo, C. Villani, Optimal Transportation and its Applications. Martina Franca, Italy 2001. Editors: L. A. Caffarelli, S. Salsa (2003)

Vol. 1814: P. Bank, F. Baudoin, H. Föllmer, L.C.G. Rogers, M. Soner, N. Touzi, Paris-Princeton Lectures on Mathematical Finance 2002 (2003)

Vol. 1815: A. M. Vershik (Ed.), Asymptotic Combinatorics with Applications to Mathematical Physics. St. Petersburg, Russia 2001 (2003)

Vol. 1816: S. Albeverio, W. Schachermayer, M. Talagrand, Lectures on Probability Theory and Statistics. Ecole d'Eté de Probabilités de Saint-Flour XXX-2000. Editor: P. Bernard (2003)

Vol. 1817: E. Koelink, W. Van Assche(Eds.), Orthogonal Polynomials and Special Functions. Leuven 2002 (2003)

Vol. 1818: M. Bildhauer, Convex Variational Problems with Linear, nearly Linear and/or Anisotropic Growth Conditions (2003)

Vol. 1819: D. Masser, Yu. V. Nesterenko, H. P. Schlickewei, W. M. Schmidt, M. Waldschmidt, Diophantine Approximation. Cetraro, Italy 2000. Editors: F. Amoroso, U. Zannier (2003)

Vol. 1820: F. Hiai, H. Kosaki, Means of Hilbert Space Operators (2003)

Vol. 1821: S. Teufel, Adiabatic Perturbation Theory in Quantum Dynamics (2003)

Vol. 1822: S.-N. Chow, R. Conti, R. Johnson, J. Mallet-Paret, R. Nussbaum, Dynamical Systems. Cetraro, Italy 2000. Editors: J. W. Macki, P. Zecca (2003)

Vol. 1823: A. M. Anile, W. Allegretto, C. Ringhofer, Mathematical Problems in Semiconductor Physics. Cetraro, Italy 1998. Editor: A. M. Anile (2003)

Vol. 1824: J. A. Navarro González, J. B. Sancho de Salas, $\mathscr{C}^\infty$ – Differentiable Spaces (2003)

Vol. 1825: J. H. Bramble, A. Cohen, W. Dahmen, Multiscale Problems and Methods in Numerical Simulations, Martina Franca, Italy 2001. Editor: C. Canuto (2003)

Vol. 1826: K. Dohmen, Improved Bonferroni Inequalities via Abstract Tubes. Inequalities and Identities of Inclusion-Exclusion Type. VIII, 113 p, 2003.

Vol. 1827: K. M. Pilgrim, Combinations of Complex Dynamical Systems. IX, 118 p, 2003.

Vol. 1828: D. J. Green, Gröbner Bases and the Computation of Group Cohomology. XII, 138 p, 2003.

Vol. 1829: E. Altman, B. Gaujal, A. Hordijk, Discrete-Event Control of Stochastic Networks: Multimodularity and Regularity. XIV, 313 p, 2003.

Vol. 1830: M. I. Gil', Operator Functions and Localization of Spectra. XIV, 256 p, 2003.

Vol. 1831: A. Connes, J. Cuntz, E. Guentner, N. Higson, J. E. Kaminker, Noncommutative Geometry, Martina Franca, Italy 2002. Editors: S. Doplicher, L. Longo (2004)

Vol. 1832: J. Azéma, M. Émery, M. Ledoux, M. Yor (Eds.), Séminaire de Probabilités XXXVII (2003)

Vol. 1833: D.-Q. Jiang, M. Qian, M.-P. Qian, Mathematical Theory of Nonequilibrium Steady States. On the Frontier of Probability and Dynamical Systems. IX, 280 p, 2004.

Vol. 1834: Yo. Yomdin, G. Comte, Tame Geometry with Application in Smooth Analysis. VIII, 186 p, 2004.

Vol. 1835: O.T. Izhboldin, B. Kahn, N.A. Karpenko, A. Vishik, Geometric Methods in the Algebraic Theory of Quadratic Forms. Summer School, Lens, 2000. Editor: J.-P. Tignol (2004)

Vol. 1836: C. Năstăsescu, F. Van Oystaeyen, Methods of Graded Rings. XIII, 304 p, 2004.

Vol. 1837: S. Tavaré, O. Zeitouni, Lectures on Probability Theory and Statistics. Ecole d'Eté de Probabilités de Saint-Flour XXXI-2001. Editor: J. Picard (2004)

Vol. 1838: A.J. Ganesh, N.W. O'Connell, D.J. Wischik, Big Queues. XII, 254 p, 2004.

Vol. 1839: R. Gohm, Noncommutative Stationary Processes. VIII, 170 p, 2004.

Vol. 1840: B. Tsirelson, W. Werner, Lectures on Probability Theory and Statistics. Ecole d'Eté de Probabilités de Saint-Flour XXXII-2002. Editor: J. Picard (2004)

Vol. 1841: W. Reichel, Uniqueness Theorems for Variational Problems by the Method of Transformation Groups (2004)
Vol. 1842: T. Johnsen, A.L. Knutsen, K3 Projective Models in Scrolls (2004)
Vol. 1843: B. Jefferies, Spectral Properties of Noncommuting Operators (2004)
Vol. 1844: K.F. Siburg, The Principle of Least Action in Geometry and Dynamics (2004)
Vol. 1845: Min Ho Lee, Mixed Automorphic Forms, Torus Bundles, and Jacobi Forms (2004)
Vol. 1846: H. Ammari, H. Kang, Reconstruction of Small Inhomogeneities from Boundary Measurements (2004)
Vol. 1847: T.R. Bielecki, T. Björk, M. Jeanblanc, M. Rutkowski, J.A. Scheinkman, W. Xiong, Paris-Princeton Lectures on Mathematical Finance 2003 (2004)
Vol. 1848: M. Abate, J. E. Fornaess, X. Huang, J. P. Rosay, A. Tumanov, Real Methods in Complex and CR Geometry, Martina Franca, Italy 2002. Editors: D. Zaitsev, G. Zampieri (2004)
Vol. 1849: Martin L. Brown, Heegner Modules and Elliptic Curves (2004)
Vol. 1850: V. D. Milman, G. Schechtman (Eds.), Geometric Aspects of Functional Analysis. Israel Seminar 2002-2003 (2004)
Vol. 1851: O. Catoni, Statistical Learning Theory and Stochastic Optimization (2004)
Vol. 1852: A.S. Kechris, B.D. Miller, Topics in Orbit Equivalence (2004)
Vol. 1853: Ch. Favre, M. Jonsson, The Valuative Tree (2004)
Vol. 1854: O. Saeki, Topology of Singular Fibers of Differential Maps (2004)
Vol. 1855: G. Da Prato, P.C. Kunstmann, I. Lasiecka, A. Lunardi, R. Schnaubelt, L. Weis, Functional Analytic Methods for Evolution Equations. Editors: M. Iannelli, R. Nagel, S. Piazzera (2004)
Vol. 1856: K. Back, T.R. Bielecki, C. Hipp, S. Peng, W. Schachermayer, Stochastic Methods in Finance, Bressanone/Brixen, Italy, 2003. Editors: M. Fritelli, W. Runggaldier (2004)
Vol. 1857: M. Émery, M. Ledoux, M. Yor (Eds.), Séminaire de Probabilités XXXVIII (2005)
Vol. 1858: A.S. Cherny, H.-J. Engelbert, Singular Stochastic Differential Equations (2005)
Vol. 1859: E. Letellier, Fourier Transforms of Invariant Functions on Finite Reductive Lie Algebras (2005)
Vol. 1860: A. Borisyuk, G.B. Ermentrout, A. Friedman, D. Terman, Tutorials in Mathematical Biosciences I. Mathematical Neurosciences (2005)
Vol. 1861: G. Benettin, J. Henrard, S. Kuksin, Hamiltonian Dynamics – Theory and Applications, Cetraro, Italy, 1999. Editor: A. Giorgilli (2005)
Vol. 1862: B. Helffer, F. Nier, Hypoelliptic Estimates and Spectral Theory for Fokker-Planck Operators and Witten Laplacians (2005)
Vol. 1863: H. Führ, Abstract Harmonic Analysis of Continuous Wavelet Transforms (2005)
Vol. 1864: K. Efstathiou, Metamorphoses of Hamiltonian Systems with Symmetries (2005)
Vol. 1865: D. Applebaum, B.V. R. Bhat, J. Kustermans, J. M. Lindsay, Quantum Independent Increment Processes I. From Classical Probability to Quantum Stochastic Calculus. Editors: M. Schürmann, U. Franz (2005)
Vol. 1866: O.E. Barndorff-Nielsen, U. Franz, R. Gohm, B. Kümmerer, S. Thorbjønsen, Quantum Independent Increment Processes II. Structure of Quantum Levy Processes, Classical Probability, and Physics. Editors: M. Schürmann, U. Franz, (2005)
Vol. 1867: J. Sneyd (Ed.), Tutorials in Mathematical Biosciences II. Mathematical Modeling of Calcium Dynamics and Signal Transduction. (2005)
Vol. 1868: J. Jorgenson, S. Lang, $Pos_n(R)$ and Eisenstein Sereies. (2005)
Vol. 1869: A. Dembo, T. Funaki, Lectures on Probability Theory and Statistics. Ecole d'Eté de Probabilités de Saint-Flour XXXIII-2003. Editor: J. Picard (2005)
Vol. 1870: V.I. Gurariy, W. Lusky, Geometry of Müntz Spaces and Related Questions. (2005)
Vol. 1871: P. Constantin, G. Gallavotti, A.V. Kazhikhov, Y. Meyer, S. Ukai, Mathematical Foundation of Turbulent Viscous Flows, Martina Franca, Italy, 2003. Editors: M. Cannone, T. Miyakawa (2006)
Vol. 1872: A. Friedman (Ed.), Tutorials in Mathematical Biosciences III. Cell Cycle, Proliferation, and Cancer (2006)
Vol. 1873: R. Mansuy, M. Yor, Random Times and Enlargements of Filtrations in a Brownian Setting (2006)
Vol. 1874: M. Yor, M. Andr Meyer - S minaire de Probabilités XXXIX (2006)
Vol. 1875: J. Pitman, Combinatorial Stochastic Processes. Ecole dŠEté de Probabilités de Saint-Flour XXXII-2002. Editor: J. Picard (2006)
Vol. 1876: H. Herrlich, Axiom of Choice (2006)
Vol. 1877: J. Steuding, Value Distributions of L-Functions(2006)
Vol. 1878: R. Cerf, The Wulff Crystal in Ising and Percolation Models, Ecole d'Et de Probabilits de Saint-Flour XXXIV-2004. Editor: Jean Picard (2006)
Vol. 1879: G. Slade, The Lace Expansion and its Appli- cations, Ecole d'Eté de Probabilités de Saint-Flour XXXIV-2004. Editor: Jean Picard (2006)
Vol. 1880: S. Attal, A. Joye, C.-A. Pillet, Open Quantum Systems I, The Hamiltonian Approach (2006)
Vol. 1881: S. Attal, A. Joye, C.-A. Pillet, Open Quantum Systems II, The Markovian Approach (2006)
Vol. 1882: S. Attal, A. Joye, C.-A. Pillet, Open Quantum Systems III, Recent Developments (2006)
Vol. 1883: W. Van Assche, F. Marcell n (Eds.), Orthogonal Polynomials and Special Functions, Computation and Application (2006)
Vol. 1884: N. Hayashi, E.I. Kaikina, P.I. Naumkin, I.A. Shishmarev, Asymptotics for Dissipative Nonlinear Equations (2006)
Vol. 1885: A. Telcs, The Art of Random Walks (2006)
Vol. 1886: S. Takamura, Splitting Deformations of Degenerations of Complex Curves (2006)
Vol. 1887: K. Habermann, L. Habermann, Introduction to Symplectic Dirac Operators (2006)
Vol. 1888: J. van der Hoeven, Transseries and Real Differential Algebra (2006)

Recent Reprints and New Editions

Vol. 1618: G. Pisier, Similarity Problems and Completely Bounded Maps. 1995 – Second, Expanded Edition (2001)
Vol. 1629: J.D. Moore, Lectures on Seiberg-Witten Invariants. 1997 – Second Edition (2001)
Vol. 1638: P. Vanhaecke, Integrable Systems in the realm of Algebraic Geometry. 1996 – Second Edition (2001)
Vol. 1702: J. Ma, J. Yong, Forward-Backward Stochastic Differential Equations and their Applications. 1999. – Corrected 3rd printing (2005)

4. Careful preparation of the manuscripts will help keep production time short besides ensuring satisfactory appearance of the finished book in print and online. After acceptance of the manuscript authors will be asked to prepare the final LaTeX source files (and also the corresponding dvi-, pdf- or zipped ps-file) together with the final printout made from these files. The LaTeX source files are essential for producing the full-text online version of the book (see http://www.springerlink.com/openurl.asp?genre=journal&issn=0075-8434 for the existing online volumes of LNM).

   The actual production of a Lecture Notes volume takes approximately 8 weeks.

5. Authors receive a total of 50 free copies of their volume, but no royalties. They are entitled to a discount of 33.3 % on the price of Springer books purchased for their personal use, if ordering directly from Springer.

6. Commitment to publish is made by letter of intent rather than by signing a formal contract. Springer-Verlag secures the copyright for each volume. Authors are free to reuse material contained in their LNM volumes in later publications: A brief written (or e-mail) request for formal permission is sufficient.

Addresses:

Professor J.-M. Morel, CMLA,
École Normale Supérieure de Cachan,
61 Avenue du Président Wilson, 94235 Cachan Cedex, France
E-mail: Jean-Michel.Morel@cmla.ens-cachan.fr

Professor F. Takens, Mathematisch Instituut,
Rijksuniversiteit Groningen, Postbus 800,
9700 AV Groningen, The Netherlands
E-mail: F.Takens@math.rug.nl

Professor B. Teissier, Institut Mathématique de Jussieu,
UMR 7586 du CNRS, Équipe "Géométrie et Dynamique",
175 rue du Chevaleret
75013 Paris, France
E-mail: teissier@math.jussieu.fr

For the "Mathematical Biosciences Subseries" of LNM:

Professor P. K. Maini, Center for Mathematical Biology,
Mathematical Institute, 24-29 St Giles,
Oxford OX1 3LP, UK
E-mail : maini@maths.ox.ac.uk

Springer, Mathematics Editorial, Tiergartenstr. 17,
69121 Heidelberg, Germany,
Tel.: +49 (6221) 487-8410
Fax: +49 (6221) 487-8355
E-mail: lnm@springer-sbm.com